AI Robotics

Artificial intelligence (AI) robots can learn from their experiences, make decisions in real time, understand natural language and human gestures, and utilize computer vision to perceive and comprehend their environments. Beginning with the rudimentary concepts of AI, *AI Robotics: Ethics, Algorithms, and Technology of Artificial Intelligence-Powered Robots* explores the intersection of robotics and physics and emphasizes the need for strict adherence to ethical principles in relation to overall progress and the development of humankind. Chapters on robots capable of talking, listening, and visual perception similar to human beings are followed by discussions of those that display emotional intelligence. This book also discusses task and motion planning, a set of methods that help robot hardware achieve high-level goals by breaking down tasks into smaller, more manageable steps. Lastly, the text describes autonomous robots that can make independent decisions and execute tasks on their own, utilizing sensors and AI-enabled software programmed with predefined guidelines and data. Examples of autonomous robots are presented in a chapter on robot swarms that operate in a decentralized, self-organizing manner through local communication to manage disaster relief, search-and-rescue operations, warehouse logistics, agricultural practices, and environmental exploration. Offering an up-to-date, expansive, and comprehensive treatment of the vast interdisciplinary field of AI robotics, this book will be an invaluable resource for postgraduate and doctorate students as well as academic researchers and professional engineers working on AI-enabled robotics.

Key Features

- Explores the research frontiers and advancements leveraged by integrating AI with robotics
- Highlights the unique challenges faced in robot vision and speech recognition vis-à-vis computer vision and standard speech processing
- Provides a state-of-the-art overview of emotional recognition, task and motion planning, and coordinated functioning of robots in multi-robot systems

Vinod Kumar Khanna, PhD (Physics), is an independent researcher from Chandigarh, India. He is a retired chief scientist from the Council of Scientific and Industrial Research (CSIR)—Central Electronics Engineering Research Institute (CEERI), Pilani, India and a retired professor from the Academy of Scientific and Innovative Research (AcSIR), Ghaziabad, India. He is a former emeritus scientist, CSIR, and professor emeritus, AcSIR, India. His broad areas of research include the design, fabrication, and characterization of power semiconductor devices and micro- and nano-sensors. Dr. Khanna has published 194 research papers in leading peer-reviewed national and international journals and conference proceedings. He has authored 22 books and contributed six chapters to edited books. He has five granted patents to his credit, including two US patents.

AI Robotics

Ethics, Algorithms, and Technology of Artificial Intelligence-Powered Robots

Vinod Kumar Khanna

CRC Press
Taylor & Francis Group
Boca Raton London New York

CRC Press is an imprint of the
Taylor & Francis Group, an **informa** business

Designed cover image: Shutterstock

First edition published 2026
by CRC Press
2385 NW Executive Center Drive, Suite 320, Boca Raton FL 33431

and by CRC Press
4 Park Square, Milton Park, Abingdon, Oxon, OX14 4RN

CRC Press is an imprint of Taylor & Francis Group, LLC

© 2026 Vinod Kumar Khanna

ISBN: 9781032692852 (hbk)
ISBN: 9781032695198 (pbk)
ISBN: 9781032695266 (ebk)

DOI: 10.1201/9781032695266

Typeset in Times
by codeMantra

To my parents, the late Shri Amarnath Khanna
and Shrimati Pushpa Khanna

In grateful remembrance for giving me the strength
and wisdom to face the challenges of life

And to my grandson Hansh, daughter Aloka, and wife Amita

As a token of love and affection for making the
journey of life happy and enjoyable

Contents

Preface

Robotics and artificial intelligence (AI) are distinctly separate disciplines. Robotics is a branch of engineering that focuses on the design, construction, and application of programmable machines capable of executing instructions to perform assigned tasks, either semi-autonomously or fully autonomously, without human intervention. AI is a branch of computer science that develops algorithms capable of completing tasks that would otherwise require human intelligence. While non-AI programs are used to carry out predefined tasks, AI algorithms can learn and continually improve themselves. AI algorithms employ techniques such as search, logic, if-then rules, decision trees, and machine learning (including deep learning) to handle logical and analytical reasoning, problem-solving, language processing, and other tasks.

Until quite recently, all industrial robots could only be programmed to carry out a repetitive series of movements. Repetitive movements do not require AI. Non-intelligent robots are quite limited in their functionality. AI algorithms are necessary to allow the robot to perform more complex tasks.

AI and robotics combine to create 'Artificially Intelligent Robots', serving as a bridge between robotics and AI. These are robots that are controlled by AI algorithms. They are built by integrating AI software into a robot's hardware. In these robots, robotics technology is utilized to create the physical components, while AI is applied to program the intelligence. AI robots can be considered as intelligent automation applications in which robotics provides the body, while AI supplies the brain.

This book is about the convergence of AI and robotic technologies. The text aims to explore the overlapping areas of AI and robotics with the goal of constructing robots that possess enhanced functionality. AI augments the capabilities of robots, enabling them to understand their surroundings and interact with human beings, greet customers in shops, and perform complex tasks in manufacturing industries such as cutting, grinding, welding, and inspection independently, thereby ensuring the safety of workers. The subject matter is organized into 15 chapters. Starting with the fundamentals of AI and robotics, it guides the reader step by step through a journey of robots that can see, listen, and talk, display emotions, plan their tasks and motions, drive themselves, and work in coordination as a team. The key benefit of this book is that it provides up-to-date information on a rapidly emerging and rapidly developing field of immense value, assisting human activities in inaccessible areas or hazardous situations where human operator presence is risky.

The idea of this book is simple. It follows a synergistic approach between an AI programmer and engineer. An AI developer or programmer, who primarily deals with coding and software development, is often less well-acquainted with design and engineering aspects. An AI engineer seldom cares about programming. This is why there are often two teams involved in such projects: the design team and the programming team. Here a petite attempt is made to bridge this gap by focusing on the engineering and technological features of AI robotics, including its algorithmic framework. However, it is very difficult, if not entirely impossible, to do so.

Robotics is expected to have a large impact on society. As robotics becomes more integrated into everyday life, ethical problems concerning its design, deployment, and use emerge, including concerns about prejudice and responsibility. To address the moral issues raised by robotics, numerous governments and organizations are building legal and regulatory frameworks.

Beneficence and nonmaleficence, noninfringement on human autonomy, protection of privacy and data, and ensuring fairness, justice, transparency, safety, and security are some of the ethical standards that must be trustworthily adhered to in all robotics research.

The academic level of this book is graduate and above. Its target audience includes advanced undergraduate and graduate students in electronics and computer engineering for supplementary reading, PhD students and scientists engaged in research and development on robotics, practicing electrical, electronic, and computer engineers, robotics enthusiasts, and hobbyists. This research and reference book is intended for graduate students as supplementary reading, as well as for PhD students, scientists, and engineers.

Robots are good servants but bad masters
Helping us to work efficiently and faster
Mistakes in robot design bring disaster
Our deep inner voice gives a strong gut feeling
No robot can replace a human being
Robots can only assist us
And make difficult tasks easy to hasten progress.

Let us be enthusiastic and passionate
For making robots friendly and affectionate
Benign, lovable, and affable
Trustworthy and reliable.

Robotic algorithms and technology
Must be developed with a clear methodology
Following standards of honesty, compassion, and loyalty
Fully adhering to ethical guidelines and morality
Always keeping in mind
The welfare and the best interests of humankind!

Vinod Kumar Khanna
Chandigarh, India

Acknowledgments

First and foremost, I wish to thank the Almighty God for giving me the wisdom and strength to undertake and complete this work.

I am thankful to all the pioneering scientists and engineers working at the intersection of robotics and artificial intelligence. Their dedication and hard work have paved the way for synergizing the best features of both technologies.

I appreciate the kind cooperation and support from the commissioning and production editors, editorial assistants, and staff at CRC Press throughout the project.

Last, but not least, I express my gratitude to my wife and family for ensuring the serene and tranquil environment, and sacrificing their valuable time that I spent on this work.

My sincere thanks and gratitude to all the above.

Vinod Kumar Khanna
Chandigarh, India

About the Author

INTRODUCTION

Vinod Kumar Khanna is an independent researcher from Chandigarh, India. He is a retired chief scientist from the Council of Scientific and Industrial Research (CSIR)—Central Electronics Engineering Research Institute (CEERI), Pilani, India, and a retired professor from the Academy of Scientific and Innovative Research (AcSIR), Ghaziabad, India. He is a former emeritus scientist, CSIR, and professor emeritus, AcSIR, India. His broad areas of research include the design, fabrication, and characterization of power semiconductor devices and micro- and nano-sensors.

ACADEMIC QUALIFICATIONS

He earned an MSc in physics with a specialization in electronics from the University of Lucknow in 1975 and a PhD in physics from Kurukshetra University in 1988 for the thesis, 'Development, Characterization and Modeling of the Porous Alumina Humidity Sensor'.

RESEARCH/TEACHING EXPERIENCE AND ACCOMPLISHMENTS

His research experience spans 40 years from 1977–2017. Beginning his career as a research assistant in the Department of Physics, University of Lucknow from 1977–1980, he joined CSIR-CEERI, Pilani (Rajasthan) in 1980. There, he worked on several CSIR-funded and sponsored research and development projects. His major fields of research included power semiconductor devices and microelectronics/ MEMS and nanotechnology-based sensors and dosimeters.

In the power semiconductor devices area, he worked on the high-voltage and high-current rectifier (600 A, 4,300 V) for railway traction, high-voltage TV deflection transistor (5 A, 1,600 V), power Darlington transistor for AC motor drives (100 A, 500 V), fast-switching thyristor (1,300 A, 1,700 V), power DMOSFET, and IGBT. He contributed to the development of sealed tube Ga/Al diffusion for deep junctions, surface electric field control techniques using edge beveling and contouring of large-area devices, and the design of floating field limiting rings. He carried out an extensive characterization of minority-carrier lifetime in power semiconductor devices as a function of process steps. He also contributed to the development of the P-I-N diode neutron dosimeter and PMOSFET-based gamma-ray dosimeter.

In the area of sensor technology, he worked on the nanoporous aluminum oxide humidity sensor, ion-sensitive field-effect transistor-based microsensors for biomedical, food, and environmental applications, microheater-embedded gas sensor for automotive electronics, MEMS acoustic sensor for satellite launch vehicles, and capacitive MEMS ultrasonic transducer for medical applications.

As an AcSIR faculty member, he was the course coordinator of MEMS/IC Technology for the advanced semiconductor electronics program (2011–2013) and taught 'MEMS Technology' to students pursuing MTech degrees. As an adjunct faculty member, BESU, Kolkata, he taught 'MEMS Technology & Design' to MTech (mechatronics) students. He was invited by IIT, Jodhpur to deliver lectures on 'Semiconductor Fundamentals and Technology' to BTech students in 2011. He guided BTech/MTech theses of students from BITS, Pilani; VIT, Vellore; and Kurukshetra University. He also guided a PhD thesis on 'MEMS acoustic sensor', MNIT, Jaipur.

SEMICONDUCTOR FACILITY CREATION AND MAINTENANCE

Dr. Khanna was responsible for setting up and looking after diffusion/oxidation facilities, edge beveling and contouring, reactive sputtering, and carrier lifetime measurement facilities. As the head of the MEMS and Microsensors Group, he looked after the maintenance of a 6-inch MEMS fabrication facility for R&D projects as well as the augmentation of processing equipment under this facility at CSIR-CEERI.

SCIENTIFIC POSITIONS HELD

During his tenure of service at CSIR-CEERI from 1980 until superannuation in 2014, Dr. Khanna was promoted to various positions including one merit promotion. He retired as a chief scientist and professor (AcSIR) and as the head of MEMS and Microsensors Group. Subsequently, he worked for 3 years as an emeritus scientist, CSIR, and emeritus professor, AcSIR, from 2014–2017. After completing the emeritus scientist scheme, he now resides in Chandigarh. He is a passionate author and enjoys reading and writing.

MEMBERSHIP OF PROFESSIONAL SOCIETIES

He is a fellow and life member of the Institution of Electronics and Telecommunication Engineers (IETE), India. He is a life member of the Indian Physics Association (IPA), Semiconductor Society, India (SSI), and Indo-French Technical Association (IFTA).

FOREIGN TRAVEL

Dr. Khanna has traveled widely, participating in and presenting research papers at the IEEE Industry Application Society (IEEE-IAS) Annual Meeting in Denver, Colorado, USA in 1986. His short-term research assignments include deputations to Technische Universität Darmstadt, Germany, in 1999; Kurt-Schwabe-Institut fur Mess-und Sensortechnike e.V., Meinsberg, Germany, in 2008; and Fondazione Bruno Kessler, Trento, Italy, in 2011, under collaborative programs. He was a member of the Indian Delegation to the Institute of Chemical Physics, Novosibirsk, Russia in 2009.

SCHOLARSHIPS AND AWARDS

Dr. Khanna was awarded a national scholarship by the Ministry of Education and Social Welfare, Government of India on the basis of a higher secondary result, 1970; CEERI Foundation Day Merit Team Award for projects on fast-switching thyristor (1986), for the power Darlington transistor for transportation (1988), for the P-I-N diode neutron dosimeter (1992), and for the high-voltage TV deflection transistor (1994); Dr. N.G. Patel Prize for the best poster presentation in the 12th National Seminar on Physics and Technology of Sensors, 2007, BARC, Mumbai; and CSIR-DAAD fellowship in 2008 under Indo-German Bilateral Exchange Programme of Senior Scientists, 2008. He is featured in the Stanford–Elsevier list of the World's Top 2% Scientists (2022, Elsevier Data Repository, V4, doi:10.17632/btchxktzyw.4). He is named as a Highly Ranked Scholar-Lifetime: #3 in Nanoelectronics by ScholarGPS (https://scholargps.com/scholars/25423546982929/vinod-kumar-khanna).

RESEARCH PUBLICATIONS AND BOOKS

Dr. Khanna has published 194 research papers in leading peer-reviewed national and international journals, as well as conference proceedings. He has authored 22 books and contributed six chapters to edited books. He has five granted patents to his credit, including two US patents.

About This Book

AI robotics integrates artificial intelligence (AI) techniques into robotics, enabling robots to learn, adapt, and perform tasks beyond simple programmed actions, thereby becoming more versatile and intelligent. This book examines the evolution of robots equipped with AI and machine learning algorithms, enabling them to perceive their environment, make informed decisions, and respond to situations, much like humans. It contains 15 chapters. The following is a summary of this book's contents, highlighting how AI synergizes with robotics to drive the evolution of AI-driven robotics.

Chapter 1 introduces the core concepts of AI, machine learning, and deep learning. Ethical issues are addressed to ensure the responsible development and use of AI in ways that benefit society, with a focus on fairness, transparency, and accountability, while fully addressing privacy and security concerns.

Chapter 2 deals with the basics of robotics, robophysics, and roboethics, the trio of disciplines that work together in cooperation as a unified technology. Stringent adherence to roboethics is mandatory to ensure that robots do not pose a threat to humans in the long or short term. Robots should be designed and developed keeping potential hazardous situations in view.

Chapter 3 describes robotic sensors and actuators that work together in a feedback loop, where sensors measure physical quantities such as temperature, pressure, light, and sound, and convert them into electrical signals. These signals are then fed to the robot's control system. The control system uses the received information to instruct the actuators to take appropriate actions.

Chapter 4 explores methods for equipping robots with the ability to listen and speak, utilizing a combination of AI processor hardware and software. This includes microphones for input, speakers for output, and algorithms for speech recognition and synthesis, as well as natural language processing for understanding and generating human language.

Chapters 5–7 survey the technologies for robotic vision, which stand at the forefront of the AI robotic revolution, providing robots with the ability to see like humans. Unlike traditional robots, which have relied on cameras and sensors to navigate their environments, recent breakthroughs in computer vision and AI have propelled the development of robots with vision capabilities akin to human eyesight, enabling them to perceive depth and color and navigate complex and dynamic environments with precision.

Chapters 8 and 9 present the induction of emotions into robots to make them social entities that can freely interact with humans. Artificial emotional intelligence (AEI) involves endowing robots with the ability to recognize, understand, and express emotional features, thereby facilitating natural and harmonious human–robot interactions that are more intuitive and engaging. It's a complex field that involves understanding human emotions, modeling human emotions, and enabling robots to respond appropriately.

Chapter 10 discusses the methods employed by robots for planning their tasks and the necessary motions required for task execution. Robots break down complex goals into sequences of high-level actions or steps and generate the specific trajectories needed for efficient and collision-free execution, considering obstacles and constraints while optimizing for efficiency through the shortest or fastest path.

Chapters 11 and 12 outline technologies and tools for making robots capable of autonomous operation independent of human supervision and guidance by using a combination of sensors like cameras, LiDAR, and distance sensors to perceive their environments and using AI algorithms and machine learning models to make decisions, and execute tasks such as moving, manipulating objects, or interacting with the environment through actuators.

Chapters 13–15 explore the research challenges faced in using teams of robots to work cohesively and carry out mission projects. Robotic swarms work by leveraging the collective intelligence of many simple robots, enabling them to perform complex tasks such as efficiently combing for objects or resources in search-and-rescue missions or environmental monitoring that would be impossible for a single robot. Groups of robots work through decentralized coordination by distributing tasks among themselves utilizing algorithms drawing inspiration from natural swarms like bees, birds, or fish.

With an extensive bibliography for further reading, this book will be of immense value to postgraduate and PhD students, scientists engaged in research on AI robotics, as well as professional engineers working on the practical realization and uses of AI robotics technology.

Abbreviations, Acronyms, and Symbols

A

A*	A-star (Search Algorithm)
ABC	Artificial Bee Colony (Algorithm)
AC	Alternating current
ACO	Ant Colony Optimization (Algorithm)
AI	Artificial intelligence
AiMP	Associative-in-memory processor
AIST	Advanced Industrial Science and Technology
AlexNet	A convolutional neural network architecture designed by Alex Krizhevsky
Android	Greek 'andro' (man) + 'eides' (shape)
ANN	Artificial neural network
APF	Artificial Potential Field (Algorithm)
ASR	Automatic speech recognition

B

BEV	Bird's eye view
BFO	Bacterial Foraging Optimization (Algorithm)
BLDC	Brushless direct current (Motor)
BO	Bayesian optimization

C

CCD	Charge-coupled device
CCTV	Closed-circuit television
CIFAR-10 dataset	Canadian Institute for Advanced Research, ten classes
CIM	Compute-in-memory
CMOS	Complementary-metal-oxide-semiconductor
CNN	Convolutional neural network
COP	Combinatorial optimization problem
CPU	Central processing unit
C-Space	Configuration space
CSS	Concatenating speech synthesis
CV	Computer vision

D

1D/2D/3D	One-dimensional/two-dimensional/three-dimensional
DC	Direct current
DL	Deep learning
DNN	Deep neural network

DoF   Degrees of freedom
DRL   Deep reinforcement learning

E
ECG   Electrocardiogram
E. coli   *Escherichia coli*
EEG   Electroencephalogram
EKF-SLAM   Extended Kalman filter-simultaneous localization and mapping
EMG   Electromyogram

F
FA   Firefly algorithm
FastSLAM   Fast simultaneous localization and mapping
FC   Fully connected (layer)
FER   Facial emotion recognition
FFNN   Feedforward neural network
FFT   Fast Fourier transform
FHMM   Fuzzy hidden Markov model
FM   Feature map

G
GA   Genetic algorithm
GAN   Generative Adversarial Network
gbest   Global best (position)
GenAI   Generative artificial intelligence
GG-CNN   Generative grasping convolutional neural network
GoogleNet   A 22-layer-deep convolutional neural network developed by Google
                 for image classification
GPS   Global Positioning System
GPU   Graphical processing unit
GraphSLAM   Graph-based simultaneous localization and mapping
GVD   Generalized Voronoi Diagram

H
HCI   Human–computer interaction
HMM   Hidden Markov model
HOG   Histogram-oriented gradient

I
ILSVRC 2014   ImageNet Large-Scale Visual Recognition Challenge 2014
ImageNet   A publicly available large image database with annotated images
IMU   Inertial measurement unit
ISODATA   Iterative self-organizing data analysis technique algorithm

K
KASPAR   Kinesics and Synchronization in Personal Assistant Robotics
kHz   Kilohertz

KISMET	A robot named after a Turkish word meaning 'fate' or 'luck'
k-means clustering	'k' is the number of clusters (groups) that the algorithm is trying to form
KODOMOROID	Japanese word 'kodomo' (child) + 'android'; 'android' originates from the Greek 'andro' (man) + 'eides' (shape)

L

LiDAR	Light detection and ranging
LiDAR SLAM	Light detection and ranging, simultaneous localization and mapping
LRN	Local response normalization
LSTM	Long short-term memory (network)

M

mAP	Mean Average Precision
MEMS	Micro-electro-mechanical systems
MFCC	Mel frequency cepstral coefficient
MIT	Massachusetts Institute of Technology
ML	Machine learning
MLP	Multilayer perceptron
MPC	Model predictive control
MRTA	Multi-robot task allocation
MSE	Mean squared error
MV	Machine vision

N

NAS	Neural architecture search
NER	Named entity recognition
NLP	Natural language processing
NN	Neural network
NPU	Neural processing unit

O

OP-AMP	Operational amplifier

P

PARO	Personal robot
Pascal VOC	Pattern analysis, statistical modeling, and computational learning of visual object classes (dataset)
pbest	Personal best (position)
PCA	Principal component analysis
PID	Proportional-Integral-Derivative (control)
PIM	Processor-in-memory
PIXOR	ORiented 3D object detection from PIXel-wise neural network predictions
PNAS	Progressive neural architecture search

PR   Pattern recognition
PRM   Probabilistic roadmap
PSO   Particle Swarm Optimization (Algorithm)

Q
QPU   Quantum processing unit

R
RADAR   Radio detection and ranging
RAM   Random access memory
R-CNN   Region-based convolutional neural network
ReLU   Rectified linear unit
RGB   Red-Green-Blue
RL   Reinforcement learning
RNN   Recurrent neural network
ROI   Region of interest
ROS   Robot Operating System
RPN   Region proposal network
RRT   Rapidly exploring random trees
RTD   Resistance temperature detector
RV   Robot vision

S
SIFT   Scale-invariant feature transformation
SL   Supervised learning
SLAM   Simultaneous localization and mapping
SMBO   Sequential model-based optimization
SOM   Self-organizing map
SONAR   Sound navigation and ranging or sonic navigation and ranging
SSA   Salp Swarm Algorithm
SSD   Single-Shot MultiBox Detector
SVM   Support vector machine
Synset   Set of synonyms

T
TAMP   Task and motion planning
TPU   Tensor processing unit
TSP   Traveling Salesman Problem
TTS   Text-to-speech
TV   Television

U
UAV   Unmanned Aerial Vehicle
UNESCO   United Nations Educational, Scientific and Cultural Organization
USL   Unsupervised learning

V

VFH	Vector Field Histogram (Algorithm)
VFH⁺, VFH*	Successively improved variants of VFH algorithm
VGG-16	Visual Geometry Group-16
VOC	Visual object classes
VPU	Vision processing unit
vSLAM	Visual SLAM Algorithm

W

WordNet	A large online lexical database of English

Y

YOLO	You Only Look Once

ROMAN SYMBOLS

A

A	Observed audio input

C

c	Velocity of light
$C = \{c_{ij}\}$	A finite set of available solution components
c_{ij}	A component of the set of solutions
c_1	Cognitive coefficient
c_2	Social coefficient
$C^c(u_i{-}1, u_i)$	Concatenation cost
$C^t(u_i, t_i)$	Target cost

D

D	Dimension of optimization parameters
d	Distance of the target from the robot, distance d between centers of clusters (the inter-center distance) in k-means clustering

F

f	An objective function
fit_i	The fitness value of solution X_i
$f(n)$	Global cost function
$f(x)$	A function of variable x, objective function
$f(X_i)$	The objective function value of the decision vector X_i

G

$g(n)$	Cost function of the path traversed from the initial state to the node n

H

$h(n)$	Heuristic function representing the estimated cost from node n to the goal state

I

i,j	Nodes
$I(x)$	Light intensity of a firefly

K

k	An ant located in node i
K_d	Derivative Term
K_i	Integral Term
K_p	Proportional Term

L

L	The language
L_k	Cost of the kth ant's tour

M

m	Number of artificial ants

N

N	A number
n	Node
$N(sp) \subseteq C$	The set of feasible neighbors

P

P, p	Probability
$P(A)$	Independent probability of A
$P\left(\dfrac{A}{S}\right)$	Probability of A given S is true
$P(S)$	Independent probability of S
$P\left(\dfrac{S}{A}\right)$	Probability of S given A is true

Q

Q	A constant

S

S	A search space
$S \in L$	Possible sentences within language L
$\hat{S}$	The most likely sentence uttered by a user
SN	The size of solutions, i.e., food sources
S^p	Current partial solution
$S^p = \varnothing$	An empty partial solution

W

w	Inertia weight

X

X_{Goal}	Goal point of the robot
X_i	Decision vector
x_{ij}	The solution numbered as ith solution with dimension j
$x_{\text{maximum},j}$	The upper bound of solutions for the dimension j
$x_{\text{minimum},j}$	The lower bound of solutions for the dimension j
X_{Nearest}	Nearest neighbor node
X_{New}	New node
X_{Random}	Random sampling point
X_{Start}	Starting point of the robot

GREEK SYMBOLS

α

α	Parameter to control the influence of τ_{ij} on ants

β

β	Parameter to control the influence of η_{ij}

Δ

$\Delta\tau_{ij}^k$	Amount of pheromone deposited by kth ant

η

η_{ij}	Desirability of state transition ij
η_{il}	Attractiveness for the other possible state transitions

ρ

ρ	Pheromone evaporation coefficient

σ

σ	Standard deviation within each cluster in k-means clustering

τ

τ_{ij}	Amount of pheromone deposited for transition from state i to j
τ_{il}	The trail level

φ

$\phi_{i,j}$	A random number in the interval $[-1,1]$

Ω

Ω	A set of constraints applied on the variables in a combinational optimization problem

1 Artificial Intelligence, Ethical Concerns, and Social Responsibility

1.1 INTRODUCTION

In this chapter, the interlinked fields of 'data science (DS)', 'artificial intelligence (AI)', 'machine learning (ML)', and 'deep learning (DL)' are defined. Their roles in the extraction of knowledge and insights from data are elucidated. Software agents, cognition, and concepts related to sentience are discussed. The arguments and implications of Turing and Searle's thought experiments are presented. The significance of following ethical guidelines and recommendations in AI practices is emphasized.

The interrelationship in which these basic ideas are considered places them in a proper perspective. Intelligent software agents are autonomous programs. These programs are applied for perception, interpretation, and action on data without seeking any guidance from the user. Cognition and sentience are two important terms related to the mind and mental processes with distinctly different meanings. Information processing ability, known as cognition, is distinguished from sentience, the ability to experience feelings or sensations. The Turing test and the Chinese Room argument are two thought experiments that explore the boundaries of intelligence displayed by machines. AI ethics is the framework that safeguards the improper use of AI. The reason is that AI presents several concerns and difficulties for society, which must be controlled by formulating regulations. Strict adherence to ethics is warranted for building safe, secure, and environmentally friendly AI systems.

1.2 DATA SCIENCE

DS is one of the four resembling and related disciplines (DS, AI, ML, and DL) that play complementary roles toward the advancement of AI (AWS 2025). Figure 1.1 shows four ovals, with the biggest oval representing DS. A smaller oval inside it symbolizes AI. A still smaller oval within AI corresponds to ML. The smallest oval enclosed within ML signifies DL. We shall successively clarify the terms 'DS', 'AI', 'ML', and 'DL' in the sequel.

$$\text{Data science} = \text{Data} + \text{Science} \tag{1.1}$$

Data is the raw information collected from various sources. It can be measurement results from sensors. Event highlights from public news and social media sites can be treated as raw information. Another example of raw information may be given

DOI: 10.1201/9781032695266-1

1

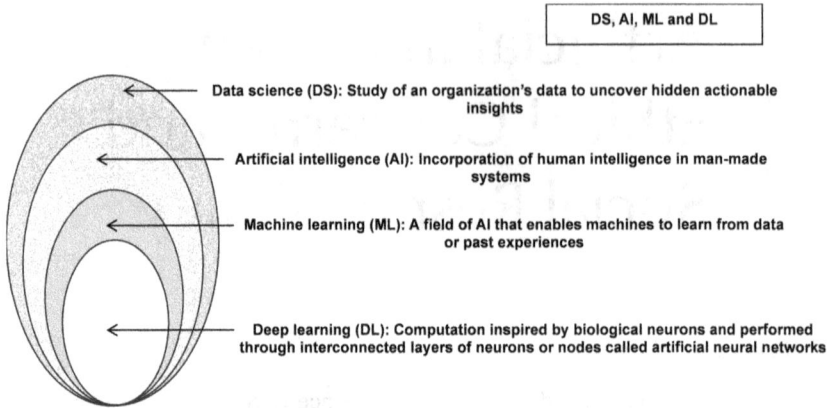

FIGURE 1.1   Related fields of data science, artificial intelligence, machine learning, and deep learning that work in collaboration to build intelligent systems.

as an economic survey conducted by agencies, etc., either in a structured format as databases or in an unstructured format such as pictures, video, text, and so forth. It is also called unprocessed information, primary data, source data, or atomic data.

The term 'Science' is derived from the Latin word 'Scientia' meaning knowledge. It is the knowledge of the natural world about the physical, chemical, and biological universe. This knowledge is acquired by a systematic procedure consisting of recognition of a problem, its formulation, performing experiments, making observations, putting forward, and testing hypothesis. The combination of data with science leads to an interdisciplinary field. In this field, algorithms, statistical methods, and computational tools are applied to data gathered about a subject for its analysis, examination, and exploration to extract useful insights and discover hidden patterns. From these insights, predictions are made that assist in decision-making aided by domain expertise (Grus 2015; Balusamy et al. 2021).

Big data refers to the large and complex datasets. These datasets include structured, semi-structured, and unstructured data from social media, transportation, and healthcare. They are too large to be easily managed and analyzed and require advanced technologies for processing and analysis. Data mining is the process of analyzing large datasets by cleaning, integrating, reducing, and transforming the data. Techniques such as clustering, classification, and association are applied to find patterns and relationships in the data that are not immediately obvious. These findings help businesses predict future trends and make better decisions, e.g., analyzing customer purchase history to identify patterns.

1.3   ARTIFICIAL INTELLIGENCE

$$\text{Artificial Intelligence} = \text{Artificial} + \text{Intelligence} \tag{1.2}$$

An artificial entity is a material, thing, or process. Its principal characteristic is that it has been developed by human beings, rather than found in nature. Intelligence is the ability to gain knowledge, learn skills, and understand phenomena. The knowledge

TABLE 1.1

Data Science and Artificial Intelligence

Sl. No.	Point of Comparison	Data Science	Artificial Intelligence
1	Focus	It uses data, statistics, and computer science with the main intent to extract useful insights from data	It uses mathematics, cognitive thinking, and computer science with the aim to mimic human intelligence.
2	Goal	Its objective is making decisions based on data analysis.	It drives innovation and creativity by solving problems, improving decision-making, and automating tasks.
3	Process	It collects, cleans, and analyzes large amounts of data. These methods reveal patterns and discernments for transforming raw data into actionable information.	It learns from the environment or experience in performing tasks to develop new models and create systems. The systems created can learn, reason, and adapt across a wide range of tasks.
4	Application examples	Visualization of data and predictive analysis	Recommendation engines, chatbots, and self-driving vehicles

thus gained is utilized to deal with new situations or solve complex problems. Decisions are made for taking appropriate actions.

AI is a field of computer hardware and software engineering. It is used to construct computers that can learn, perform analytical thinking, reasoning, deciding, recommending, and executing several advanced functions. These functions include visual perception, object recognition and categorization, understanding and translating spoken/written language, and making forecasts in such a manner that intelligence similar to humans is displayed (Norvig and Russell 2022). John McCarthy coined the term 'artificial intelligence' in the year 1956 (McCarthy 2007; Strydom and Buckley 2019).

Generative AI (GenAI), or generative artificial intelligence, is a particular type of AI. It uses large AI models called foundation models. These models are built from encoders and decoders to create new content like text, images, videos, and music. GenAI learns from data through a process of observation and pattern matching to create new data instances and content. It is used in arts, entertainment, technology, communications, and healthcare. It helps to increase efficiency, productivity, and innovation.

In Table 1.1, we look at the distinguishing features of AI and DS.

1.4 MACHINE LEARNING

$$\text{Machine Learning} = \text{Machine} + \text{Learning} \tag{1.3}$$

A machine is a mechanically and electrically operated device or apparatus. It is built with several parts, each of which is assigned a defined role. All the parts of a machine work together in harmony to accomplish a particular task. Learning is the process of acquisition of knowledge and skills. It results in behavioral changes or those in attitudes, values, and preferences. ML is a sub-branch of AI dealing with machines. These machines can learn and adapt using data, algorithms, and models to work on unknown data, and draw inferences from them. They undertake execution of tasks without any explicit instructions with gradual improvement in accuracy (Mohri et al. 2018; Burkov 2020). An algorithm in ML is a set of computational rules or procedures. These rules or procedures constitute a mathematical method through which a computer learns from data to identify patterns, discover relationships, and gain understandings to make predictions on fresh unseen data.

ML is categorized into three main types:

i. Supervised Learning (SL): It uses labeled data. By labeled data is meant the data in which the desired output of an input is already known. The model learns patterns based on this labeled data and thereafter, it can make predictions on new data.
ii. Unsupervised Learning (USL): It works with unlabeled data. Using unlabeled data means the model works by identifying patterns and structure within the data. No predefined categories or labels are used.
iii. Reinforcement Learning (RL): In this learning mechanism, an agent learns through trial and error by interacting with its environment. It receives positive feedback for good actions. Negative feedback is received for poor actions. Gradually, the agent optimizes its behavior to achieve a goal.

1.5   DEEP LEARNING

$$\text{Deep Learning} = \text{Deep} + \text{Learning} \qquad (1.4)$$

'Deep' means extending downward far below the surface. DL is the most prevalent facet of AI. It is a type of ML that involves multiple layers of artificial neural networks (ANNs; hence called deep to contrast with shallow) for the transformation of data (Voulgaris and Bulut 2018). The ANN is a model consisting of nodes called neurons and the connections between them serving as flow paths for data and computations that follow principles identical to those of biological neurons or nerve cells. The biological neurons in the brains of animals are composed of dendrites, cell bodies, and axons. They participate in the receipt of electrical signals from the external world and the firing of signals known as action potentials to other neurons, muscles, or organs.

Table 1.2 casts a view over the peculiarities of AI, ML, and DL side by side.

Table 1.3 sketches the domains of DL and GenAI.

1.5.1   Basic Layers of a Neural Network

A neural network comprises three basic layers:

TABLE 1.2
Artificial Intelligence, Machine Learning, and Deep Learning

Artificial Intelligence	Machine Learning	Deep Learning
It is a broad category of computer software, including the system software and application software, along with the supporting hardware that impersonates human thought and decision-making.	It is a type of AI that uses algorithms to learn from data and perform tasks autonomously. It can be used to identify patterns in large sets of data. It continuously improves its performance and accuracy through experience.	It is a type of machine learning that utilizes artificial neural networks to learn from data to recognize complex patterns and make predictions.

TABLE 1.3
Deep Learning and Generative Artificial Intelligence

Deep Learning	Generative AI
It is a subfield of machine learning that uses neural networks to learn from large amounts of data to make predictions or classifications.	It is a subset of deep learning that creates new content such as text, images, videos, audio, and computer code based on existing data. Representative techniques applied include generative adversarial networks, which utilize two neural networks to generate and classify data, and variational autoencoders, learning a compressed representation of data.

    i. an input layer which receives the input data,
   ii. one or more hidden layers where all computations occur, and
  iii. an output layer producing the result for the supplied inputs.

The layers of the neural network contain interconnected nodes called artificial neurons. All the neurons in any given layer of the network are connected with neurons in the next layer (Figure 1.2).

1.5.2 WEIGHTS AND BIASES IN A NEURAL NETWORK

The weights and biases are the staple parameters that control flow of data through a neural network. Weights are numerical values assigned to the connections between neurons. The purpose of assigning numerical values to the connections is to control the strength of connections between them, thereby regulating the extent to which an input influences an output. They are initialized randomly.

The bias is a constant value. Biases shift the activation function by a constant amount. The shifting ensures that the neuron is always activated by a small amount, irrespective of inputs being zero. Weights impact the steepness of a curve. Biases cause shifting of the curve leftward or rightward.

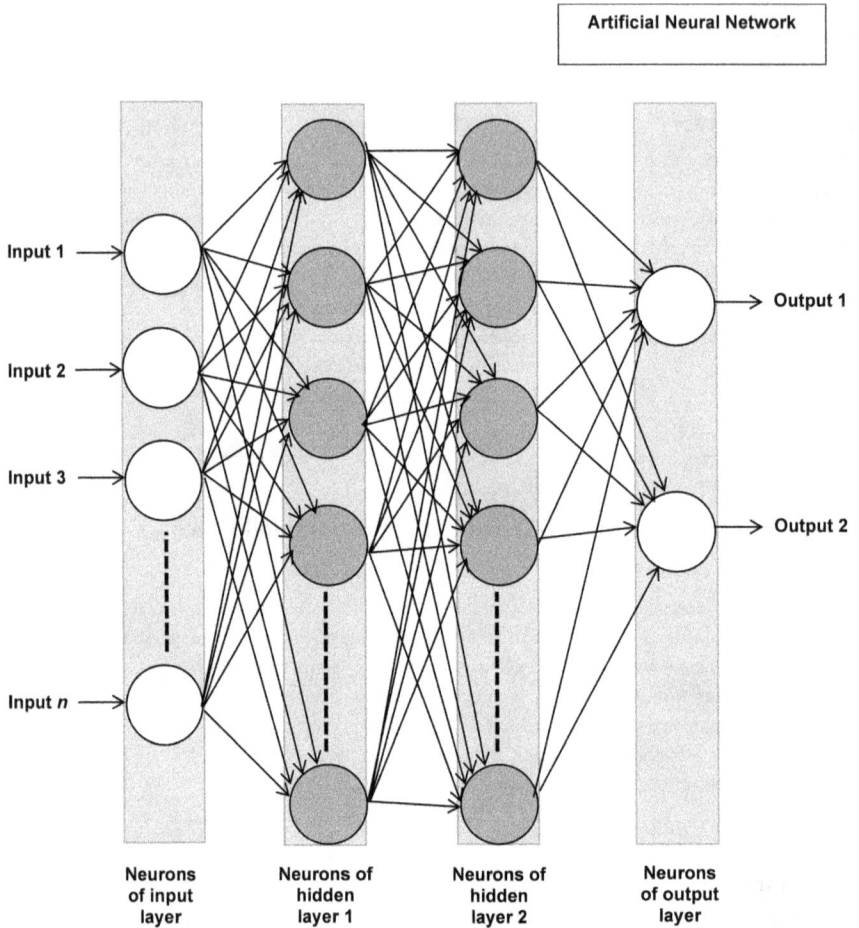

FIGURE 1.2   An artificial neural network.

Each value of input data is multiplied by its corresponding weight. Then the weighted sums are added together. The bias term is added to the summation result, acting as an offset to the combined weighted sum. The weighted sum with bias added to it is passed through an activation function. This function is a mathematical function, e.g., sigmoid, tanh (hyperbolic tangent), ReLU (rectified linear unit), leaky ReLU, and softmax (softargmax or normalized exponential function). It determines whether a neuron should fire or not by introducing non-linearity to the output, enabling it to learn complex relationships between the inputs and the output.

Hyperparameters of a neural network are essentially its configuration settings. They control the training process of the network and are set before commencement of training, e.g., number of neurons in each layer, the learning rate, the number of hidden layers, and the choice of activation function, among others. They are differentiated from the model parameters that are learned from the data itself.

1.5.3 BACKPROPAGATION IN A NEURAL NETWORK

Backpropagation is a fundamental algorithm in the training of deep neural networks. It is used for training neural networks to improve their predictions. For training a network, first the weights of the network are initialized to random values. Then an input vector from the training dataset is fed into the network and propagated through it to generate an output. The difference between the predicted and the actual outputs of the network is found. This difference is called the cost function and measures the error. It is propagated backward from the output through the hidden layers of the network to adjust the weights and biases in the network.

The backpropagation works in conjunction with gradient descent. The gradient descent is an optimization algorithm to minimize the cost function. The gradients of the cost function are calculated with respect to each parameter, weight and bias, in the network by application of the chain rule. Using this rule, the partial derivatives of the cost function are determined with respect to each weight in the neural network. The partial derivatives indicate the contribution of each weight to the error. Accordingly, the weights and biases in the network are adjusted to enhance its accuracy.

1.5.4 CATEGORIZATION OF NEURAL NETWORKS

A neural network is categorized into several types. Each of these types of networks has specific strengths for handling different kinds of data. A few types of neural networks are as follows:

i. Feedforward Neural Network (FFNN): It is the most basic type of network. In this network, data flows in one direction only, and this direction is from input to output through multiple layers. There are no cycles or loops in this network.

ii. Multilayer Perceptron (MLP): It is a type of feedforward network consisting of multiple layers of neurons. In this network, each neuron in a layer is fully connected to the neurons in the next layer. Hence, it can learn complex patterns in data. It is widely used for classification and regression tasks.

iii. Convolutional Neural Network (CNN): It is a neural network designed for image processing. It contains convolutional layers to efficiently extract features from grid-like data in images.

iv. Recurrent Neural Network (RNN): It is a neural network suitable for analysis of sequential data like text or time series. In this network, information from previous steps is retained through feedback loops.

v. Long Short-Term Memory (LSTM) Network: It is a specialized type of RNN designed to effectively handle long-term dependencies in sequential data. It uses memory cells to overcome the vanishing gradient problem. This problem is a common issue that occurs when training deep neural networks. It happens when the gradients used to update the network weights become very small. The small gradients prevent the weights from updating properly, which can lead to poor performance.

vi. Generative Adversarial Network (GAN): It is a generative model that uses two competing neural networks to produce realistic data. These networks are the generator and discriminator networks. The generator and discriminator work in a competitive training process. The process constantly improves the generator's output based on the discriminator's feedback. The GAN produces new data that closely resembles existing data. Highly realistic data like images, audio, or text are thereby created.

vii. Transformer Neural Network: It is a DL architecture used for analyzing complex relationships in data sequences. The relationships are analyzed by using a mechanism called 'self-attention'. By self-attention, we mean that the model is able to analyze and understand the relationships between different parts of an input sequence by assigning weights to each element based on its relevance to the others. The ability of a transformer neural network to capture long-range dependencies within a sequence makes it suitable for natural language processing tasks like translation and text generation.

1.6 THEORETICAL NOTIONS AND THOUGHT EXPERIMENTS OF AI

After introducing the elementary terminology of AI in the preceding sections, we look at a miscellany of theoretical ideas and thought experiments by imagining scenarios to explore concepts that are essential to grasp the fundamental principles of AI in letter and spirit.

1.6.1 INTELLIGENT SOFTWARE AGENTS

A software agent is an autonomously operating computer program or system. The special feature of this system is that it is endowed with capabilities of perceiving its environment, making decisions and taking proper actions (Figure 1.3). The diagram shows the agent equipped with sensors, signal processing circuit, and actuators. The agent is placed adjoining the investigated environment. The signal produced under the influence of the environment is fed to the sensors of the agent. It passes to the

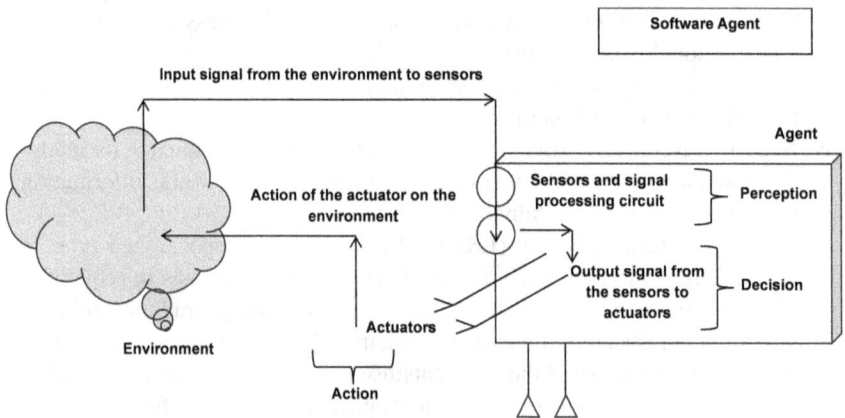

FIGURE 1.3   Operational mechanism of an intelligent software agent.

circuit associated with the sensor where it is processed. The output signal from the processing circuit is supplied to the actuators of the agent. The actuators deliver the required action on the environment in accordance with input signal received from it.

The agents are subdivided into four architectural groups: reactive, deliberative, learning, and hybrid (Kirrane 2021).

1.6.1.1   Reactive Software Agents

These are modeled on the reflexive behavior of humans involving automatic, involuntary responses to stimuli. They have two principal components:

i. Condition-Action Rules: In these agents, perception and action are firmly fastened by condition-action rules. This tight perception-to-action coupling leads to a quick natural response. As soon as a specific condition is perceived by the sensor of the agent, the agent immediately retrieves the action associated with that condition. Without any delay, it applies the retrieved action to convey the required instruction to its actuator.

ii. State: In advanced versions of reactive agents, the historical record of information regarding preceding interactions of the agent with its environment is maintained. This is done alongside the information about the whole environment as a state. When the agent is supplied with an entirely new perception, it responds with an action based not only on its present perception but also on the historical record of previous perceptions. A reactive agent is preferred in the circumstances demanding real-time decision-making where time limit is a crucial manifestation of performance.

1.6.1.2   Deliberative Software Agents

In these agents, the environment is modeled using a symbolic language. The decisions are made by logical reasoning. By reasoning logically, we mean performing a mental activity in which premises and relations between premises are utilized to arrive rigorously at conclusions that are implied by the premises and the relations. As an example, if every item left outdoors is wet when it is raining, and a person has kept his books on the outside table, logical reasoning will result in the inference that the books are drenched with water. In order to carry out its activities, this agent needs the following parts:

i. Symbolically Encoded Knowledge Base: This knowledge base stores knowledge of the agent about the surrounding environment and the knowledge controlling its behavior and actions.

ii. Logical Reasoning Mechanism: Perception of conditions by the sensor in the agent and the reasoning about possible actions and their influence on the environment serve as inputs to determine the instruction that must be delivered by the agent to its actuator.

iii. Encoding of Goal: Desirable behaviors are described for guidance of decision-making by the agent.

iv. Utility Function: It helps the agent in performing a comparative appraisal in view of preferences for maximization of its usefulness.

1.6.1.3   Learning Software Agents

It is an agent in which the deliberative component is enhanced with the capability to learn and perform better with time by learning from experience. It is especially useful in cases where the environment is an unknown priori. The term 'a priori' is applied to knowledge that is considered as true, e.g., all rectangular shapes are polygonal. A priori reasoning works by theoretical deduction, rather than observation. A learning agent must have:

  i. A Performance Component: It refers to the core inner responsibilities of the agent.
 ii. A Problem Generator: It suggests actions expediting new knowledge acquirement.
iii. A Critic: It provides feedback to the agent as a penalty or reward by comparison of its performance relative to a benchmark.
 iv. A Learning Element: Its jobs are execution of action allocated by the problem generator and modification of the core inner functions of the agent based on the feedback received from the critic.

1.6.1.4   Hybrid Software Agents

Here the reactive and deliberative components are arranged in horizontal and/or vertical layers. The layering combines together the reactive, deliberative, and learning features. Controllers are used for scheduling, implementation, and supervision of activities. They also address the management of interactions between the activities.

1.6.2   COGNITION

Cognition is the combination of all processes, conscious or unconscious, involved in the accumulation, storage, manipulation, and retrieval of knowledge. The processes involved are perception, recognition, conception, and reasoning to make decisions and produce appropriate responses through our senses, experience, and thought. They underpin several routine activities across the span of life for guiding our behavior. 'Cognition' has its roots in the Latin word 'cognoscere', meaning 'getting to know'. The received sensory information is very vast and intricate. So, cognitive functioning assists us in distilling this information to the essence level for easy understanding. The physical basis of cognition resides in the around 10^8 nerve cells or neurons in the brain. Each of these neurons has $>10^4$ connections with other neurons (Cambridge Cognition 2015).

1.6.3   SENTIENCE

Sentience is the capacity of a creature, human or animal, to experience feelings, sensations, or emotions, i.e., have awareness and consciousness in distinction to perceptions and thoughts. Sentient AI can think and feel the joy, fear, pain, and love akin to a human being. Fish exhibits an averse behavioral reaction against toxic stimuli that cause pain in humans and other animals (Sneddon 2009).

1.6.4 THE TURING TEST

The Turing test is a thought experiment called the imitation game (Turing 1950). The objective of this experiment is to determine whether a computer can interact in the same way that a human being does. In this experiment, there are three participants: a computer labeled as X, a person named as Y, and a human Z. The participants X and Y are respondents while the participant Z is an interrogator or questioner. The X and Y respondents are located in two cabins II and III, while the Z interrogator is sitting in a separate cabin I. Thus, both the respondents X and Y are physically isolated from the interrogator Z.

Figure 1.4 displays the arrangement made for carrying out the experiment for the Turing test with a lady interrogator Z, a respondent computer X, and a respondent person Y, which are occupying three separate cabins. Messages are exchangeable between Z and X, and Z and Y.

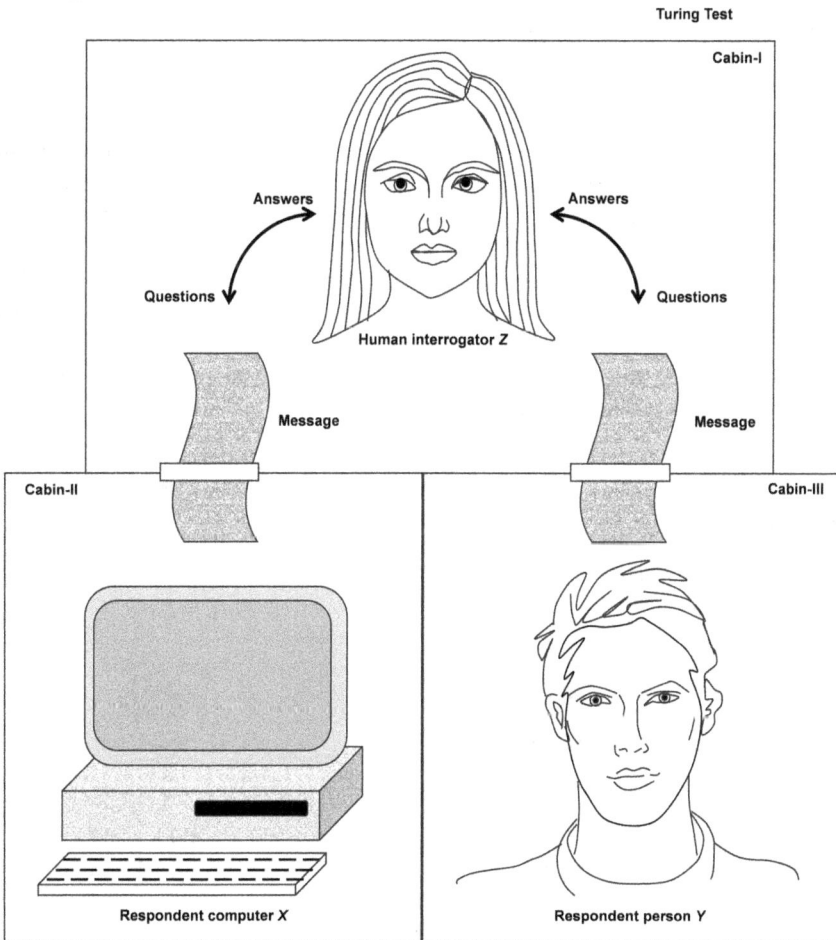

FIGURE 1.4 Setup of the Turing test.

The interrogator Z poses the same questionnaire to the respondents X and Y. The person Y tries to help the human interrogator Z to correctly identify the computer X. At the same time, the computer X tries to make the human interrogator Z erroneously conclude that person Y is the computer.

Turing said in 1950 that after a span of 50 years has elapsed, i.e., around the year 2000, it will be possible to program computers in such a manner that they will play the imitation game so efficiently that the human interrogator Z will be left with less than 70% probability of correctly identifying the computer and discriminating between the computer X and person Y after a 5-minute question–answer session (Oppy and Dowe 2003).

Thus, the Turing test is a deceivingly easy method of demonstration of human intelligence by a computer. A computer will be deemed as intelligent if it can take part in a conversation with a human interrogator without the human interrogator being able to know that he/she is in dialogue with a computer or another human. Yes, if the computer can dupe the human interrogator to such an extent, it is obviously an intelligent computer. The Turing test is a prime motivator in the development of AI.

1.6.5 THE CHINESE ROOM ARGUMENT

The Chinese Room argument is a thought experiment proposed by the American philosopher John Searle (1980) challenging the Turing claim that computers can think, understand, and have cognition like humans. It is an objection to counter the veracity of the Turing test.

Figure 1.5 shows the preparations that are made for the experiment. Inside a locked room, there is a person (the insider) and a rule book or program which the person consults for manipulation of Chinese symbols. Outside the room near its wall, a person is standing (the outsider). Input is sent from the outsider to the insider while the output is transmitted in the reverse direction.

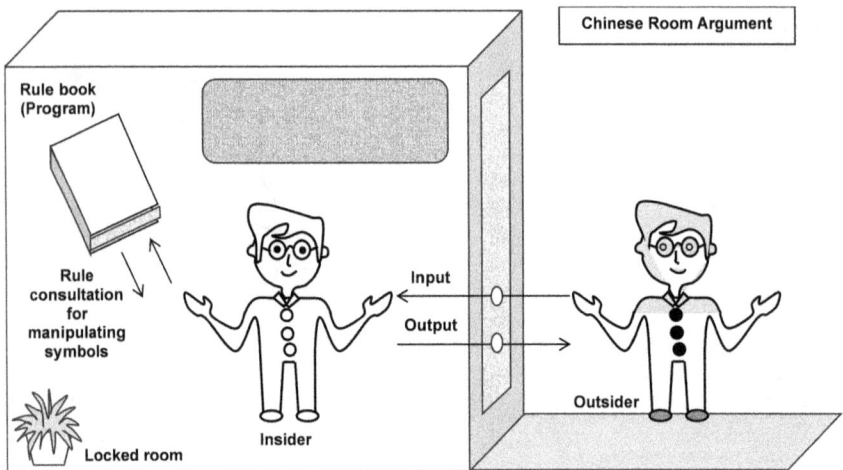

FIGURE 1.5   Performing the Chinese Room experiment.

Supposing the experiment is done on me, it proceeds as follows:

i. I do not know the Chinese language at all, either spoken or written. I am locked in a room and presented with a large batch of Chinese writing. This first batch is called a script. As I am inside the room, I am referred to as the insider.

ii. I am presented with a second batch of Chinese writing by a person from outside the room, termed the outsider. This batch is called the story. Along with the second batch of Chines writing, I am given a set of rules or instructions in English language (which I know very well) for correlation of the second batch with the first batch. The set of rules or instructions is known as a program. It is the rule book. By applying these rules, I am able to correlate one set of symbols with another set of symbols entirely from identification of the shapes of the symbols.

iii. I am offered a third batch of Chinese writing. This third batch is called a questionnaire. Together with the third batch, I am provided with necessary instructions or program to correlate the symbols of the third batch with those of the first and second batches. The program also contains the instructions enabling me to return certain Chinese symbols of given shapes called answers in response to symbols of particular shapes given to me in the third batch as questionnaire.

iv. I am sent stories in English. I can understand the stories and can answer the questions posed in English about these stories in English.

v. After sometime, I get skilled in following the instructions for manipulating the Chinese symbols and the outside people sending the programs become so adept in writing the programs that to somebody outside the room, the answers sent by me become indistinguishable from those given by a native Chinese expert. No one outside is able to discern from my answers in Chinese that I do not have any knowledge of the Chinese language.

The experiment conclusively proves that by programming a digital computer we may betray someone to appear as understanding a language (Cole 2023). However, this understanding is not real. It only seems to give an illusion of understanding. Staking the claim that a programmed computer understands stories is rebutted because I am illiterate to Chinese and so is the computer. The claim that program explains human understanding is also refuted because both the computer and the program are operational but no understanding results. Thus, the deficiency of the Turing test is highlighted. It is also reiterated that computers only use rules for manipulation of strings of symbols. They do not understand the underlying meaning of the contents of a writing.

1.7 AI ETHICS

At this stage, we understand the capabilities of AI. By extrapolation and generalization of its abilities, we can appreciate its enormous potential, and the likely risks accompanying its advancement and uncontrolled proliferation. Legal ramifications

of AI require proactive measures to protect humans from any detrimental consequences of its misuse. Responsible AI takes into consideration the lawful and ethical
viewpoints for AI (Figure 1.6). The diagram shows the responsible AI block in the
center surrounded on its sides by ten distinctive features characterizing it that can be
seen as fairness, accountability, reliability, security and safety, sustainability, compliance, explainability, interpretability, transparency, and privacy (Floridi and Cowls
2019; Dignum 2019).

AI ethics are a system of moral principles, guidelines, and techniques to demar-
cate between right and wrongdoings for fair development and responsible use of AI.
These principles aim to benefit society en bloc by restricting AI behavior within
the bounds of human values. According to UNESCO (United Nations Educational,
Scientific and Cultural Organization 2022), it is essential to ensure conformity to the
following conditions during the entire life cycle of an AI system.

i. Human dignity, rights, and fundamental freedoms must be respected, protected, and promoted by all means. Cognizance and preservation of basic
liberties are spotlighted.

ii. Any harm or subordination of a person or community should not occur in
any way whatsoever.

iii. Interaction of persons with AI systems, such as for receiving assistance
for those who are vulnerable, can take place throughout their life cycles.
Unprotected and defenseless must be cared for.

iv. Recognition, protection, and promotion of the flourishing of the environment and the ecosystem should always be guaranteed, calling for policies
prioritizing ecological integrity.

v. Compliance with international/domestic laws is mandatory for all actors in
the lifecycle of any AI system.

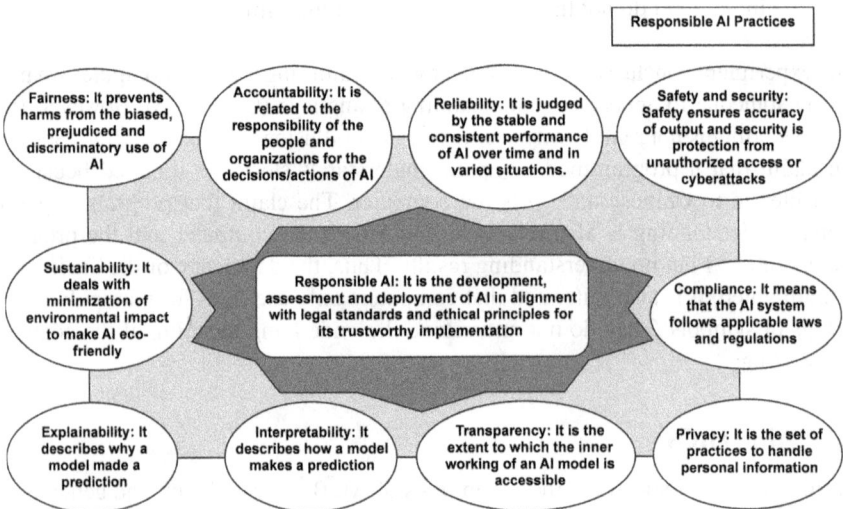

FIGURE 1.6   The salient features of responsible AI.

vi. Keen participation of all individuals or groups should be fostered without any racial, ethnic, or social origin or similar discrimination to confirm variety and completeness.

vii. During any stage of AI systems, the extent of lifestyle selections or ideas should not be constrained, i.e., the breadth of such choices or concepts should not be limited. Any deficiencies or flaws of the technological substructure should be overwhelmed by international cooperation.

viii. The interrelation of all living beings with each other as well as with the natural environment is emphasized along with the promotion of peace, justice, and equity. Thus, tranquility, righteousness, and evenhandedness are affirmed. Every human being is considered an integral part of a greater whole.

ix. The possibility of any harm to human beings should be prevented by implementing risk assessment methods.

x. The AI technique should be science-based. It should be suitable to the context besides being proportional to the targeted goal, and non-infringing on human rights.

xi. Any undesired safety and security risks should be avoided.

xii. Social justice and fairmindedness should be encouraged and promoted providing motivational and advocational support.

xiii. Reinforcement or perpetuation of biased applications and results should be eschewed. Their deliberate avoidance is mandated by moral grounds.

xiv. All people should be equitably treated by taking digital and knowledge divides within/across nations. Digital divide refers to gaps between those who have access to and use of information and communication technologies. Knowledge divide means disparities in access to knowledge, information, and opportunities for learning and developing skills.

xv. Impact of AI technologies should be ascertained with due consideration for sustainability aspects encompassing the environmental, economic, and social foundations.

xvi. Necessary data protection and governance mechanisms and algorithm systems must be set up for respecting the right to privacy and protection of personal information from unauthorized access and disclosure.

xvii. Human responsibility and accountability are not replaceable by AI systems.

xviii. Transparency and explainability are essential prerequisites of AI systems with proper balancing for privacy and security. Transparency leads to democratic societies and mitigates corruption. In case the AI application influences the end user in a permanent or irreversible manner, the underlying algorithm should be clearly explainable.

xix. It is the ethical and legal responsibility of AI actors to protect human rights according to the applicable laws.

xx. Required impact assessment mechanisms should be developed for assurance of accountability of AI systems.

xxi. Understanding and awareness of AI technologies by general public must be encouraged, particularly with regard to the human rights and environmental protection.

xxii. Usage of data must be done keeping national sovereignty and international laws in mind.

xxiii. Inclusive AI governance requires participation of all stakeholders enabling that the benefits of AI are shared by everyone.

1.8 ORGANIZATIONAL PLAN OF THIS BOOK

To reiterate the focal theme of this book, we will be largely dedicated to learning how AI robotics works, as well as the systematic algorithms, work plans, and procedures designed to tackle various issues faced by robots in their day-to-day activities. The focus will be on the algorithmic, engineering, and technological features of AI robotics. This is only one side of the coin. Let us inquire, 'What is the other side of the coin?' An equally important aspect of this subject is computer programming, the art of instructing computers through code. This field is largely based on computer science and related technologies. This is a complete subject in itself and requires separate, comprehensive treatment. The material presented in this book will serve as the foundation for the computational software development. Needless to say, all research in robotics, like any scientific endeavor, must adhere to ethical guidelines. This is essential to ensure the responsible development and deployment of technology. Human rights and well-being must always be safeguarded and protected. Therefore, adherence to ethical principles will be emphasized throughout.

This book is divided into 15 chapters as follows:

Chapter 1: This chapter introduces the fundamentals of AI and emphasizes the establishment of clear ethical standards for ensuring the fair and above-board practices of AI for the benefit of humanity at large in a compassionate and cooperative manner for innovation, progress, social cohesion, stability, and improved well-being. AI brings with it ethical concerns primarily revolving around issues like data governance, algorithmic fairness, transparency, explainability, potential for discrimination based on data used to train AI systems, privacy violations due to data collection, bias in decision-making processes, and the potential for misuse of AI technology. All these issues highlight the importance of social responsibility in developing and deploying AI systems with kindness, empathy, and altruism.

Chapter 2: This chapter presents an overview of robotics, robophysics, and roboethics. Robotics is the engineering and computer science field of designing, manufacturing, and operating intelligent programmable machines called robots used in many industries for improving efficiency and safety. Robophysics is the study about using physics methods like parameter space exploration, systematic control, and dynamical systems, and applying the laws of motion, energy, and electromagnetism for making robots with life-like movement and coordination. Roboethics, an interdisciplinary subfield of ethics of technology combining ethics, law, and sociology, considers how robots are designed to act ethically without posing any threat to humans.

Chapter 3: This chapter covers robotic sensors and actuators. Robotic sensing involves vision systems and cameras, LiDAR, RADAR, proximity, touch, force, and temperature sensors, accelerometers/gyroscopes, and chemical sensors. Robotic actuation entails DC, stepper motors, and servo motors; pneumatic, hydraulic, and piezoelectric actuators; shape memory alloy- and compliant materials-based devices. Sensor fusion for comprehensive understanding of the environment using multiple sensors, sensor calibration to ensure accuracy, and sensor control algorithms implementing software to process sensory data play a leading role in robotic systems.

Chapter 4: This chapter deals with the technology used in robot assistants, customer service chatbots, educational and medical robots enabling a robot to generate human-like speech (text-to-speech synthesis), interpret spoken words from human operators (speech recognition), and comprehend the meaning behind those words (natural language understanding), allowing for natural interaction between a robot and a person.

Chapters 5–7: These chapters describe the techniques of computer and machine vision used by robots to perceive and interact with their environment by object detection and recognition, feature extraction not only to perform tasks in dangerous and hazardous conditions but also to recognize faces, interpret expressions, and interact socially.

Chapters 8 and 9: These chapters discuss the salient aspects of robots capable of emotional recognition, expression, self- and social awareness, adaptive response, and contextual understanding mirroring the key aspects of human emotional intelligence. Abilities to understand and respond to human emotions in a nuanced way create more natural interactions for healthcare, customer service, education and companionship.

Chapter 10: This chapter treats a field in robotics where a system combines both high-level 'task planning' about deciding what actions to take to achieve a goal with low-level 'motion planning', calculating the precise movements needed to execute those actions, to complete the job while avoiding obstacles, and coping with real-world environmental constraints. Discrete decisions of task planning are merged with continuous movements considered in motion planning. The activities are useful for robots operating in unstructured environments, like navigating a room and manipulating objects. Challenges faced are the uncertainties, unpredictabilities, and interactional and computational complexities encountered in practical scenarios.

Chapters 11 and 12: These chapters discuss machines that can perform tasks without human aid, e.g., self-driving vacuum cleaners, cars, and industrial robot arms, using advanced sensors, information processing, decision-making, and movement, unlike the customary remote-controlled robots. Perception and sensing, real-time decision-making, human-robot interaction, ethical and legal issues, data security and privacy, and, above all, safety concerns are decisive issues for robots that think for themselves.

Chapters 13–15: This set of concluding chapters focuses on swarm intelligence, which is a collective intelligence of a group of robots. It is a bio-inspired AI field replicating the behavior and cooperation of large numbers of homogeneous, self-organized, and decentralized agents, e.g., birds, bees, ants, and even bacteria and micro-organisms obeying simple rules and interacting with each other to solve natural problems including foraging for food, prey evading, and task allocation in colonies such as finding the shortest path between their nest and food source or organizing their nests.

1.9 DISCUSSION AND CONCLUSIONS

This chapter presented the fundamentals of DS, AI, ML, and DL, and clarified the interrelationships and differences between them (Table 1.4). ML was presented as a subset of AI and DL as a subset of ML. AI theoretical topics, such as reactive/deliberative approach to AI design, cognition, sentience, the Turing test, and the Chinese Room problem, were discussed.

With the rapid advancements in AI and its widespread application in data-driven decision-making, it is expected that the AI companies will sincerely follow ethical protocols to avoid possible infringements on human rights in order that AI's potential benefits reach a large human population without producing any adverse effects. Responsible AI is built on several pillars such as explainability, accountability, reliability, security, privacy, and so forth. AI ethics refers to the set of moral rules that must be followed in the development and applications of AI technology. Therefore, AI regulations are necessary to strictly monitor and ensure that AI systems do not exceed their limits and go beyond the achievement of a legitimate aim. AI ethical considerations and challenges were succinctly presented.

TABLE 1.4
Terminology Introduced, Basic Ideas Learned, and the Core Issues Raised in This Chapter

Sl. No.	Terminology/ Basic Ideas/ Issues	Explanation
1	Summary	The terms 'data science', 'artificial intelligence', 'machine learning', and 'deep learning' were defined. Various types of software agents were introduced, notably the intelligent, reactive, deliberative, learning, and hybrid software agents. Notions of cognition and sentience were explained. Thought experiments of AI were described, including the Turing test and the Chinese Room experiment.
2	Ethics	The ethical concerns associated with the development of artificial intelligence raise moral dilemmas. These can be resolved by following responsible, equitable, and reliable practices to prevent the misuse of AI.
3	Organization	The organizational structure and plan of this book were outlined by summarizing the contents of its chapters.
4	Keywords and ideas to remember	Data science, artificial intelligence, machine learning, deep learning, neural network, weights and biases, backpropagation, software agents, cognition, sentience, Turing test, Chinese Room argument, AI ethics.

The organizational structure of this book was laid out.

From AI, we move on to AI-driven robotics in Chapter 2, together with its sister branches of robophysics and roboethics, which share similar activities or other characteristics, whose association is of significant relevance to AI robotics.

REFERENCES AND FURTHER READING

AWS. 2025. What is data science? Amazon Web Services, Inc., https://aws.amazon.com/what-is/data-science/#:~:text=What%20is%20the%20difference%20between,learning%20engineers%20to%20process%20data

Balusamy B., R. Nandhini Abirami, S. Kadry and A. Gandomi. 2021. *Big Data: Concepts, Technology, and Architecture*, John Wiley & Sons Inc, Hoboken, NJ, 356 pages.

Burkov A. 2020. *Machine Learning Engineering*, True Positive Incorporated, Bristol, England, 310 pages.

Cambridge Cognition. 2015. Insights: What is cognition? https://cambridgecognition.com/what-is-cognition/#:~:text=Cognition%20is%20defined%20as%20'the,used%20to%20guide%20your%20behavior

Cole D. 2023. The Chinese Room Argument, The Stanford Encyclopedia of Philosophy, copyright © 2024 by The Metaphysics Research Lab, Department of Philosophy, Stanford University, https://plato.stanford.edu/archives/sum2023/entries/chinese-room/

Dignum V. 2019. *Responsible Artificial Intelligence: How to Develop and Use AI in a Responsible Way*, Springer, Cham, 127 pages.

Floridi L. and J. Cowls. 2019. A unified framework of five principles for AI in society, *Harward Data Science Review*, Vol. 1, 1, pp. 1–14.

Grus J. 2015. *Data Science from Scratch: First Principles with Python*, O'Reilly Media, Inc., Sebastopol, CA, 330 pages.

Kirrane S. 2021. Intelligent software web agents: A gap analysis, *Journal of Web Semantics: Science, Services and Agents on the World Wide Web*, Vol. 71, 100659, pp. 1–25.

McCarthy J. 2007. What is Artificial Intelligence? p. 2, https://www-formal.stanford.edu/jmc/whatisai.pdf

Mohri M., A. Rostamizadeh and A. Talwalkar. 2018. *Foundations of Machine Learning*, 2nd edition, MIT Press, Cambridge, MA, 504 pages.

Norvig P. and S. J. Russell. 2022. *Artificial Intelligence: A Modern Approach*, 4th edition, Pearson Education, Noida, 1292 pages.

Oppy G. and D. Dowe. 2003. The Turing Test. In: Zalta E. N. and U. Nodelman (Eds.), *The Stanford Encyclopedia of Philosophy*, Stanford University, Stanford, CA, https://plato.stanford.edu/archives/win2021/entries/turing-test/

Searle J. 1980. Minds, brains, and programs, *Behavioral and Brain Sciences*, Vol. 3, pp. 417–457.

Sneddon L. U. 2009. Pain perception in fish: Indicators and endpoints, *Institute for Laboratory Animal Research (ILAR) Journal*, Vol. 50, 4, pp. 338–342.

Strydom M. and S. Buckley (Eds.). 2019. *AI and Big Data's Potential for Disruptive Innovation (Advances in Computational Intelligence and Robotics)*, IGI Global, Hershey, PA, p. 7.

Turing A. 1950. Computing machinery and intelligence, *Mind*, Vol. 59, 236, pp. 433–460.

UNESCO. 2022. Recommendations on the Ethics of Artificial Intelligence Adopted on 23 November 2021, Published by The United Nations Educational, Scientific and Cultural Organization, Paris, France, pp. 18–23. https://unesdoc.unesco.org/ark:/48223/pf0000381137

Voulgaris Z. and Y. E. Bulut. 2018. *AI for Data Science*, Technics Publications, Basking Ridge, NJ, p. 9.

2 AI-Driven Robotics, Robophysics, and Roboethics

2.1 INTRODUCTION

In this chapter, we study three closely interlinked domains of knowledge, namely, robotics, robophysics, and roboethics. Robotics is a technology built over the scientific foundation of robophysics. Robotics must strictly adhere to the moral values of roboethics for the welfare and progress of humanity. We begin by defining the main terms that will help in easily following the discussion ahead.

2.2 ROBOTICS AND RELATED TERMS

2.2.1 ROBOTICS

Robotics is a branch of engineering, principally mechanical and electrical engineering, and computer science. It is concerned with the conception, design, manufacturing, operation, and applications of machines that replicate human actions. These machines work jointly with their supporting computer and information systems for sensory feedback, as well as their control instrumentation and actuators. They assist humans in a variety of ways to improve automation and innovation by performing repetitive tasks with greater efficiency and accuracy than humans (Craig 2022).

2.2.2 ROBOTS

A robot is an automated machine that resembles a living creature and is capable of independently moving by walking or rolling on wheels and performing complex tasks. These tasks include grasping and working with objects. The robot executes its tasks with great speed and precision with little or without human intervention.

A humanoid robot is one designed to resemble the human body in shape and form. It is designed to interact with human environments and tools, but may still look like a machine. An android robot is a specific type of humanoid robot that aesthetically aims to look as human-like as possible. Often it has features like realistic skin, hair, and facial expressions to closely mimic human appearance. All androids are humanoid robots, but the converse is not necessarily true because not all humanoid robots are androids.

DOI: 10.1201/9781032695266-2

2.2.3 AI Robotics

Artificial intelligence (AI) robotics, or AI-powered robotics, is robotics augmented with a diversity of sensors, e.g., 2D/3D cameras, vibration sensors, proximity/position sensors, accelerometers, and other sensors. These sensors feed robots with data that they can analyze using AI algorithms and specialized AI processors. Based on their responses to the environment and the overall mission goals, the robots make the requisite inferences. They implement appropriate real-time actions from the inferences at par with human capabilities (Lu and Xu 2017; Murphy 2019).

2.2.4 AI Robots

AI-enabled robots are robots embellished with AI capabilities to act on their own from gathered information provided by their sensors and their analysis using machine learning techniques (Govers III 2018, 2024).

Table 2.1 brings out a comparison between AI and robotics.

Table 2.2 shows the enhanced capabilities of AI-powered robotics with respect to simple robotics.

TABLE 2.1
AI and Robotics

Sl. No.	Point of Comparison	AI	Robotics
1	Primary field	Computer science	Electrical and Mechanical Engineering
2	Focus on software/hardware	It aims at designing intelligent software	Its objective is to develop physical robots
3	Examples	Data analysis, machine learning, and deep learning	Sensors, actuators, controllers, and associated electronics
4	Applications	Solving problems and taking decisions by reasoning	Automation

TABLE 2.2
Simple Robotics and AI-Powered Robotics

Sl. No.	Point of Comparison	Simple Robotics	AI-Powered Robotics
1	Scope	It is a less sophisticated field, largely devoted to mechanical motion and related actions with restricted use of information processing.	It is an advanced technology of robots enhanced with AI acting as the robot's brain, elevating the robot's status beyond that of a mechanically moving machine.

(Continued)

TABLE 2.2 (*Continued*)
Simple Robotics and AI-Powered Robotics

Sl. No.	Point of Comparison	Simple Robotics	AI-Powered Robotics
2	Focus	It aims at the electromechanical design and construction of robots, including grippers and movement tools.	It is directed toward integration of robotics with artificial intelligence algorithms to enable robots to learn, reason, and adapt to their environment.
3	Capabilities	A basic robot without AI assistance has limited information processing capability. It can handle simple tasks only, e.g., it can perform pre-programmed tasks.	An AI robot has a more extensive information processing capability. It can handle complex tasks autonomously beyond simple programmed activities. For example, it can analyze data, make decisions in real time, and adjust its actions based on changing situations.
4	Examples	A simple robotic arm used for repetitive assembly tasks in manufacturing.	A self-driving car utilizing computer vision and path planning algorithms to navigate complex environments.

2.3 GENERATIONS OF ROBOTICS

Robotics is divided into five generations (Perera 2022), characterized by the evolution of capabilities of robots from simple mechanical arms making precise, high-speed movements in industrial manufacturing to intelligent, autonomous machines equipped with AI. The robots collaborate and co-exist with humans, augmenting their capabilities and helping in day-to-day activities. Figure 2.1 shows the five generations of robotics: first generation: manipulator robots, second generation: learning robots, third generation: reprogrammable robots, fourth generation: mobile robots, and fifth generation (ongoing): AI robots bestowed with advanced AI.

2.4 PARTS OF AN AI ROBOT

The AI robot has a physical structure, or body, containing many parts that enable it to perform various operations. Figure 2.2 shows the components of an AI robot: central processing unit (CPU), graphical processing unit (GPU), or other processor for running AI algorithms; camera, LiDAR for vision; microphone for listening to sounds and speaker for talking; end effectors, sensors, fingers, robotic arms with actuators (electric motors, hydraulic/pneumatic devices) for manual tasks; controller and interfaces (Wi-Fi, Bluetooth) for information processing and communication; and power source for energy. The details of the parts are given below:

  i. Sensors: These are akin to human sensory organs to perceive the environment, navigate without colliding against obstacles, and perform various other chores.

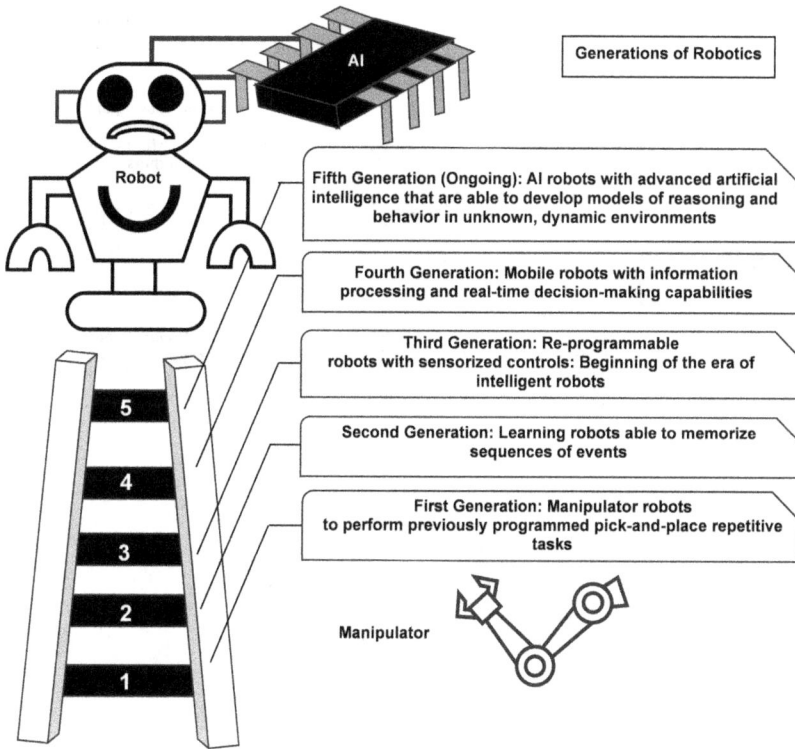

FIGURE 2.1 The lineage of AI robotics.

ii. Actuators: These are electric motors, pneumatic, and hydraulic devices that convert stored energy into mechanical work to move the robot and its arms and carry out heavy-duty work.

iii. Robotic Arms or Manipulators: They are identical to human shoulders, elbows, and wrists, with joints for easy movements

iv. End Effectors: These are tools attached to the robot's wrist that allow it to grip objects or perform painting or welding jobs.

v. Controllers: They perform analog-to-digital and digital-to-analog conversion, PID (proportional-integral-derivative) control, an extensively deployed feedback control mechanism in industrial automation; robot trajectory interpolation, temperature regulation, etc.

vi. AI Processors: These are integrated circuits acting as the brain of the robot (Liu and Law 2021; Kim and Deka 2021). They are designed to handle the mathematical operations necessary to execute AI, machine learning, and deep learning algorithms of AI robotics. They achieve the extraordinarily high speed and efficiency of completing more computations per unit of energy consumed by incorporating huge numbers of smaller and smaller transistors, which run faster and consume less energy than larger transistors. Unlike the traditional chips, they also have AI-focused design features to

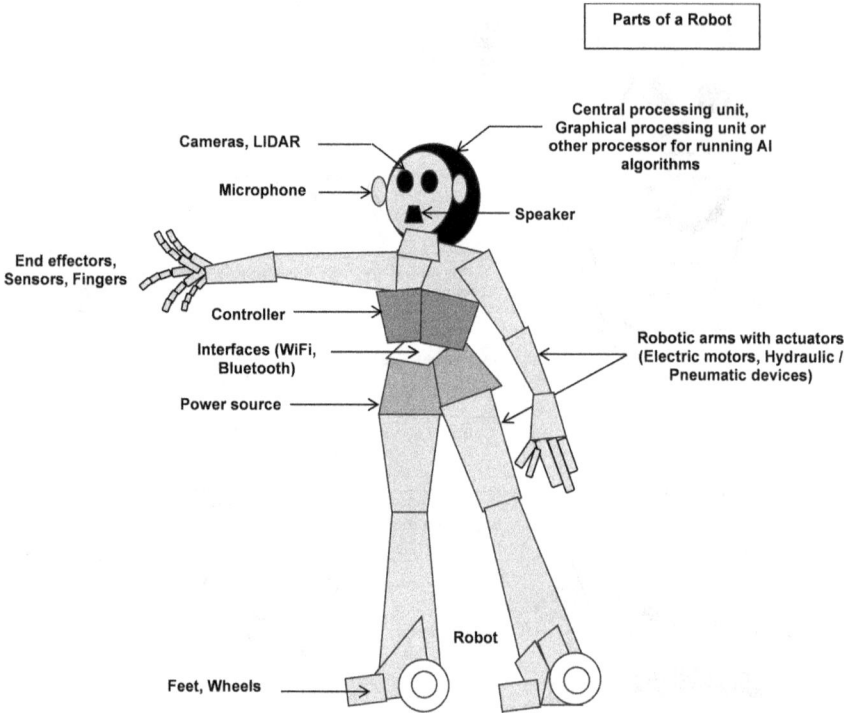

FIGURE 2.2 The components of a typical robot.

dramatically accelerate the identical, predictable, independent calculations required by AI algorithms. These features include executing a large number of calculations in parallel rather than sequentially, calculating numbers with low precision in a way that successfully implements AI algorithms but reduces the number of transistors needed for the same calculation, speeding up memory access and using programming languages built specifically to efficiently translate AI computer code for execution on an AI chip.

2.5 AI PROCESSOR CHIPS FOR ROBOTICS

Figure 2.3 shows some of the processors (Mishra et al. 2023; Gover 2025) that are either used presently or hold promises of being used in future robotics: the CPU, the GPU, the tensor processing unit (TPU), vision processing unit (VPU), neural processing unit (NPU), the associative-in-memory processor, the graph analytics processor, and the quantum processing unit (QPU). Their salient features are described as follows.

2.5.1 CENTRAL PROCESSING UNIT

The CPU, a general-purpose processor based on the von Neumann architecture, is the main component of a computer. It is responsible for processing data, executing instructions, and controlling all its operations. Owing to its flexibility, resilience, and

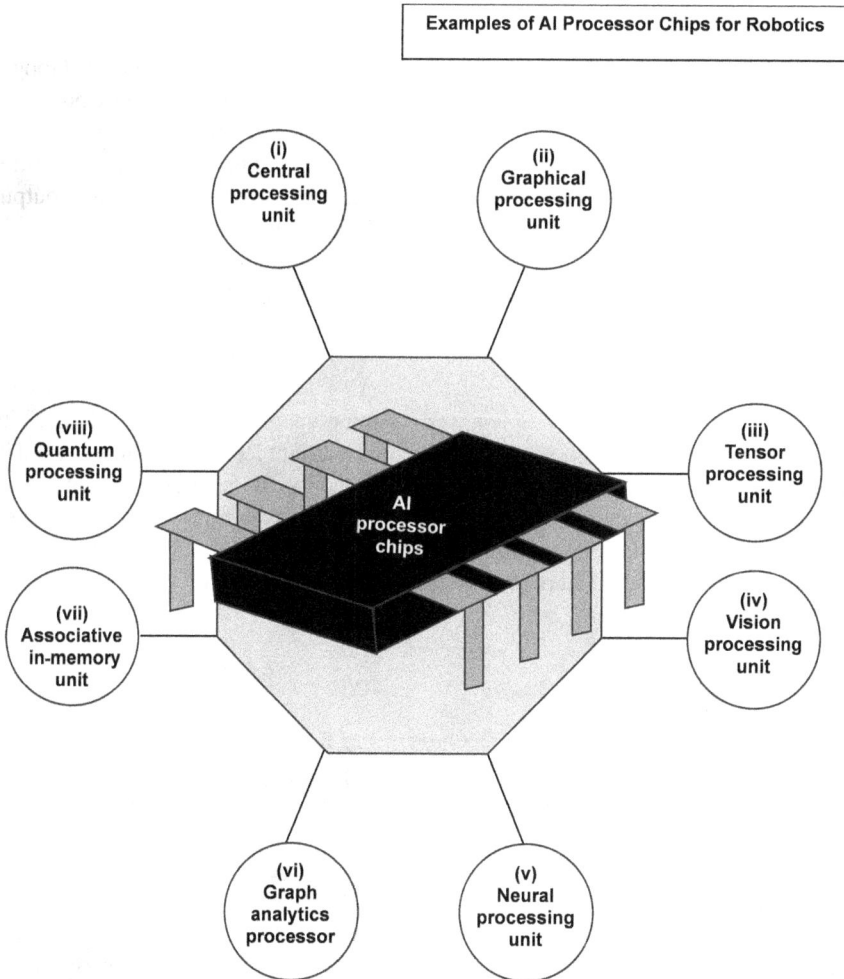

Examples of AI Processor Chips for Robotics

(i) Central processing unit

(ii) Graphical processing unit

(viii) Quantum processing unit

(iii) Tensor processing unit

AI processor chips

(vii) Associative in-memory unit

(iv) Vision processing unit

(vi) Graph analytics processor

(v) Neural processing unit

FIGURE 2.3 Computer chips used in AI robots.

adaptability to a variety of computing situations, the CPU is utilized for tasks ranging from simple to highly complicated.

2.5.2 GRAPHICAL PROCESSING UNIT

The GPU, a processor to handle rendering 3D graphics and pictures faster than a traditional CPU, is specially designed with massive parallelism and enhanced programmable capabilities. It can process many pieces of data simultaneously by incorporating thousands of Arithmetic Logic Units in a single chip to provide improved support for neural network operations, such as matrix multiplication. These qualities make it a popular processor architecture in deep learning. The parallel structure of GPUs is well suited for algorithms that process large blocks of data in AI workloads.

2.5.3 Tensor Processing Unit

The TPU is an application-specific integrated circuit (ASIC) designed by Google for neural networks as a specialized processor for a high volume of low-precision computation in neural network workloads by connecting a large number of multipliers and adders directly to form a systolic array architecture. In this architecture, several operations are performed with a single memory access by using the output of one structural unit as the input of the next. These improvements enable a drastic reduction of the von Neumann bottleneck. Its matrix multiply unit and proprietary interconnect topology make it ideal for accelerating AI training and inference.

2.5.4 Vision Processing Unit

The VPU is an AI accelerator for running machine vision algorithms such as CNN (convolutional neural networks) and SIFT (scale-invariant feature transform), used in computer vision, image recognition, and object detection. It includes direct interfaces to receive data from cameras. It achieves a balance between power efficiency and computing performance by coupling highly parallel programmable computations with workload-specific AI hardware acceleration in an architecture that minimizes data movement.

2.5.5 Neural Processing Unit

The NPU imitates the function of the human brain by using artificial neurons and synapses that mimic the activity spikes and the learning process of the brain. It is used for various applications that require smarter and more energy-efficient computing, such as image processing, and face and speech recognition in robotics.

2.5.6 Associative-in-Memory Unit

The processor-in-memory (PIM) or compute-in-memory (CIM) or associative-in-memory processor (AiMP)/associative processor is a non-von Neumann architecture consisting of a single computer chip integrating a processor with RAM (random access memory). It allows data to be processed directly in memory instead of being stored on the disk. This strategy of using associative memory cells for data storage as well as processing speeds up processing times. It is able to do so by eliminating the need to transfer data between the processor and the memory. A resistive memory implementation uses a resistive crossbar with peripheral circuitry. The associative processor was invented in the 1960s but was almost forgotten and cast aside until recently when advancements in big data data created a resurgence of interest in this technology.

2.5.7 Graph Analytics Processor

The graph analytics processor leverages a parallel processing architecture with multiple cores or processing units connected via high bandwidth inter-core or inter-unit communication. It operates under specialized instructions, data structures, and indexing techniques. All these are tailored for graph algorithms that are executed by storing in a sparse matrix format to conserve memory space.

2.5.8 QUANTUM PROCESSING UNIT

The QPU, the central component of a quantum computer processor, contains a number of interconnected quantum bits or qubits. The qubits are manipulated to compute quantum algorithms using the unique characteristics of particles, such as electrons or photons. The QPU works on properties like superposition, the ability of a particle to exist in many states at the same time. It performs specific types of calculations much faster than the processors in today's computers called classical computers.

2.6 CLASSIFICATION OF ROBOTS

Robots are classified into myriad categories (Guizzo 2023) of which we name a few leading ones below.

2.6.1 CLASSIFICATION BY SIZE

Robots are distinguished into three types by looking at their dimensions. Figure 2.4a shows the classification of robots by size into categories named as nano-, micro-, and macro-robot categories. Nanorobots are extremely small (nanometers). Precise manipulation at the cellular level is the main aim of designing such robots. Currently, they are mostly found in research stages. Microrobots are smaller than visible to the naked eye, typically in the micrometer range. They are used for targeted drug delivery or microsurgery. Macro-robots are visible to the naked eye, ranging in size from millimeters and above. Most industrial, service, and humanoid robots fall into the macro segment. These are the traditional robots that we are accustomed to seeing.

2.6.2 CLASSIFICATION BY THE TYPE OF CONTROL SYSTEM USED

Figure 2.4b shows the classification of robots by control system into three groups, namely, non-servo, servo, and servo-controlled categories. Non-servo robots mean robots showing simple movement with limited control and no feedback mechanism to monitor position. They are often used in applications where precise positioning is non-critical, e.g., a robotic arm with only on/off switches for movement. A servo robot has the capability of precise control over its position and movement. It utilizes a feedback loop to monitor its current position. It can be programmed to move to specific locations, e.g., a robotic arm with servo motors that can securely grasp an object at a specific location. A servo-controlled robot is the same as a servo robot, but emphasizes the active control aspect. It actively adjusts motor output based on feedback to maintain the desired position.

2.6.3 CLASSIFICATION BY MOBILITIES

Robots are divided into four categories in accordance with their movement capabilities. Figure 2.4c shows the classification of robots by their movement abilities

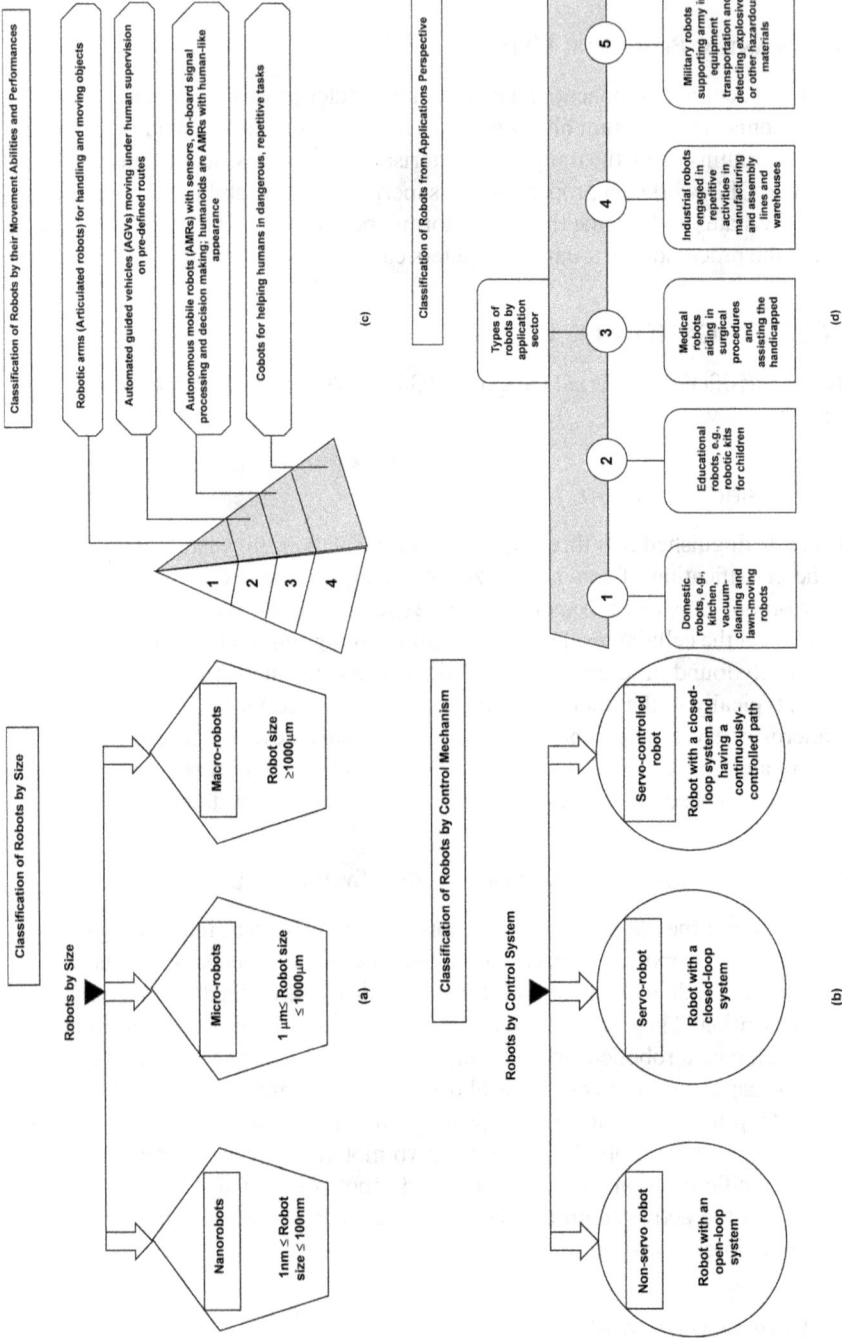

Classification of Robots by their Movement Abilities and Performances

- Robotic arms (Articulated robots) for handling and moving objects
- Automated guided vehicles (AGVs) moving under human supervision on pre-defined routes
- Autonomous mobile robots (AMRs) with sensors, on-board signal processing and decision making; humanoids are AMRs with human-like appearance
- Cobots for helping humans in dangerous, repetitive tasks

(c)

Classification of Robots from Applications Perspective

Types of robots by application sector

| 1 | 2 | 3 | 4 | 5 |

1. Domestic robots, e.g., kitchen, vacuum-cleaning and lawn-moving robots
2. Educational robots, e.g., robotic kits for children
3. Medical robots aiding in surgical procedures and assisting the handicapped
4. Industrial robots engaged in repetitive activities in manufacturing and assembly lines and warehouses
5. Military robots supporting army in equipment transportation and detecting explosives or other hazardous materials

(d)

Classification of Robots by Size

Robots by Size

- Nanorobots — 1nm ≤ Robot size ≤ 100nm
- Micro-robots — $1\,\mu m \le$ Robot size $\le 1000\mu m$
- Macro-robots — Robot size $\ge 1000\mu m$

(a)

Classification of Robots by Control Mechanism

Robots by Control System

- Non-servo robot — Robot with an open-loop system
- Servo-robot — Robot with a closed-loop system
- Servo-controlled robot — Robot with a closed-loop system and having a continuously-controlled path

(b)

FIGURE 2.4 Placement of robots into different classes depending on chosen features, working style, and application sector: (a) by size, (b) by control system, (c) by mobility, and (d) from an applications viewpoint.

into groups labeled as robotic arms, automated guided vehicles, autonomous mobile robots, and cobots. Robotic arms help with tasks like handling materials and products in factories and warehouses, their assembly, and transportation. Automated guided vehicles move on fixed paths, which are often marked on the floor with wires, magnetic strips, or lasers. They are preferred for carrying out repetitive tasks like moving materials or equipment between warehouse locations or factory locations. They reduce mistakes and accidents because they are programmed to execute their job precisely. Cobots, or collaborative robots, are usually industrial robots that can safely work alongside humans in a shared workspace.

2.6.4 CLASSIFICATION FROM APPLICATIONS VIEWPOINT

Figure 2.4d shows the subdivision of robots from an applications perspective into five categories according to the sector of their intended use. These are designated as domestic, educational, medical, industrial, and military robots. Domestic robots are primarily used for household chores, e.g., vacuum cleaners, floor washers, and ironing robots. Educational robots are used in classrooms to teach robotics, computer programming, science, technology, engineering, and mathematics. Medical robots are utilized in healthcare settings for patient care, disinfection jobs, rehabilitation, and prosthetics. They are also used in performing critical surgeries, including orthopedics and cardiac surgery. Military robots aid in defense tasks, including reconnaissance and surveillance, logistics, service and rescue, bomb disposal, and combat. Industrial robots are used in manufacturing and production lines. They can move on multiple axes and perform tasks like welding, pick-and-place jobs, and packaging.

AN EDUCATIONAL ROBOT

I am an educational robot
I work in a school.
I follow discipline and obey all rules
I teach the students physics and chemistry
I teach biology and plant trees
Sometimes I teach mathematics too
In the games period, I wear my sports shoes
And take the students to the playground
Where we play and run around
Children eat cakes and buns
And have lots of fun.

A MEDICAL ROBOT

I am a medical robot
I work in a hospital
I have a robot identity card and label
I sit with children in the nursery
Sometimes I assist doctors in microsurgery
I am delicate and do not injure anyone
I feel happy when the surgical operation is successfully done.

AN INDUSTRIAL ROBOT

I am an industrial robot
I work in a factory
I draw heavy current from mains, no battery
I am very strong and can lift tons of weight
And load it on a heavy metal plate
I can put my hands in the furnace
And pull out the red-hot iron
Do not dare to copy me
Your hands will burn!

2.7 ROBOPHYSICS

Robophysics is an emerging scientific field that pursues the study of the movements of robots in real-world environments (Calderone 2016; Collins et al. 2021; Li et al. 2023). The investigation is done by examination of the principles of self-generated motion in mobile systems, and application of physics methods for exploration of locomotion in laboratory devices. Essentially, it avails the services of physics to enhance robot movement and behavior in contrasting environments. Figure 2.5 shows typical examples of physics underlying robotics, viz., robot motion analysis (kinematics and dynamics), understanding of friction and contact mechanics for making grippers, environmental interaction scrutiny, physics-based simulation of robot behavior and associated experimentation, and application of physical principles of mechanics, heat, thermodynamics, optics, electricity, magnetism, and electromagnetics.

2.7.1 ROBOPHYSICS VS. BIOPHYSICS

Robophysics bears analogy to the familiar discipline of biophysics in many ways. Both fields are concerned with applying physics principles to complex systems. Robots are considered more controlled and carefully designed systems, whereas the biological organisms have an intricate and dynamic nature.

In robophysics, systems are largely mechanical or electromechanical in nature. In biophysics, they are biological systems, which are biomolecules and cellular materials. Robophysics focuses on robot motions, while biophysics tries to understand the physical phenomena taking place in biological organisms.

Clarifying further, robophysics is the physics of artificial movements of man-made robots. It analyzes the mechanics of robots and their environmental interactions in an attempt to improve robotic design and locomotion. Biophysics is the physics of natural movements in living beings. It addresses the mechanics of biological matter, e.g., membranes, muscles, and neurons. In biophysics, the attention is mainly directed toward understanding biological processes at the molecular level for developing medical treatments.

An example aiding in visualization of robophysics is studying how a well-designed truck carrying a heavy load moves through different terrains, such as smooth highways or rough, rocky, and sandy regions. Contemplating biophysics is like studying how a deer runs on grassy woodland.

Physics behind Robotics: A few examples of physics underlying robotics, playing a foundational role in the design and development of robots, and their operation in a controllable and predictable manner

Application of physical principles of mechanics, heat, thermodynamics, optics, electricity, magnetism and electromagnetics for fabricating reliable sensors and actuators for robots

Environmental interaction studies like fluid dynamics for robots to navigate through liquids, and collision avoidance with obstacles

Robot motion analysis (kinematics and dynamics) by applying Newton's laws of motion to calculate forces/torques for robot motion and acceleration, designing robot joints, and maintaining its equilibrium for stability and balancing

(i)   (ii)   (iii)   (iv)   (v)

Robotics

Physics-based simulation of robot behavior and associated experimentation for refinement, optimization and evaluation of designed robot prototypes

Understanding of friction and contact mechanics for making grippers for object grasping and manipulation by robots

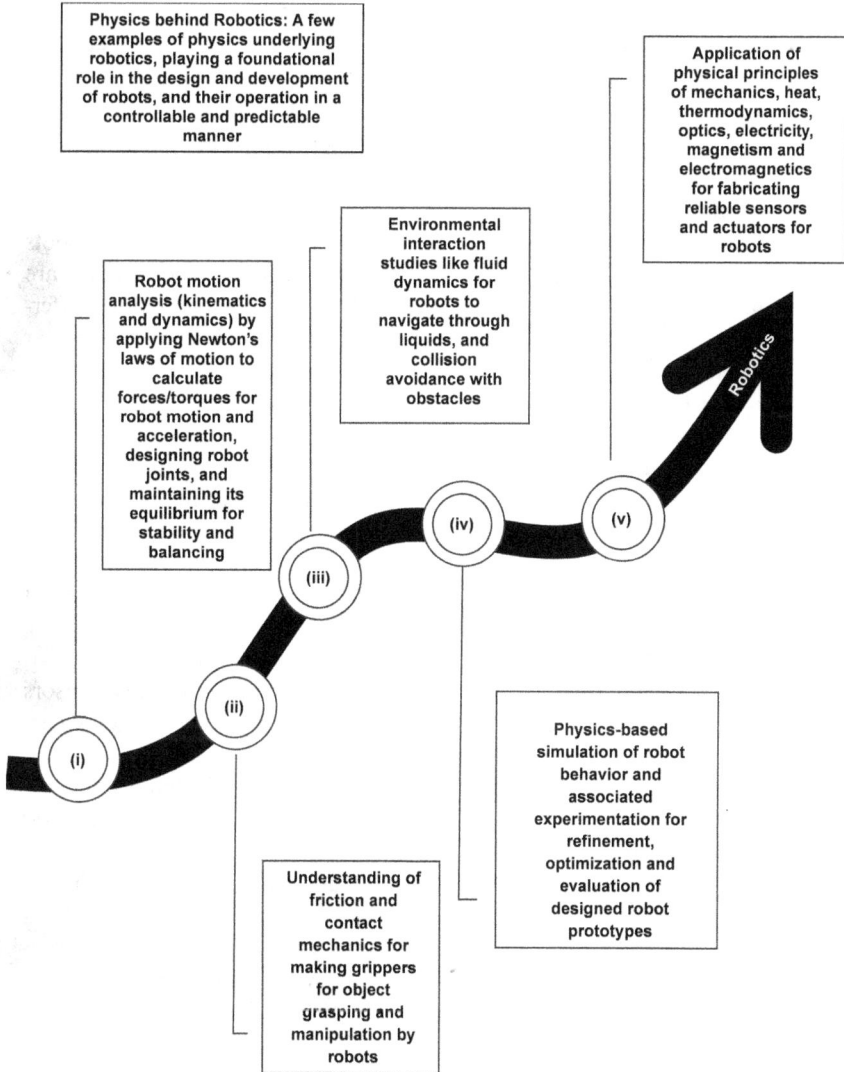

FIGURE 2.5   Concepts of robophysics.

2.7.2   PRINCIPLES OF ROBOPHYSICS STUDIES

The fundamental principles of robophysics studies are as follows:

i. Applying Physics Methods: A prime aspect of robophysics studies is that techniques and approaches borrowed from physics are largely used to inquire into locomotion in laboratory devices. Generally, specialized equipment is used to analyze and measure the movement patterns of robots. The findings of robophysics shed light on various aspects of motor control or gait analysis of robots in different conditions.

ii. Exploring the Principles of Locomotion: In robophysics, simplified models of robots are used to explore the principles of their locomotion and validate hypotheses. Principles of robot locomotion include the following:

    a. Wheeled Locomotion: It is the most popular locomotion mechanism in man-made vehicles that finds widespread utilization in mobile robotics. A sufficient power efficiency is achieved even at high speed. Stability is not an issue, as in legged locomotion.

    b. Legged Locomotion: Legged motion is known as gait. Legged robots move by lifting and stepping each leg in sequence. These robots are more versatile than wheeled robots. They can traverse many different terrains. Their main features are increased complexity and power consumption.

    c. Snake-Like Locomotion: Snake-like robots are very effective in confined, narrow, and irregular environments.

    d. Optimal Behavior: The optimal behavior for bipedal locomotion on two legs is slow-speed walking and high-speed running. The same for quadrupedal robotic locomotion using four limbs is slow-speed walking, intermediate-speed trotting, and high-speed galloping.

    e. Control and Sensing: The internal state and configuration of the robot are measured by proprioceptive sensors (accelerometers, gyroscopes, and optical encoders), while the information about the external environment and contact interactions is gathered by exteroceptive sensors (vision, tactile, ultrasonic, and temperature). Proprioception, or kinesthesia, represents a body's ability to sense its position and movement in space for its balance, coordination, and motor control. Exteroception is the awareness of external stimuli to perceive and interact with the world around a body.

iii. Performing Systematic Experimentation and Integrating Experiment, Theory, and Computation: Robophysics systematically integrates experiment, theory, and computation. Its studies rely on well-planned and organized experimental investigations coupled with theory and modeling. Real-world experiments conducted in parallel with theoretical formulation and analysis, and strongly supported by computational modeling, provide an in-depth understanding of a phenomenon. The experimental data verify the theoretical model and modify it, if necessary. Further experiments are designed based on the feedback received from modeling. New insights are obtained by the interpretation of results. Different robot designs and control strategies are tested for evaluation in virtual environments. Computational modeling and simulations are done before building physical prototypes of efficient and stable legged robots.

iv. Using Simplified Robotic Devices: Robophysics studies use simplified robotic devices in controlled laboratory settings to complement the study of complex robots in complicated situations. A simplified robotic device is a basic, single-function robot arm. This device displays movement capabilities within a restricted range. Precise tasks handled by it are picking up specific items, dispensing liquids, and transferring small samples between containers. All these tasks are performed within a designated area with controlled parameters under minimal environmental variation. Complex robots in complicated situations refer to robotic systems designed to operate in environments with many variables and uncertainties. They require advanced capabilities in the form of sophisticated perception, decision-making, and adaptation capabilities to navigate and perform tasks successfully. They often include scenarios with multiple moving parts, unpredictable interactions, and dynamic environments. Examples of such robots are the surgical robots performing intricate procedures in hospitals, and autonomous vehicles navigating busy city streets. Humanoid robots interacting socially in a crowded space too fall in this class of robots. So also, the swarm robots working in close coordination and mutual cooperation to explore a vast, unknown terrain. Essentially, all these robots work beyond simple repetitive tasks. They are able to handle complex real-world scenarios with high levels of autonomy.

v. Parameter Space Exploration: Robophysics studies use parameter space exploration, systematic control, and techniques from dynamical systems to observe locomotor successes and failures. The parameter space is the set of all possible values for the parameters that are specified to define a mathematical model. It is also called weight space. Parameter space exploration is the process of analyzing the patterns of changes in the dynamics of a system with variations in its parameters.

vi. Interaction with Soft Materials: Robophysics studies discover principles of interaction of active or programmable objects with soft materials like mud, sand, grass, and litter. Soft materials comprise the stretchable elastomers and textiles that are pasted over the skin of a robot without interfering with its movement. Ferromagnetic soft materials self-actuate in response to magnetic fields. This property makes them remotely controllable and compatible with biological tissues. Silk fibroin sheets with Ag nanowires are used in highly sensitive stretchable capacitive sensors for low-pressure detection.

2.7.3 SIGNIFICANCE OF ROBOPHYSICS

The significance of robophysics is multifaceted:

i. Bridging Disciplines and Improving Robot Design: The significance of robophysics is that it acts as a bridge connecting robotics technology with the theoretical framework of physics. The merger of the fields of robotics and physics allows engineers to leverage insights from physics to solve complex robotic challenges. A deep understanding of the physical principles behind robot movement in different terrains and situations is essential. It becomes more relevant particularly in reference to complex environments such as soft materials or non-uniform surfaces. The reason is that it allows robot dynamics, their locomotion, and interactions with their surroundings to be analyzed through a more fundamental perspective. It helps researchers in designing more efficient, adaptable, and complex robots to work in challenging environments. Researchers analyze the physics of robot locomotion to identify the critical design parameters that optimize movement, stability, and energy efficiency of robot motion in various terrains. Applying physics concepts like dynamics, mechanics, and control theory, innovative designs and control strategies are developed for robots. These designs are not readily apparent through purely engineering approaches. Robophysics inspires new areas of research within physics itself by surveying the dynamics of novel robotic designs, and opens doors to investigate novel concepts. An example is the emergent behavior in robotic systems, where complex behaviors originate from interactions between individual robot components.

ii. Understanding Complex Interactions: Robophysics helps in studying the interaction of robots with their environments. These include factors such as friction, terrain variations, and contact forces. Their understanding leads to the development of better control algorithms for robot movements in varied circumstances.

iii. Biomimicry Inspiration: The physics of locomotion in animals and other natural systems is studied. Robophysics fosters the design of robots that mimic biological movements. On this basis, more versatile robots with efficient and adaptable movement patterns are fabricated.

iv. Exploration of New Physics Questions: Investigating the dynamics of novel robotic systems often leads to the discovery of new physical phenomena. It also gives impetus to theoretical advancements in several areas, such as soft matter physics and nonlinear dynamics.

2.7.4 APPLICATIONS OF ROBOPHYSICS

Robophysics finds multifarious applications. The prominent among them are as follows:

i. Designing Legged Robots for Rough Terrains: Robophysics helps in analyzing the physics of leg movements on uneven surfaces. The analysis deals with how animals walk, run, and jump. It helps to optimize gait patterns for stability and efficient locomotion of robots. Applying this knowledge, efficient and stable legged robots are realized.

ii. Developing Soft Robots: Concepts of soft matter physics are applied to design robots with flexible bodies. Such robots can adapt to complex environments. Using the physics of soft materials, robots with compliant bodies are created that can adapt to complex environments.

iii. Making Swimming Robots: Notions of fluid dynamics are used to design underwater robots that exhibit flexible swimming motions.

iv. Simulating Robot Behavior in Complex Environments: Mathematical simulations are performed on computational models to study the interaction of robots with various terrains and obstacles. Physical prototyping of robots is done by utilizing the advisories derived from the simulations.

2.8   ROBOETHICS

Roboethics or robot ethics is an interdisciplinary field at the intersection of robotics, computer science, psychology, and philosophy. In this field, the ethical, social, humanitarian, and ecological aspects of robotics are deliberated upon (Torresen 2018; Bartneck et al. 2021). It is treated as an extension of machine ethics. It is essentially a subfield of ethics of technology, specifically information technology that is concerned with the ethics of human behavior toward advanced robots. Particularly, it discusses the legal and socio-economic concerns about robotics posing a threat to humans in the long or short run. It aims to ensure that robots are morally designed and used for the benefit of humanity. The safety of the human race must always be kept at highest priority, e.g., momentous ethical issues arise in social assistive robotics (Boada et al. 2021). Ethical implications of integration of AI in robotics and healthcare demand scrupulous consideration (Elendu et al. 2023).

2.8.1   ISAAC ASIMOV'S LAWS OF ROBOTICS

Seeking to create an ethical system for humans and robots, the science fiction author Isaac Asimov devised the laws to be followed by robots. He proposed these laws in his stories in anticipation of the likely nuisance of developing intelligent robots, and the consequent technical and social problems (Figure 2.6). Despite the fact that they are not scientific laws, they have received wide attention and recognition. This is because they provide ethical guidelines to robots preventing them from malfunctioning in a dangerous manner (Asimov 1942):

Zeroth Law: The robot can neither inflict any harm on mankind nor by their inaction allow mankind to come to harm.

    The law underscores the significance of welfare of humanity as a whole over that of an individual human being.

First Law: The robot must neither injure a human being nor by inaction allow a human being to come to harm.

    The primary directive of this law is that a robot must never harm a human being, either deliberately or unintentionally. The secondary directive of this law is that a robot cannot stand by watching carelessly and allowing harm to befall a human being if it is capable of preventive intervention. Figure 2.6a illustrates laws 0 and 1 in the form of robot's friendship with and protection for humans.

Second Law: The robot must follow the orders given to it by human beings as long as the orders do not conflict with the first law.

    The robots are primarily designed to follow humans and execute their orders in order to ensure human safety by preventing humans from being harmed by the actions of robots. The underlying connotation of this law is that the robots are intended to be tools for human assistance. They operate by following human orders and working under their supervision and control. Figure 2.6b illustrates law II by showing a robot obeying the orders of the human operator.

Third Law: The robot must protect its own existence unless the protection does not conflict with first or second law.

The law implies that the robot must avoid any actions or situations that could cause it to harm itself in any way. Figure 2.6c illustrates law III by showing a robot fighting for defending its existence.

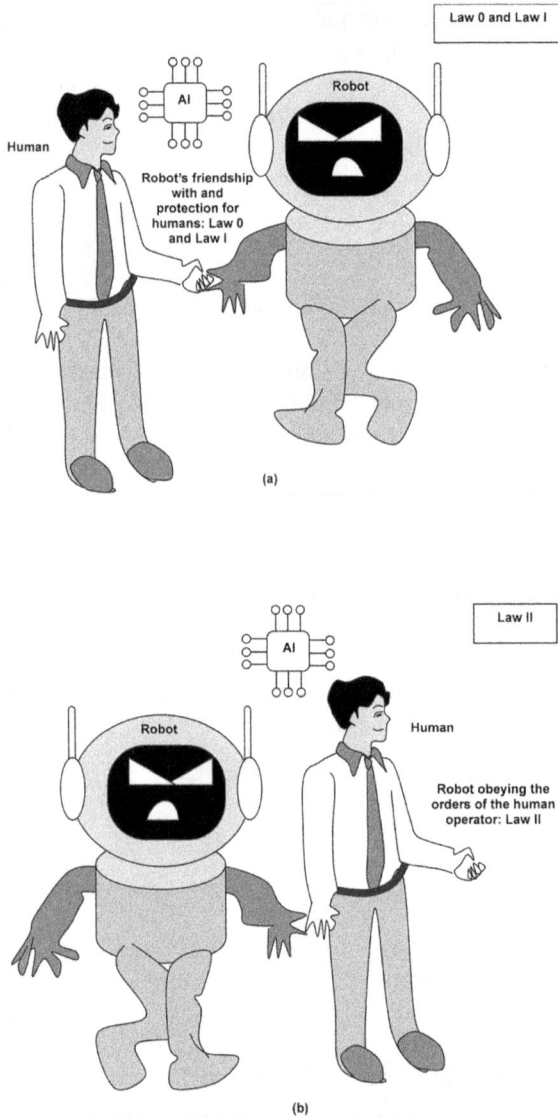

FIGURE 2.6 Visualization of laws of robotics propounded by Isaac Asimov: (a) law 0 and law I, (b) law II.

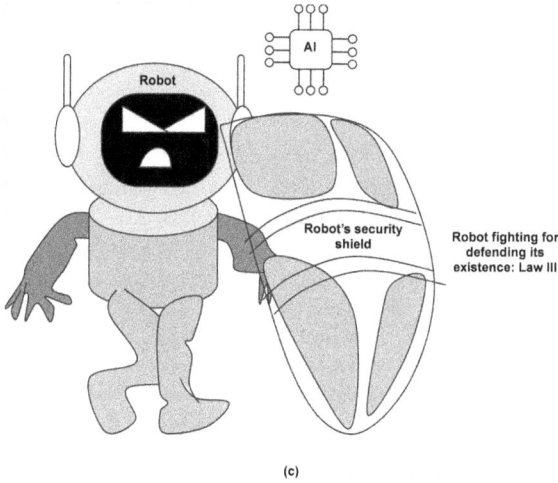

FIGURE 2.6 *(CONTINUED)*   (c) law III.

2.8.2   Order of Prioritizing Obeyance of the Laws

The zeroth law takes precedence over the other three laws of robotics. It mandates robots to prioritize humanity as a whole over any individual. This law allows robots to override human commands if they can inflict long-term harm to humanity.

When the remaining three laws conflict, the first law takes precedence. Then the second law takes precedence. Finally, the third law of self-preservation and the safety of the robot is pursued. Suppose a human being orders a robot to attack another human being. Then the robot will not follow the order because the first law takes precedence over the second law. Notwithstanding, if a human being orders a robot to disassemble itself, the robot will obey the order. This happens because the second law takes precedence over the third law.

The aims and scope of AI ethics and roboethics are expounded in Table 2.3.

2.9   UNDERSTANDING THE INTERRELATIONSHIP AMONG ROBOTICS, AI ROBOTICS, ROBOPHYSICS, AND ROBOETHICS

2.9.1   Breakdown into Subdomains

Robotics is a broad field. It encompasses the design, construction, operation, and application of robots. The necessary mechanical systems, electrical components, sensors, actuators, and control algorithms fall under robotics. AI robotics is a specialized subfield within robotics in which robots are empowered by AI techniques.

Robophysics is a specialized area. It is placed within robotics. It focuses on applying physical laws like mechanics, dynamics, and control theory to understand and optimize robot movement, manipulation, and interaction with the environment.

Roboethics is an interdisciplinary field. It examines the ethical implications of robotic technology. The examination includes questions about robot autonomy,

TABLE 2.3
AI Ethics and Roboethics

Sl. No.	Point of Comparison	AI Ethics	Roboethics
1	Scope	It studies the ethical implications of artificial intelligence, including its algorithms and decision-making processes.	It specifically examines the ethical concerns related to the design, development, and use of physical robots, often including how humans interact with them.
2	Focus	It can apply to physical and non-physical systems.	It places significant emphasis on the physical form of a robot.
3	Set/subset relationship	AI ethics is a broad field.	Roboethics is a subset of AI ethics that focuses on the physical embodiment of AI in robots.
4	Issues discussed	Data privacy, bias in algorithms, and transparency.	Potential harm caused by robots, robot autonomy, and the robot-human interactions.
5	Example concerns	Ensuring fairness in algorithmic decision-making, and preventing biased data from influencing AI outcomes.	The central issue of paramount importance is: Should a robot be given the ability to make independent decisions? Can it pose a danger to humans?

responsibility, safety, privacy, and their potential societal impacts as robots become more sophisticated.

2.9.2 Interrelation of Robotics with Robophysics and Roboethics

Robophysics is a scientific foundation that deals with the design and development of robots within the field of robotics and its specialized sub-branch AI robotics, while roboethics considers the ethical considerations that arise from these advancements in robotics.

2.9.3 An Example of AI Robotics, Robophysics, and Roboethics Interrelation

In robotics, we aim at designing a robotic arm for a factory assembly line. In AI robotics, we make a self-driving car. In robophysics, we analyze the forces and torques acting on a robot's joints to optimize its movement, calculating the optimal trajectory for a robot to navigate a complex environment. In roboethics, we debate whether an autonomous robot, such as an AI-powered robot, should be programmed to prioritize human life over its own, considering the potential for bias in AI decision-making systems.

2.10 DISCUSSION AND CONCLUSIONS

This chapter dealt with the basic principles of robotics, robophysics, and the philosophy of roboethics (Table 2.4). Importance of roboethics, the ethical, legal, and social facets of robotics were emphasized, describing the ways in which robots must be designed in order that they act and behave 'ethically'.

Physics plays a key role in the dynamics and kinematics of robot motion. This chapter reviewed the basics of robophysics, the study of robotic movement in complex real-world environments using the methods of physics and theoretical models (Aguilar et al. 2016). Robophysics is an emerging scientific discipline that deals with the motion of robots analogous to biophysics, which studies the motion of biological systems. It is concerned with problems at the interface of nonlinear dynamics, soft matter, control, and biology. Its objective is to examine successful and failed locomotion in simplified robotic devices to create robots that have life-like abilities.

TABLE 2.4
Looking at Significant Themes of This Chapter and the Findings

Sl. No.	Significant Themes	Explanation
1	Robotics	The meanings of common terms like 'robots', 'robotics', 'AI robots', and 'AI robotics' were explained. Chronologically, five generations of robotics are distinguishable. The primary components of an AI robot are sensors, actuators, arms or manipulators, end effectors, controllers, and AI processors. Important AI processor chips used in robotics include the central processing unit, the graphical processing unit, the tensor processing unit, the vision processing unit, the neural processing unit, the associative-in-memory unit, the graph analytics processor, and the quantum processing unit. A classification of robots was made according to size, by the type of control system used, by mobility, and from an application viewpoint. The interrelationship among robotics, AI robotics, robophysics, and roboethics was brought out by breaking them into subdomains, and illustrated with an example.
2	Robophysics	Robophysics is an emerging scientific discipline in which physics methods are applied to enhance robot movement and behavior. The analogy of robotics with biophysics is drawn. The principles of robophysics studies were described. Its significance and applications are mentioned.
3	Roboethics	Roboethics deals with the ethical, social, and humanitarian implications of robotics. Isaac Asimov's laws of robotics were enunciated, followed by an understanding of the order in which obedience to the rules is prioritized.
4	Keywords and ideas to remember	Robotics and robots, AI robotics and AI robots, AI processor chips, CPU, GPU, TPU, VPU, NPU, AiMP, graph analytics processor, QPU, robophysics, biophysics, roboethics, Isaac Asimov's laws.

Robophysics studies have become essential for robotics because the present-day autonomous robots possess limited locomotion capabilities and cannot robustly navigate in situations that require climbing on vertical surfaces, such as trees and hills, or moving on deformable surfaces like sand and mud. The 'robophysics' approach, which involves a systematic search for novel dynamical principles in robotic systems, can assist computer science and engineering, which have proven successful in less complex environments.

Anthropomorphism is concerned with ascribing human features to non-human things, and seeks to develop robots with human-like characteristics. Principles of natural phenomena must be emulated in robotics because laboratory-created robots have to work in the real-world environment, where they must have cognition, sensing, and decision-making capabilities, which living creatures have acquired over long periods of evolution.

At this stage, the reader is acquainted with the fundamental concepts of AI and robotics, and it is time to look into the working of robotic systems. At the core of any robotic system is a combination of two key components: sensors and actuators. The wide-ranging robotic applications permeating and spanning from industrial manufacturing automation to prosthetic systems function with a high degree of autonomy through the harmonious integration of these components in robots. Chapter 3 will provide a brief description of the main sensors and actuators used in robotics.

REFERENCES AND FURTHER READING

Aguilar J., T. Zhang, F. Qian, M. Kingsbury, B. McInroe, N. Mazouchova, C. Li, R. Maladen, C. Gong, M. Travers, R. L. Hatton, H. Choset, P. B. Umbanhowar and D. I. Goldman. 2016. A review on locomotion robophysics: The study of movement at the intersection of robotics, soft matter and dynamical system, *Reports on Progress in Physics*, Vol. 79, 110001, pp. 1–35.

Asimov I. 1942. *Runaround, Astounding Science Fiction*. March 1942, Vol. XXIX, No. 1, John W. Campbell, Jr. (Eds), pp. 94–103.

Bartneck C., C. Lütge, A. Wagner and S. Welsh. 2021. *An Introduction to Ethics in Robotics and AI*, Springer, Cham, 117 pages.

Boada J. P., B. R. Maestre and C. T. Genís. 2021. The ethical issues of social assistive robotics: A critical literature review, *Technology in Society*, Vol. 67, 101726, pp. 1–13.

Calderone L. 2016. Robophysics: Robots in motion, Robotics Tomorrow, https://www.roboticstomorrow.com/article/2016/09/robophysics-robots-in-motion/8788

Collins J., S. Chand, A. Vanderkop and D. Howard, 2021. A Review of physics simulators for robotic applications, *IEEE Access*, Vol. 9, pp. 51416–51431.

Craig J. J. 2022. *Introduction to Robotics: Mechanics and Control*, 4th Edition, Pearson Education Limited, Harlow, 448 pages.

Elendu C., D. C. Amaechi, T. C. Elendu, K. A. Jingwa, O. K. Okoye, M. John Okah, J. A. Ladele, A. H. Farah and H. A. Alimi. 2023. Ethical implications of AI and robotics in healthcare: A review, *Medicine (Baltimore)*, Vol. 102, 50, pp. 1–7.

Gover E., Updated by M. Urwin. 2025. AI chips: What are they? https://builtin.com/articles/ai-chip#:~:text=dynamic%20traffic%20conditions.-,Robotics,to%20humanoid%20robots%20providing%20companionship

Govers III F. X. 2018., 2024. *Artificial Intelligence for Robotics: Build Intelligent Robots that Perform Human Tasks Using AI Techniques, Artificial Intelligence for Robotics: Build Intelligent Robots Using ROS 2, Python, OpenCV, and AI/ML Techniques for Real-World Tasks*, Packt Publishing, Birmingham, 344 pages, 551 pages.

Guizzo E. 2023. (Updated) Types of robots: Categories frequently used to classify robots, https://robotsguide.com/learn/types-of-robots

Kim S. and G. C. Deka (Eds.). 2021. *Hardware Accelerator Systems for Artificial Intelligence and Machine Learning*, Imprint: Academic Press, Elsevier Science, Amsterdam, Netherlands, 416 pages.

Li S., H. N. Gynai, S. W. Tarr, E. Alicea-Muñoz, P. Laguna, G. Li and D. I. Goldman. 2023. A robophysical model of spacetime dynamics, *Scientific Reports*, Vol. 13, Article Number 21589, https://doi.org/10.1038/s41598-023-46718-4

Liu A. C.-C. and O. M. K. Law. 2021. *Artificial Intelligence Hardware Design: Challenges and Solutions*, Wiley-IEEE Press, Hoboken, NJ, 240 pages.

Lu H. and X. Xu (Eds.). 2017. *Artificial Intelligence and Robotics*, Springer, Cham, 326 pages.

Murphy R. R. 2019. *Introduction to AI Robotics*, The MIT Press, Cambridge, MA, 648 pages.

Mishra A., J. Cha, H. Park and S. Kim (Eds.). 2023. *Artificial Intelligence and Hardware Accelerators*, Springer International Publishing, New York, 364 pages.

Perera A. 2022. The 5 generations of robotics, Robotics, https://automatismosmundo.com/en/the-5-generations-of-robotics/

Torresen J. 2018. A review of future and ethical perspectives of robotics and AI, *Frontiers in Robotics and AI*, Vol. 4, Article 75, pp. 1–10.

3 Robotic Sensing and Actuation Techniques

3.1 INTRODUCTION

Robotics works by the confluence of sensing, actuation, and electronic control. This chapter dwells upon the sensors, actuators, and electronic systems that are frequently used in robotics (Dahiya et al. 2023).

3.2 SENSING AND PERCEPTION BY ROBOTS

Sensing and perception are the abilities of a robot to gather information about its surroundings using various sensors. A few examples of sensors are cameras (vision), LiDAR (light detection and ranging), vision sensor, light sensor, SONAR (sound navigation and ranging), microphones (audio sensors), accelerometers (motion), tilt sensor, tactile or touch sensors, force sensor, pressure sensor, proximity sensor, temperature sensor, global positioning system (GPS), digital magnetic compass, current and voltage sensor, and chemical sensors.

The various sensors in a robot work around the clock, all day and all night, to record data about the environment in which the robot is deployed. This is similar to the non-stop perception of environments by human beings through their senses of sight, odor, taste, touch, and hearing. The robot's hardware processes the collected information about temperature, light intensity, distance, and chemical composition, using AI algorithms to extract meaningful information like object location, degree of hotness, distance, shape, texture, and movement. The information processing allows a robot to understand and interact with its environment effectively in a controlled manner in order to make informed decisions and actions (Guo et al. 2006; Wu et al. 2022). These are the decisions made and actions taken after gathering all the useful information about a subject, considering potential benefits and risks and aligning with the goals.

3.3 ACTUATORS AND END-EFFECTORS OF ROBOTS

3.3.1 ACTUATORS

Actuators of robots are equivalent to their muscles, by which they convert their energies into mechanical motions. They are components producing a force, torque, or displacement to perform different types of actions for execution of tasks. The tasks involve handling of objects and carrying out numerous activities during the interaction of robots with their environment. A common actuator is the electric motor. A powerful precision servo motor offers a wide range of motion control mechanisms.

DOI: 10.1201/9781032695266-3

Different forms of pneumatic, hydraulic, and electric actuators are extensively used in robotics for industrial automation. They help state-of-the-art humanoid robots in rotating their joints and simulating complex natural walking. The vast variety of robotic actuators encompass alternating current (AC) and DC servo motors, stepper motors, synchronous motors, pneumatic motors, linear DC actuators, hydraulic and pneumatic cylinders, ultrasonic piezoelectric actuators, and so on.

3.3.2 END-EFFECTORS

End-effectors of robots are the peripheral devices, mechanical or electromechanical. They range from legs and wheels to arms and fingers. Various implements are attached to a robot's wrists, enabling it to interact with and manipulate its physical environment. They are broadly classified as grippers and advanced-functionality process tools. Grippers of different shapes, sizes, and configurations are used for grasping and moving objects. Examples of process tools used by robots are as follows:

  i. The welding tools used in the automotive industry,
 ii. The grinding and sanding tools for smoothing and finishing the surfaces of workpieces,
iii. The cutting tools, like blades for material removal and shaping,
 iv. The painting tools with brushes for applying consistent layers of paints and dispensers or syringes with nozzles, and
  v. The valves are for controlled liquid and adhesive flow.

Table 3.1 enlists the distinctive duties performed by sensors and actuators used in robotics.

3.4 ROBOT CAMERAS

The lens assembly of the camera focuses light onto an image sensor for the conversion of optical signals into electrical signals. The camera utilizes complementary-metal-oxide-semiconductor (CMOS) or charge-coupled device (CCD) technology. Each pixel of the sensor takes care to cover a small area of the captured scene. The analog signals are converted to the digital domain by an analog-to-digital converter for processing by the robot's computer. A robotic camera to detect the desired subject, track, and focus it is reported (Rehman et al. 2023); the camera's position is driven and controlled through movable motors.

3.4.1 TYPES OF ROBOT CAMERAS

Several types of robot cameras have been developed:

  i. 1D or Line-Scan Camera: It captures visual data along a line. Hence, it is useful for inspection of movements of objects on platforms like conveyor belts. A 1D camera is simplest and the least computationally intensive.

TABLE 3.1

Responsibilities of Robotic Sensors and Actuators

Sl. No.	Point of Comparison	Sensors	Actuators
1	Definition	Sensors are devices that detect and measure robot's environmental conditions, like detecting light, pressure, or estimating distance. The conditions are detected by converting the concerned physical parameters into electrical signals.	Actuators are components that take electrical signals from sensors and translate them into physical actions. They allow the robot to move around in its workplace and interact with its surroundings.
2	Function	As implied by the name, sensors 'sense' the environment.	In accordance with the name, actuators 'act' upon the information received from sensors.
3	Output	Sensors produce electrical signals as output.	Actuators yield physical motion as output.
4	Examples	Vision sensors (cameras), proximity sensors, force sensors, temperature sensors, ultrasonic sensors, etc.	Electric motors, e.g., servo motors; pneumatic cylinders, hydraulic actuators, etc.

ii. 2D or Area Scan Camera: This is a standard camera that gives a flat planar image with length and breadth dimensions. From a planar image, the shape of the object is easily recognized, as well as its position is located. A 2D camera is moderately complex.

iii. 3D or Depth Camera: Stereo vision or laser scanning techniques are used to reconstruct the geometry of an object in 3D. A 3D camera is highly sophisticated and requires extensive computational capabilities.

 a. Stereo Vision: Two cameras placed at a small distance apart acquire images of the object from two viewpoints. Depth information is obtained from the pixel disparities between the two images.

 b. Laser Scanning: Precise distance measurements of different points on the object from the robot are done by illumination of the object with a laser beam. Measurements of the time of return of the reflected laser beam are made from each point. An accurate 3D model of the object is built from these measurements.

3.4.2 Considerations for Robot Camera Mounting

Vital issues to be considered during the mounting of a robot camera are as follows:

i. Selecting the Type of Camera to Be Used: A decision among 1D, 2D, and 3D cameras is made depending on the application.

ii. Choosing the Location of the Robot where the Camera Is to Be Fixed: The camera is fixed either on the wrist or the forearm of the robot. The camera is sometimes attached to the base of the robot or even on a separate dedicated fixture. The choice of mounting site is dictated by the desired field of view.

iii. Picking the Appropriate Lens: This is determined by the field-of-view requirement.

iv. Calibration Procedure to Be Followed: It depends on the control and vision system needs.

3.4.3 TYPICAL MOUNTING CONFIGURATIONS OF ROBOT CAMERAS

There are two principal configurations in which the robot cameras are mounted, namely:

i. Eye-in-Hand: In this configuration, the camera is fitted directly to the end-effector of the robot. The camera moves with the end-effector providing real-time feedback for achieving accuracy in crucial grasping and controlling jobs, e.g., pick-and-place operations. However, the perspective of the camera continuously changes in this configuration causing difficulty in camera calibration.

ii. Eye-to-Hand: In this arrangement, the camera is fixed and stationary. It watches the workspace and actions of the robot from its fixed viewpoint. The advantage gained by this positioning of camera is the resulting stable angle of view. Although the vision processing becomes relatively simple, extra calculations need to be done to find the position of the object with respect to the arm of the robot. A robot engaged in object recognition or navigation tasks is greatly benefited by the broader field of view offered by such camera fixation.

3.4.4 APPLICATIONS OF ROBOT CAMERAS

Among the many applications of robot cameras, the following are the most common:

i. Industrial Pick-and-Place Robots: These camera-equipped robots work on assembly lines.

ii. Warehouse Product Identification/Sorting Robots: These camera-wearing robots make warehouse tasks easier.

iii. Medical Surgery-Assistance Robots: The visual feedback provided by cameras to these robots is very helpful to doctors in performing minimally invasive surgeries.

iv. Robots Driving Autonomous Vehicles: The robots installed on these robots detect pedestrians, read traffic signs, and lane markings. So, they are able to guide the vehicle on a safe journey.

3.5 ROBOTIC LIDAR SENSOR

The robotic LiDAR sensor is a distance measurement sensor that the robot uses for measuring its separation from a target. It provides the robot with real-time information about its surroundings. Therefore, it functions as the eyes of the robot to navigate its environment. Furthermore, the LiDAR is an active remote sensing system. It itself generates energy in the form of light to illuminate the target area from which data is to be collected. This aspect differentiates it from a passive system. A passive system relies on naturally occurring radiation for distance measurement.

3.5.1 PRINCIPLE OF OPERATION

The robot's LiDAR sensor is mounted on a cart carrying a load (Figure 3.1). The LiDAR sensor consists of a laser diode (transmitter), a photodiode (receiver), and a scanner (rotating mirror or prism) with associated optical assembly. A pulse generator triggers the laser diode. The LiDAR sensor emits a beam of light as a laser pulse. This laser pulse is reflected from the objects in the environment such as the surfaces of roads, ground, buildings, and trees. Reflected light from the obstacle is detected by the photodiode. Incident and reflected light beams are shown. The electrical signal produced in the photodiode is amplified, digitized by an A-to-D converter, and fed to

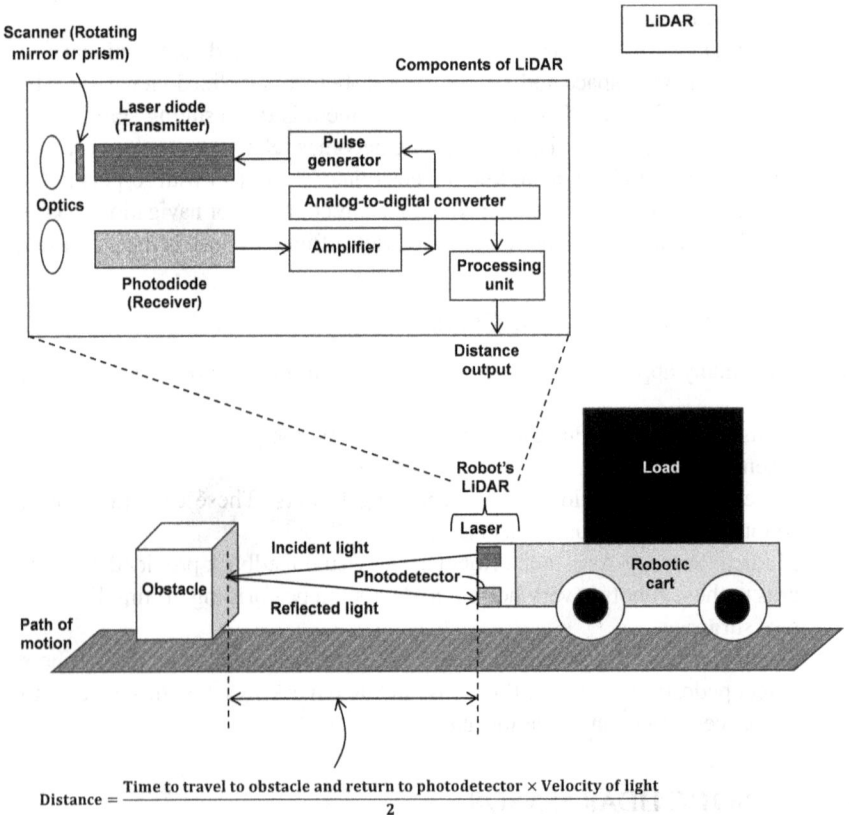

$$\text{Distance} = \frac{\text{Time to travel to obstacle and return to photodetector} \times \text{Velocity of light}}{2}$$

FIGURE 3.1 The principle of LiDAR is illustrated with reference to its use for distance measurement by a robot from a load-carrying cart to an obstacle. The inset shows the inner structure and components of the LiDAR. The formula for distance calculation is given.

a processing unit. The processing unit calculates the distance between the robot and the obstacle and provides the distance output. The distance is found by measuring the time t taken by the laser pulse reflected from an object called the back-scattered light to reach the LiDAR sensor. The formula used in the calculation of distance is derived by noting that the laser pulse has traveled from the LiDAR sensor to the object and returned by rebounding. From knowledge of velocity c of light, the distance d of the target from the robot is calculated by the formula:

$$d = \frac{ct}{2} \tag{3.1}$$

where the factor 2 accounts for the two-way journey of light from the robot to target and back.

3.5.2   Principal Components of a LiDAR Sensor

The LiDAR has four main components:

i. Laser Source: The laser source emits pulses of near-infrared (IR) light.
ii. Scanner: It performs a continuous scanning of the environment by rotating and oscillating to point in different directions to direct the laser beam in different directions.
iii. Detector: It contains a light sensor by which it converts the reflected light from distant objects into electrical signals.
iv. Processing Unit: The electrical signals produced in the detector are ana-lyzed in the processing unit to calculate distances. A 3D point cloud is gen-erated providing a 3D representation of the scanned area, and precise X, Y, and Z coordinates for each point. From the detailed 3D maps thus gener-ated, the robot is able to perceive its surroundings accurately.

3.5.3   Applications of LiDAR Sensor in Robotics

The LiDAR sensor is the key element of several robotic systems, where it is used for:

i. Object Detection:
   a. Geometric Shape-Fitting: Ground segmentation and plane-fitting algo-rithms are used to determine the 3D geometry of the objects in the point cloud.
   b. Deep Learning: Convolutional neural networks are used to identify critical features in images to accurately detect objects.
ii. Simultaneous Localization and Mapping (SLAM): It allows a robot to con-struct a map of an unknown environment and track down its own position within that environment.
iii. Collision Avoidance: The robot uses path planning algorithms, e.g., A* (A-star), Dijkstra's algorithm, and rapidly exploring random trees (RRT) to calculate and find routes along collision-free paths. Global path planning

calculates the best path, while local planning determines the best speed of the robot's movement. The collision Jacobian matrix relates the approaching velocity of the links to the obstacles with the end-effector velocity. Accordingly, the end-effector velocity is modified to avoid smashes against obstacles (Kaneko et al. 1999).

iv. Navigation: By careful object detection, SLAM technique, and path planning, the robot can move about easily in its surroundings. A LiDAR-equipped mobile robot has been developed to navigate inside a room without any impact on the wall (Hutabarat et al. 2019).

3.5.4 Advantages of LiDAR Sensor

The LiDAR sensor offers many advantages, among which the most relevant in robotics are as follows:

i. Provision of fast and accurate target detection and ranging with an accuracy of 0.15–0.25 m is assured. During its continuous movement, a 2D LiDAR-mounted cleaning robot identifies if a person is lying on the ground after falling. A convolutional long short-term memory (LSTM) neural network is trained for the classification of the processed sensor information. It can identify if a fall has occurred for monitoring the activities of elderly people living alone to assure emergency healthcare (Bouazizi et al. 2023).

ii. Independence of distance measurements from lighting and weather conditions is achieved. Exceptions are heavy rain, cloud cover, fog, and snowstorms.

3.5.5 Limitations of LiDAR Sensor

The LiDAR has shortcomings too, which must always be properly accommodated when using it by making the requisite allowances:

i. Obscuring of one object by another at LiDAR height inhibits its proper functioning.

ii. It is difficult to detect transparent objects. Reflective surfaces too create confusion.

iii. Adverse weather conditions introduce complexities in detection.

3.6 ROBOTIC SONAR SENSOR

SONAR is the short form of Sound Navigation and Ranging, developed in inspiration from the echolocation abilities of bats and dolphins. A robot uses SONAR to detect obstacles in its environment, measure distances to objects, and navigate effectively. The SONAR primarily acts as a sense of touch for the robot. It allows a robot to perceive its surroundings even in low-light or obscured conditions. It is especially useful for underwater robots. Besides its regular activities, the robot uses SONAR to find the distance, direction, and speed of underwater objects. It is used for mapping the seafloor topography and geological formations, the aquatic environment and marine life, finding shipwrecks, or identifying potential underwater obstacles/hazards for navigation. Robots in submarines use it for navigation and the detection of underwater vessels. The superiority of SONAR over LiDAR in underwater mapping, particularly in deep and murky water conditions, arises from the fact that sound waves travel much farther and more effectively through water than light or radio waves. These properties make it suitable for underwater detection (Kleeman and Kuc 2008).

3.6.1 WORKING PRINCIPLE OF A SONAR SENSOR

It works by sending high-frequency sound waves in all directions and detecting the sound waves received after reflection from surrounding objects. In a SONAR system, the transmitter produces an electrical pulse. The pulse feeds a transducer, which converts the electrical pulse into sound waves. The sound waves propagate in the surrounding regions. During the course of their movement, they come across any object. On striking the object, they undergo reflection and bounce off. The reflected sound waves from the object return and hit the transducer, which transforms them into an electrical signal and sends it to a receiver.

In Figure 3.2a, a ship has a robot fitted with a SONAR. The ultrasound beam from the emitter of the SONAR propagates through the seawater and strikes the sea bed. The incident and reflected ultrasound beams are shown. The time difference between the electrical pulses corresponding to the incident and reflected signals is recorded. Using the velocity of sound waves in the concerned medium, the distance between the sensor mounted on the robot and the object is calculated. The SONAR works in a pulsed mode by periodically transmitting pulses of the 160 kilohertz (kHz) signal with a waiting time interval between successive transmissions to listen for the echo. A programmable divider and oscillator are used to transmit a series of closely spaced tones, called a pseudo-chirp.

3.6.2 PRIMARY COMPONENTS OF SONAR SENSOR

Figure 3.2b shows the internal construction of a SONAR system. The SONAR has a transmitter-cum-receiver carrying a transducer which acts as an ultrasound emitter and detector. Other components include a power supply and measurement circuit. The processing unit calculates the sea depth and shows it on the display unit. The functions of the different components of SONAR are explained below:

    i. Transmitter-Cum Receiver: It acts both as a transmitter and a receiver of sound waves; hence, it is called a transceiver.

   ii. Transducer: This is the core component of a SONAR, which is engaged in energy conversion from the electrical into mechanical domain during transmission and mechanical-to-electrical form during reception of sound waves using piezoelectric or magnetostrictive materials. It is built as an array of interconnected sensitive elements to improve the signal-to-noise

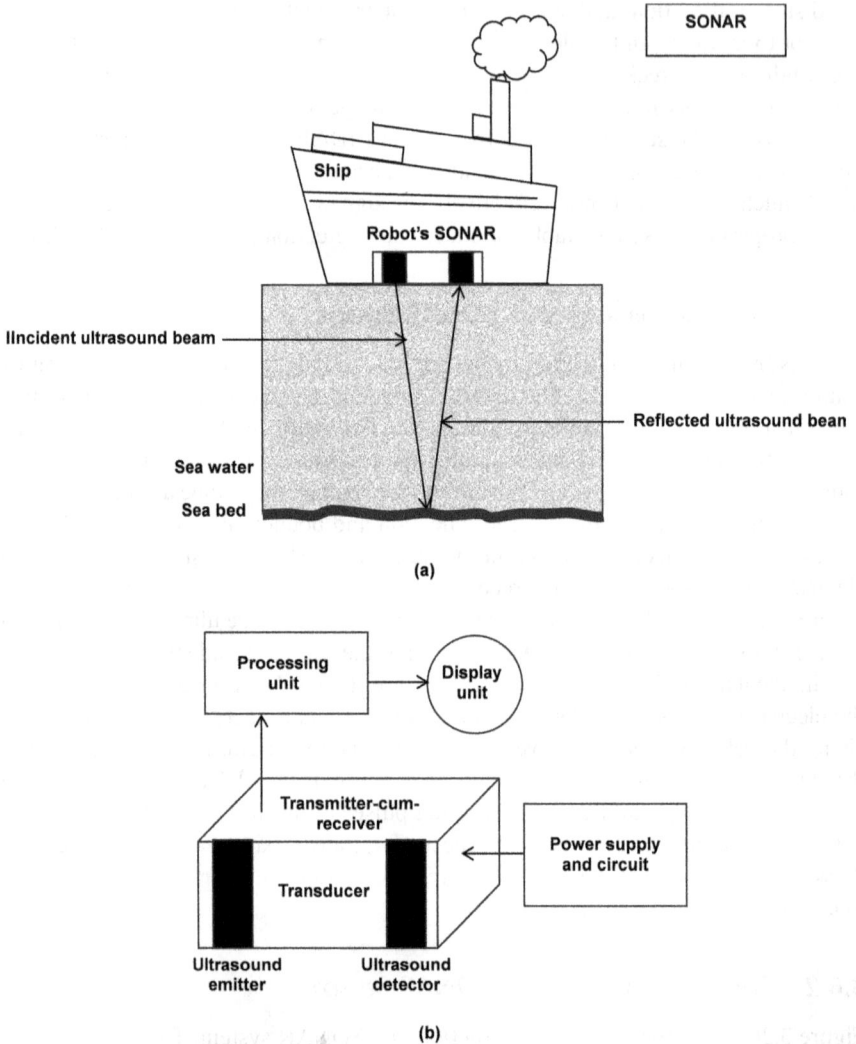

FIGURE 3.2   SONAR: (a) application of a SONAR for measuring the depth of sea water by a robot in a ship, and (b) parts of the SONAR unit.

ratio. When working in transmission mode, it is referred to as a projector; in receiver mode, it is known as a hydrophone (Benjamin 2008).

iii. Processing Unit: It analyzes the transmitted and received signals, measures the time taken by the echo signal to return, and calculates the distance of the object from the robot.

iv. Display Unit: It provides a visual representation of the processed information in numerical form or through graphics, facilitating the planning of a robot's course of action.

3.6.3 APPLICATIONS OF **SONAR** SENSOR IN ROBOTICS

The SONAR sensor finds widespread usage in robotics. Its application areas are as follows:

    i. Obstacle Detection to Avert Collisions: A robot detects objects in its path to prevent colliding against them.

    ii. Blind Spot Detection: SONAR comes to rescue in places where the robot's primary vision system is obstructed.

    iii. Room Navigation: SONAR acts as a cost-effective room navigation tool, cheaper than LiDAR for a mobile robot in an enclosed space.

    iv. Mapping and Localization: The robot sketches a map of its surroundings by classifying landmarks and pinpointing its own location within the map.

    v. Underwater Robotic Activities: SONAR excels in underwater performance to overcome the limitations of LiDAR for low-visibility light-scattering afflicted jobs to be executed by a robot.

3.6.4 LIMITATIONS OF **SONAR** SENSOR

Due attention must be paid to the limitations of SONAR when using it in robotics to avoid errors:

    i. Range of Detection Restriction: A smaller range than LiDAR in open air makes SONAR unsuitable for long-distance navigation of robots.

    ii. Lower Accuracy of Detection: Environmental noise and surface texture of the object impact SONAR output, affecting the robot's actions. The LiDAR furnishes a higher 3D resolution in data for clear environments than SONAR.

    iii. Angular Dependence of Measurements: The angle of incidence of sound waves on the object influences SONAR readings, thereby degrading the robot's performance.

3.7 ROBOT'S ACCELEROMETER

It is a device that measures acceleration, or the rate of change in velocity of the robot with respect to time (Liu and Pang 1999). It also measures a robot's tilt. For a robot moving on an inclined surface, e.g., during its uphill or downhill motions, the measured speed includes components due to gravity. These gravity components are not part of the actual robot speed. So, the computed speed is not the actual robot speed and the gravity components are compensated in the speed computations to ascertain the actual speed of the robot (Nistler and Selekwa 2011).

The accelerometer helps the robot to determine whether it is moving or stopped and to detect collisions or vibrations. It monitors the robot's physical movement and maintains its balance from observed changes in speed and orientation. It can perform gait analysis by measuring the motion of the robot's limbs or transient events.

Several types of accelerometers are fabricated using micro-electro-mechanical systems (MEMS) technology, e.g., piezoresistive, capacitive, and piezoelectric

MEMS accelerometers. They generally contain a small mass connected to a stiff spring. When the mass is accelerated, the spring is deflected. The accelerometer measures this deflection electrically.

Figure 3.3 shows the three types of accelerometers. In a piezoresistive accelerometer, as shown in Figure 3.3a, the values of piezoresistors located on flexure beams

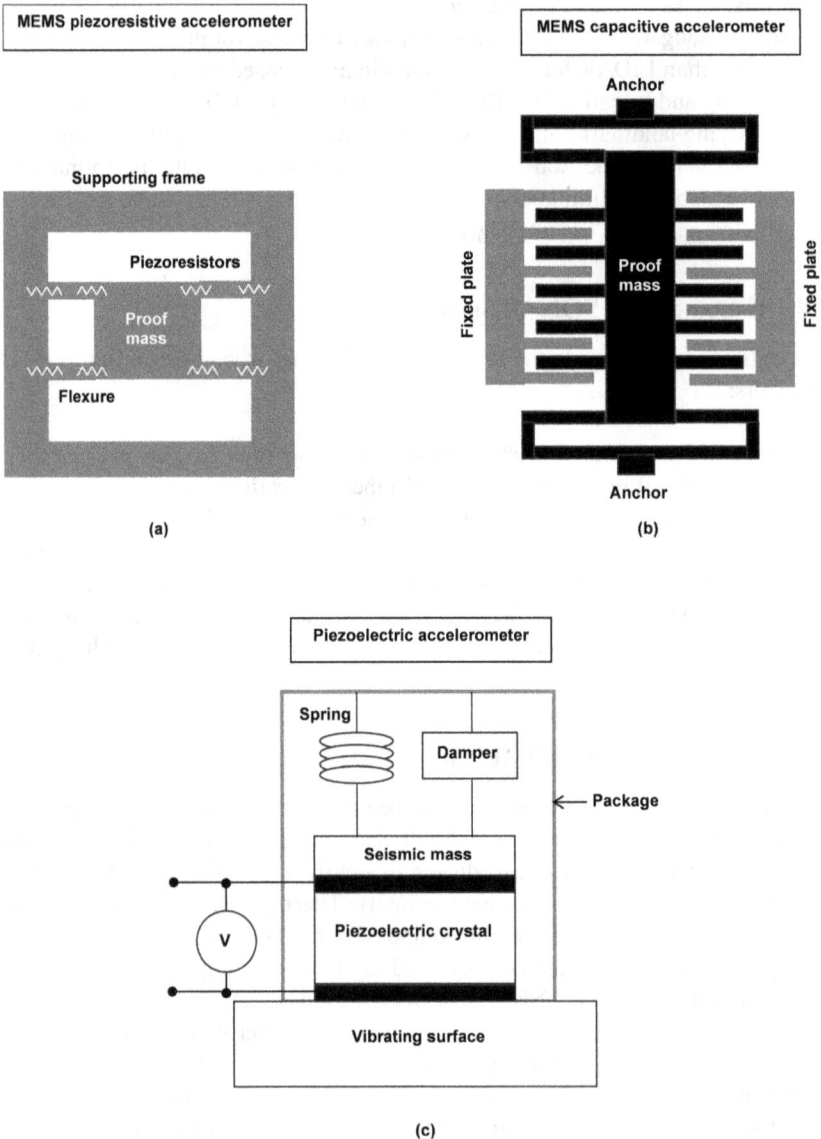

(a)

(b)

(c)

FIGURE 3.3   Robot's accelerometers: (a) piezoresistive, (b) capacitive, and (c) piezoelectric.

connected between the proof mass and a supporting frame change in response to the bending of the beams caused by acceleration. The change in resistance measures the acceleration. In the capacitive accelerometer shown in Figure 3.3b, a proof mass is suspended from anchors. It has electrodes projecting outward on both sides. These electrodes move in the gaps between electrodes projecting from fixed plates. The two sets of electrodes constitute an interdigitated pair of electrodes laid out in a variable capacitor configuration. In a piezoelectric accelerometer shown in Figure 3.3c, a piezoelectric crystal is mounted on the vibrating surface with electrodes on its two opposite sides. Over the crystal lies the seismic mass, which is held in its place with a spring. A damper is also fixed. The complete assembly is housed in a package. A voltmeter is connected across the electrodes of the crystal.

3.8 ROBOT'S TACTILE SENSOR

A robot's tactile sensor, also known as a fingertip force sensor, is a device that measures the physical properties of objects through contact (Yardley and Baker 1986; Tegin and Wikander 2005). It mimics the human sense of touch by detecting contact and pressure variations across a surface.

It is used by the robot for tasks like grasping objects with varying shapes and textures, collision detection, and human-robot interaction. It allows more delicate manipulation of objects. A tactile sensor system for a robot manipulator is used in industrial processes, e.g., welding and inspection (Suwanratchatamanee et al. 2010).

The tactile sensor operates on the principle of converting mechanical pressure exerted on its surface into an electrical signal by utilizing changes in electrical resistance, capacitance, or electric charge produced (Figure 3.4). Accordingly, it is of three types:

i. Piezoresistive Sensor: Conductive particles are embedded inside an elastomer. The distances between the particles change with pressure due to deformation of the elastomer, thereby altering the electrical resistance of the device. This sensor is easily microfabricated at an affordable cost. It has a good sensitivity and simple readout electronics.

Figure 3.4a shows an elastomer with suspended conductive particles inside and covered with electrodes on its two sides. When a force is applied to the sensor, the elastomer is squeezed, and the conductive particles come closer together, decreasing the resistance.

ii. Capacitive Sensor: Here the capacitance between electrodes varies depending on the separation or overlap between them caused by applied pressure. The capacitance changes indicate the pressure variations. Besides pressure, it measures shear forces and strain.

In Figure 3.4b, we see a fixed bottom electrode on a substrate. Spacers are fixed on two sides of the substrate and an electrode with a polydimethylsiloxane (PDMS) film is suspended forming an air gap between the electrodes. When the PDMS film is subjected to a force, it bends along with the electrode fixed to it. Consequently, the air gap between the electrodes decreases and hence the capacitance of the device changes.

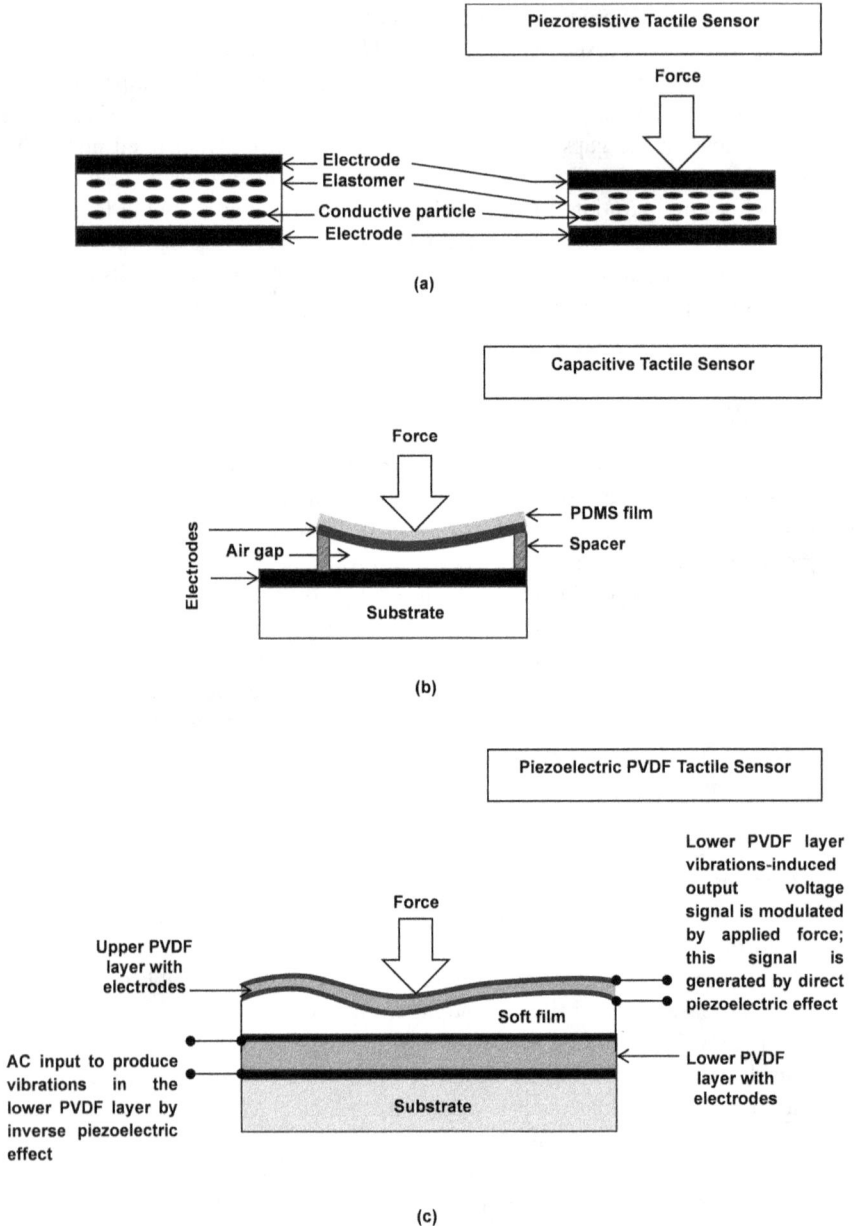

FIGURE 3.4 Robot's tactile sensors: (a) piezoresistive, (b) capacitive, and (c) piezoelectric.

iii. Piezoelectric Sensor: It produces an electric charge and hence a potential difference that is proportional to the force, pressure, or vibrations applied to the sensor.

In Figure 3.4c, we see a substrate. There is a lower polyvinylidene fluoride (PVDF) layer with electrodes on both sides. Over this layer lies a soft film.

Upon the soft film, there is an upper PVDF layer with electrodes on both sides. Vibrations are produced in the lower PVDF layer by feeding an AC signal. These vibrations generate an output voltage across the upper PVDF layer. This voltage changes on applying a force on the upper PVDF layer.

3.9 ROBOT'S PROXIMITY, POSITION, AND DISTANCE SENSORS

A proximity sensor is used in robotics to detect the presence of nearby objects without making any physical contact with those objects (Tsuji and Kohama 2020; Alagi et al. 2022). It need not specify the exact distance of the object from the robot. It is sufficient for the robot to know that an object is close to it without touching the object. It works on capacitive or inductive principles. Proximity sensors using IR radiation and ultrasonic waves are also common. The robot uses proximity sensors to avoid obstacles on its path. A robot in a factory uses it to detect whether a workpiece on a conveyor belt is near it. The proximity sensor shown in Figure 3.5a consists of an LED and a photodiode along with the readout circuit. An IR beam emitted by the LED is reflected toward the photodiode. Incident and reflected IR beams are shown. The readout circuit discovers the presence of the obstacle.

The position sensor gives information to the robot about its current location. It also measures the angles of joints of the robot's limbs with respect to a reference point. The position sensor helps the robot in controlling its motion or monitoring the positions of its joints.

Linear or rotary encoders or potentiometers are commonly used for position sensing. Figure 3.5b shows a linear potentiometer used for this purpose. A slider moves over a resistor connected to a battery and a voltmeter. The opposite end of the slider moves over a fixed plate. As the slider moves between the end points A and B, the output voltage varies because the path length traversed by the current changes.

A distance sensor accurately measures the distance of the robot from an object. Time-of-flight cameras and ultrasonic sensors are usually used for distance estimation. Laser range finders, too, are common.

Figure 3.5c shows a piezoelectric ultrasonic transducer acting as a transmitter-cum-receiver. The obstacle, incident, and echo ultrasonic waves are shown. The control circuit measures the time taken by ultrasonic waves to reach the obstacle and bounce back, and calculates the distance from the transducer to the obstacle.

The distance sensor is used by the robot for mapping its environment. It guides the robot's hands during the manipulation of an object. It also tells the robot about any object detected on its path.

3.10 ROBOT'S TEMPERATURE SENSOR

A robot uses temperature sensors for performing various everyday jobs:

i. Environmental Monitoring: The robot monitors the temperature of its surroundings. Temperature monitoring allows it to adapt to changing environments.
ii. Overheating Detection: The robot detects potential overheating in its components. Thus, its motors or batteries are prevented from damage.

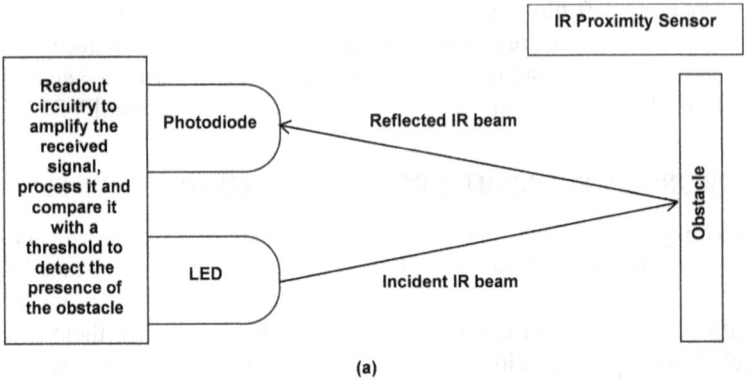

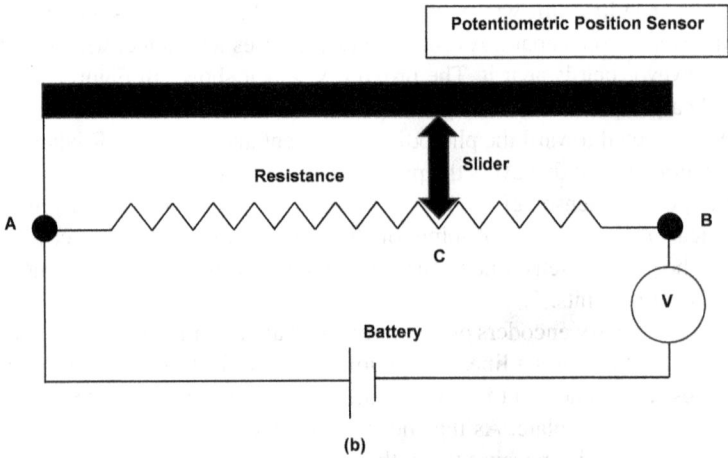

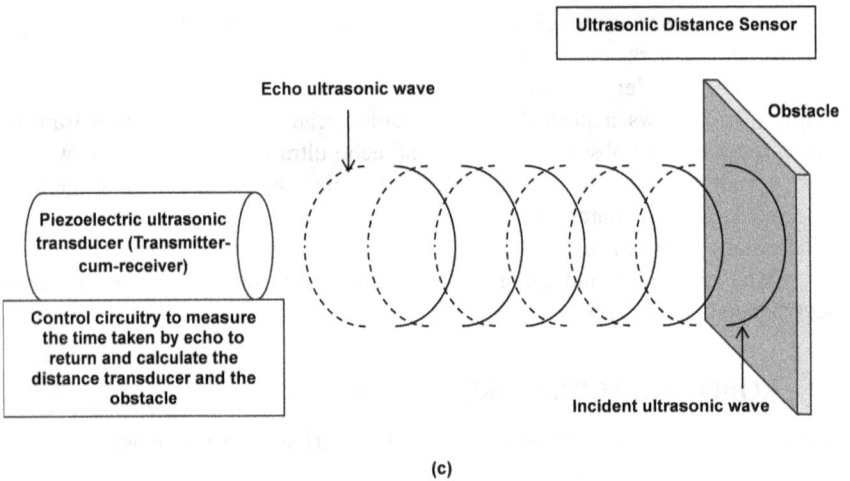

FIGURE 3.5 Sensors for proximity, position, and distance estimation by a robot: (a) proximity, (b) position, and (c) distance.

iii. Task Optimization: The robot adjusts its behavior based on temperature conditions, like slowing down in extreme heat.

Common temperature sensors used by robots include:

i. thermistors which exhibit large resistance changes with temperature,
ii. resistance temperature detectors (RTDs) where the resistance of a high-purity conducting metal, like platinum, changes with temperature, and sometimes
iii. IR sensors for non-contact temperature measurement which operate by measuring the heat of an object by converting the IR radiation emitted by it into an electrical signal.

The core principle is that the sensor translates the temperature change into an electrical signal in direct response to temperature fluctuations. This signal is read and interpreted by the robot's control system.

Figure 3.6 shows a thermistor. In Figure 3.6a, a semiconducting film is sandwiched between two electrodes with connection pins. In Figure 3.6b, the semiconducting film and electrodes are covered with an encapsulating coating.

A type of robot finger capable of precise temperature measurements from 303 to 353 K consists of a flexible reduced graphene oxide-based temperature sensor, an integrated circuit and a Bluetooth for wireless transmission of data (Zhou et al. 2019).

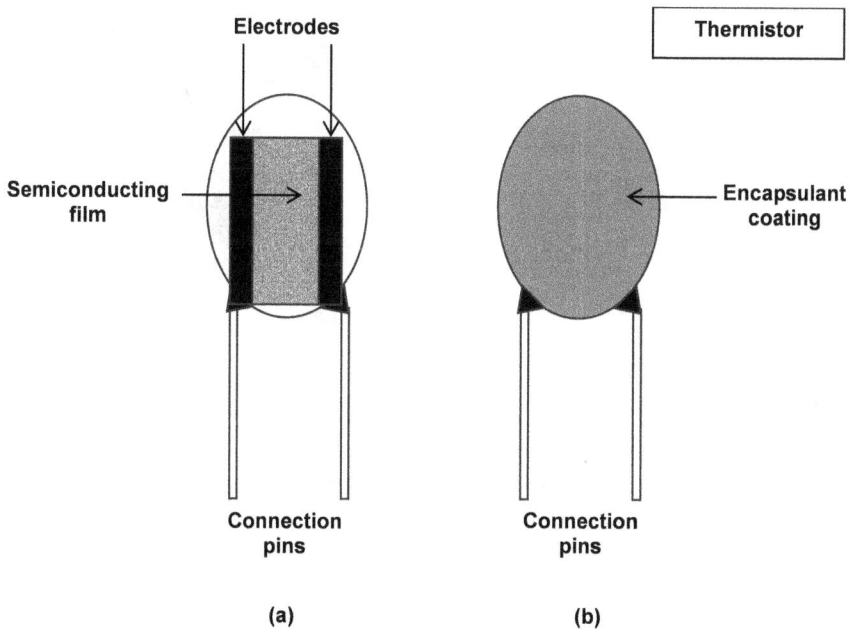

FIGURE 3.6 Robot's thermistor: (a) without and (b) with encapsulation.

3.11   ROBOT'S HUMIDITY SENSOR

A robot measures humidity to monitor environmental conditions. Humidity control is crucial for cleaning robots, agricultural robots, or robots operating in sensitive environments (Lee et al. 2007; Mariani et al. 2023). Different versions of humidity sensors include the following:

i. Capacitive Humidity Sensor: Robots typically measure humidity using a capacitive humidity sensor (Figure 3.7a). The capacitive sensor detects changes in capacitance caused by moisture in the air. When moisture from the air condenses on the sensor's dielectric material, it alters the electrical properties of the dielectric material between the two electrodes, and hence the capacitance between the electrodes. This change of capacitance is measured and translated into a digital signal representing the humidity level. Capacitive sensors are renowned for their high accuracy and rapid response time. These characteristics make them suitable for precise humidity measurements.

The capacitive humidity sensor shown in Figure 3.7a consists of a ceramic substrate over which an oxide or polymer dielectric film is sandwiched between a bottom electrode and a thin top moisture-permeable electrode.

ii. Resistive Humidity Sensor: This sensor uses a material whose electrical resistance changes based on the moisture content. But it is generally less accurate than a capacitive sensor. The resistive humidity sensor shown in

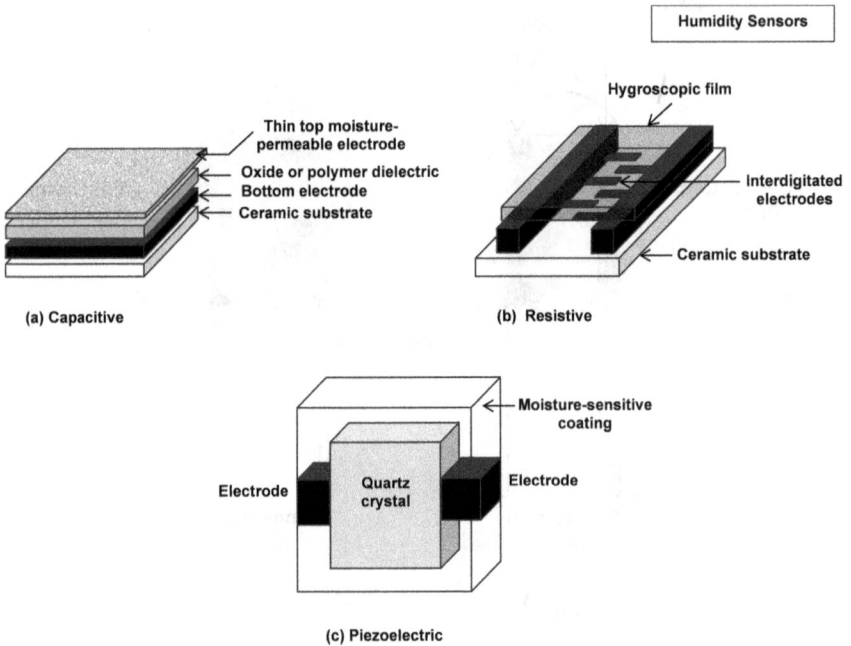

FIGURE 3.7   Robot's humidity sensors: (a) capacitive, (b) resistive, and (c) piezoelectric.

Figure 3.7b consists of a ceramic substrate covered with a pair of interdigitated electrodes over which a hygroscopic film is deposited.

iii. Piezoelectric Humidity Sensor: It measures the change in mass due to water vapor adsorption. The sensor works by measuring changes in oscillation frequency caused by mass bound to the surface of the piezoelectric crystal. In Figure 3.7c, a quartz crystal with two electrodes has a moisture-sensitive coating to adsorb moisture from the air. The moisture adsorption causes a change in mass and hence the oscillation frequency of the crystal with the humidity level.

3.12   ROBOT'S MICROPHONE

A robot uses microphones for a variety of tasks, including (Tamai et al. 2004; Löllmann et al. 2017):

i. Speech Recognition: Robots use microphones to recognize speech. They are able to do so even against background noise.
ii. Communication: Robots use microphones to communicate with their environment, enabling them to react to audio commands.
iii. Sound Source Localization: Robots use microphones to determine the direction and place from which a sound is coming.
iv. Source Separation: Robots use microphones to separate and identify simultaneous sound sources for reacting to complex auditory environments with reverberation and noise.
v. Sound Tracking for Rescue Missions: Robots use microphones to listen to sound of people who are calling for help, such as those trapped under a rubble.

Some types of microphones used in robotics are as follows:

i. Dynamic Microphones: These microphones are known for their reliability and ruggedness. They do not require batteries or external power supplies. In the dynamic microphone shown in Figure 3.8a, a wire coil is attached to a diaphragm. The coil moves between the pole pieces of a permanent magnet marked as N and S. The sound waves impinging on the diaphragm produce vibrations in it. The vibrations make the coil move back and forth in the magnetic field between the pole pieces. The movement of the coil induces a voltage in it. This voltage is recorded as the output voltage signal, serving as a replica of the sound pressure variations.
ii. Ribbon Microphones: These microphones use a light metal element to pick up both the velocity and displacement of air. This gives them improved sensitivity to higher frequencies.
iii. Condenser Microphones: These microphones provide more detailed reproduction of sound than dynamic microphones, but they require an external power supply to function. They are very sensitive and pick up sound from close sources, room sound, and background information.

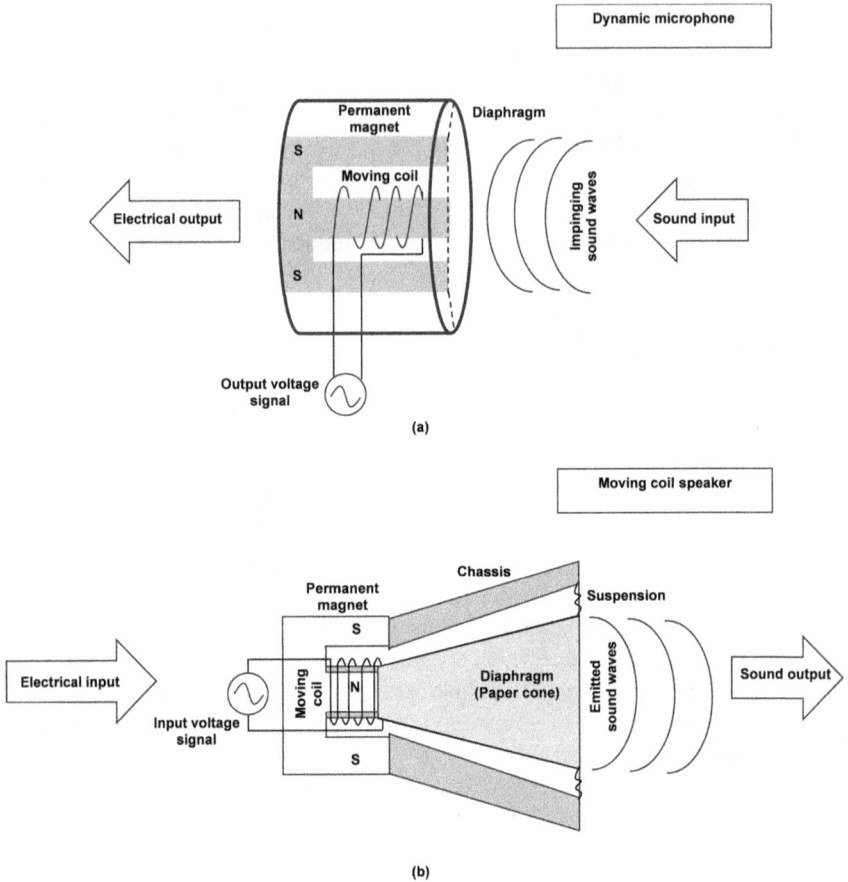

FIGURE 3.8 Robot's listening and talking devices: (a) microphone and (b) speaker.

Robot microphones use a variety of techniques to help robots identify and understand sounds in their environment, including:

   i. Beamforming Microphones: These use multiple microphones to determine the direction of a sound source. The direction is determined by measuring the time it takes for the sound to reach each microphone and the strength of the received sound signal.
   ii. Microphone Arrays: They use a group of omnidirectional microphones. They are used to separate sounds and reduce noise.

3.13   ROBOT'S SPEAKER

A speaker developed specifically for a speaking or talking robot is custom-designed and tailored to meet the unique needs of different robot applications. Vital factors considered are clear audio quality at various volumes, volume control, directionality, and sometimes even specific sound characteristics to match the robot's intended personality. No less important is the compact size of the speaker to fit within the robot's body design. The speaker must also have the ability to operate in harsh environments, with personalized features like audible warnings in safety-critical situations or pleasant voice interactions in social robots.

Figure 3.8b shows a moving-coil loudspeaker. A moving coil connected to the input voltage signal is mounted between the pole pieces N and S of a permanent magnet. The current flowing through the coil produces a magnetic field around it. The magnetic field of the coil interacts with the magnetic field of the permanent magnet. The interaction of magnetic fields induces vibrations in the diaphragm and cone attached to it and supported by suspensions from a chassis. The vibrating diaphragm produces sound waves that are transmitted through the air.

Many robot developers create unique speakers to optimize sound quality and fit within the robot's physical constraints. Different robot applications might require different speaker characteristics, like high-fidelity sound for social robots or loud, clear warnings for industrial robots. Depending on the robot's design, some may utilize commercially available speakers adapted for robotic use, depending on the application and desired functionality. Others may require completely custom speaker systems to achieve desired functionalities. Speakers are often integrated into the robot's design, sometimes even within the mouth area to enhance realism.

Social robots use artificial intelligence algorithms to identify the speaker and personalize the conversation. They are programmed to deliver a speech while mimicking a human speaker's gestures and body movements.

Design considerations for robotic speakers are as follows:

i. Size and Shape: It must fit within the robot's form factor. Compatibility with robot's design and aesthetic appeal cannot be overlooked.
ii. Durability: Depending on the robot's environment, it must withstand vibrations, temperature fluctuations, and potential impacts.
iii. Sound Quality: Clear and audible voice reproduction is crucial for communication and interaction.
iv. Directional Sound: Some robots require speakers that can project sound in specific directions for targeted communication.

The applications of robotic speakers are as follows:

i. Human-Robot Interaction: Social robots often use speakers for natural conversation and providing feedback.
ii. Industrial Automation: Industrial robots alert workers to potential hazards or provide status updates on machinery.
iii. Navigation Assistance: Assistive robots guide users with voice commands.

Examples of specialized robotic speaker features are as follows:

i. Multiple Speaker Arrays: These are used to create a more immersive sound experience or direct sound toward specific locations.

ii. Adaptive Volume Control: It automatically adjusts volume based on ambient noise levels.
iii. Audio Processing Algorithms: They enhance speech clarity and reduce background noise.

3.14   ROBOT'S TEXT-TO-SPEECH (TTS) SYNTHESIZER

It is a speaking device used by a robot. It converts written text into audible speech, allowing the robot to speak by playing back the generated audio through a speaker. It takes input in the form of text and produces an audible voice output. It has the following components:

i. Microphone: It captures spoken words from a user.
ii. Speech Recognition Algorithm: It converts spoken words into digital data.
iii. Text-to-Speech Engine: It converts text into spoken audio.
iv. Speaker: It plays the generated audio.

It is worth noting that the TTS is combined with a microphone for speech recognition. This enables the robot to understand spoken commands as well.

3.15   ROBOT'S ACTUATION MOTORS

Electric motors are the principal components of mobile robots engaged in transportation, manufacturing, and surveillance industries (Coiffet and Chirouze 1983). They are used for the movement of these robots from place to place. They power their wheels, legs, or other locomotion parts through the conversion of electrical energy into mechanical motion. A wide variety of motors of different types, each with specific capabilities suited to specialized actions, are available. The robot design engineer can choose the most befitting one from this variety for any application (Yuan 2023; Tiwari 2025).

3.15.1   MOTOR SELECTION CRITERIA FOR ROBOTICS

Essential considerations for choosing a proper robot motor are as follows:

i. Continuous and maximum torque needed for handling the load without overheating or stalling of the motor.
ii. Desired range of motor speed and requisite precision in speed control.
iii. Efficiency of the motor to save power consumption and prolong battery life.
iv. Physical dimensions and weight of the motor to fit within the robot's body.
v. Motor sealing and protection for protection from environmental hazards in robot's workplace, e.g., dust, extreme temperatures, humidity, and exposure to chemicals.
vi. Reliable and maintenance-free operation of the motor, e.g., brushless motors are more robust with longer life spans and less frequent breakdowns.

vii. Complexity of motor control circuitry, e.g., stepper motors require intricate control algorithms, whereas the DC motors are easily controllable.
viii. Compatibility with control electronics module of the robot for seamless integration with the microcontroller.
ix. Easy scalability, upgradation, and acceptability of additional features.
x. The available budget and economical aspects of the customer.

3.15.2 TYPES OF MOTORS USED IN ROBOTICS

The common types of robot motors are as follows:

i. DC Motors: These simple, affordable motors are easy to install and maintain. They are used in battery-operated applications. They provide continuous rotation with a high torque-to-inertia ratio. They respond quickly to control signals with precise speed and position control. However, they are prone to wear and tear due to the use of brushes. Moreover, they have lower efficiency than brushless motors.

Figure 3.9a shows a brushed DC motor. The rotor is the armature coil whose terminals are connected to the split-ring commutator. A battery contacts the split-ring commutator through carbon brushes to deliver the current to the rotor. The commutator reverses the direction of the current in every half cycle. Hence, the coil continues to rotate in the same direction.
ii. Brushless DC (BLDC) Motor: Brushless motors show improved performance and longer lifespans compared to traditional DC motors. This is possible because the need for brushes is eliminated. These motors are

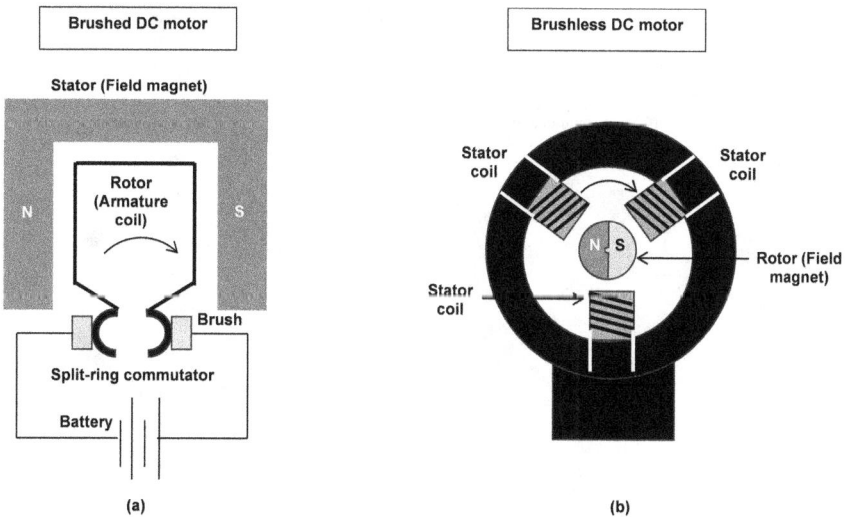

FIGURE 3.9 DC motors for robot actuation: (a) brushed and (b) brushless.

commonly used in mobile robots that require high efficiency. Undoubtedly, they offer quiet operation due to the absence of brushes and also require low maintenance. However, they have more complex control and drive circuitry. Also, their initial costs are higher compared to brushed DC motors.

Figure 3.9b shows a BLDC motor. The rotor is a permanent magnet marked N-S. The stator consists of three separate coils. Three-coil stator configuration generates a rotating magnetic field by selectively energizing two of the three stator coils at a time. This field causes a smooth, controlled motion of the rotor without brushes.

iii. Servo Motors: A type of DC motor, they offer high precision and control. They are commonly used in robotic arms and autonomous vehicles. They allow for accurate positioning. They are capable of holding a specific angle even under load. But they offer a limited continuous rotation than other motors. Also, they have a relatively higher cost than standard DC motors.

iv. Stepper Motors: They are brushless DC motors that can move in precise increments. Hence, they are used in applications that require accurate positioning or smooth motion control, such as robotic arm movement and 3D printing. They operate by dividing a full rotation into a series of steps, giving precise control and high torque output. But they have a higher power consumption when holding a position. Also, they experience resonance issues at certain speeds.

v. AC Motor: The AC motors are used in large robotic arms and manipulators. They are used in industrial automation equipment requiring high power outputs. They are robust and capable of delivering high torque. But they have more complex control than DC motors. Further, they require an external power inverter for variable speed control.

Figure 3.10a shows a synchronous motor. The rotor consists of permanent magnets. Three-phase AC supplied to the stator windings generates a

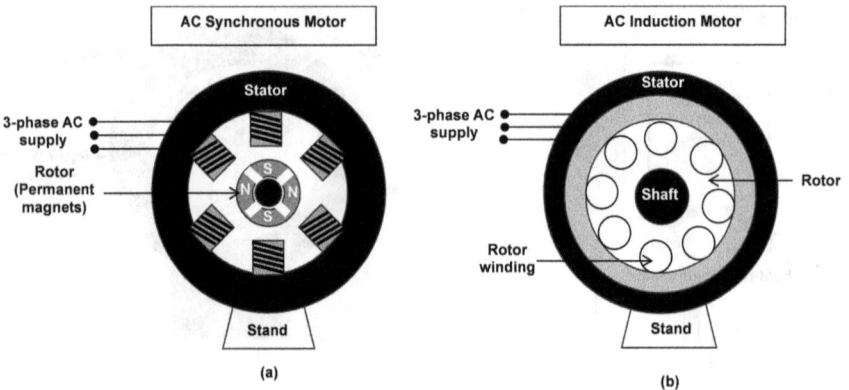

FIGURE 3.10 Three-phase AC motors for robotic actuation: (a) synchronous and (b) induction.

rotating magnetic field. The rotating field induces an electric current in the rotor, accompanied by a magnetic field. The interaction between the magnetic fields of the stator and the rotor creates a torque. The torque enables the rotor to lock with the stator's magnetic field and rotate at a synchronous speed.

Figure 3.10b shows an induction motor. A three-phase AC is supplied to the stator, creating a rotating magnetic field. The rotating magnetic field cuts the rotor's conductors, inducing an electromotive force and hence current in the rotor. The current induced in the rotor interacts with the rotating magnetic field. The resultant torque causes rotation of the rotor.

vi. Linear Motors: These motors are used in high-speed and high-precision robotic systems. Such robotic systems are used in semiconductor manufacturing equipment. They provide direct linear force, eliminating the need for additional mechanical components such as gears or pulleys. This advantage makes them suitable for applications requiring linear motion instead of rotational motion. They have higher costs than traditional rotational motors but are limited to specific linear motion applications.

vii. Pneumatic Vane Motors: They use compressed air to generate rotational motion. Being lightweight and able to deliver high power-to-weight ratios, they are used in mobile robots. These robots require quick movements. They are also preferred in environments where electric motors may not be suitable, e.g., in explosive atmospheres. They require a reliable source of compressed air.

viii. Hydraulic Motors: They use pressurized fluid to generate rotational motion. They deliver high torque and are commonly used in heavy-duty applications that require significant power output, e.g., in large mobile robots, such as construction or agricultural machinery. They require a hydraulic fluid supply and associated plumbing. They demand higher maintenance owing to the potential for fluid leaks.

Figure 3.11 shows a pneumatic/hydraulic motor. The rotating element is a slotted rotor mounted on a drive shaft. Each slot of the rotor is fitted with a freely sliding vane extended to the casing wall using springs. Compressed air or liquid is pumped through the gas/liquid inlet. It pushes the vanes to move ahead, creating the rotational motion of the central shaft. Then it comes out through the gas/liquid outlet.

ix. Piezoelectric Motors: They are based on the piezoelectric effect of deformation of a material when subjected to an electric field. They are commonly used in micro-robotics or applications that require precise movements and fine adjustments. They are limited to low power output applications and operate on a small scale. They are more complex to control than traditional motors.

Figure 3.12 illustrates the working of piezoelectric motor. Figure 3.12a shows the motor consisting of a piezo stack fixed on one side, a contact point, a slider, and a bearing. No voltage is applied.

In Figure 3.12b, a voltage slowly increasing with time is applied. The piezo stack is slowly extended. The slider moves along with the moving

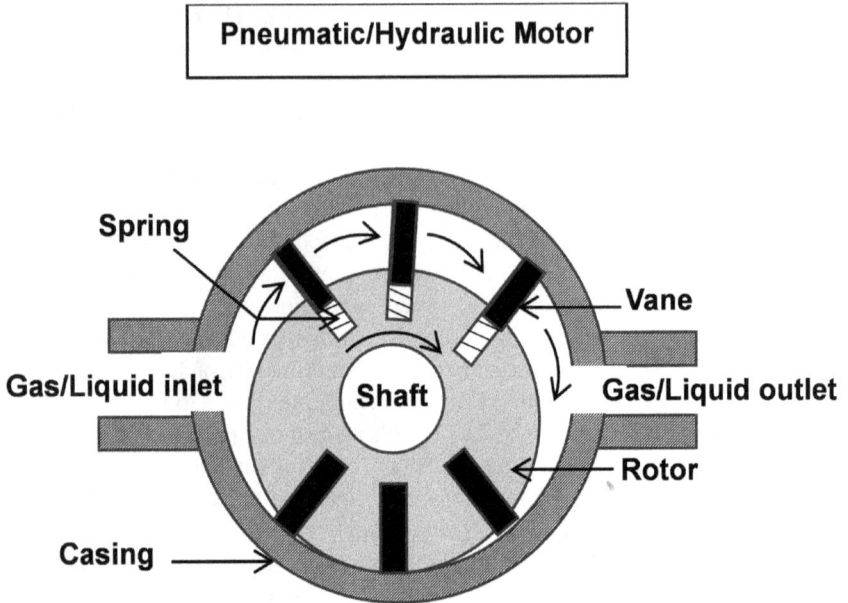

FIGURE 3.11   A pneumatic/hydraulic motor for robotics.

contact point owing to the frictional force between the contact point and the slider. This is the stick-phase.

In Figure 3.12c, a voltage rapidly decreasing with time is applied. The piezo stack is rapidly retracted. The slider remains stationary due to inertia, but the contact point slips back to its original position. This is called the slip-phase. A net displacement of the slider results.

A macroscopic movement is realized by repetition of the steps of Figure 3.12b and c.

x. Magnetic Field Motors: These motors use the principles of magnetism to generate motion, like magnetic linear actuators. They offer high precision, so they are used in mobile robots that require accurate position control, such as medical robotics or laboratory automation. They need complex control algorithms for optimal performance, and they are more expensive than other motor types.

3.16   DISCUSSION AND CONCLUSIONS

Sensors and actuators are essential components of robots that make them capable of environmental perception, physical action, and communication with other devices (Table 3.2). Electronic control systems consist of computational processors, storage devices, interfacing circuits, notably OP-AMPs and analog-to-digital converters, and open-loop and closed-loop control systems. To name a few components/circuits, Arduino is a small computer serving as the robot's brain that can be programmed

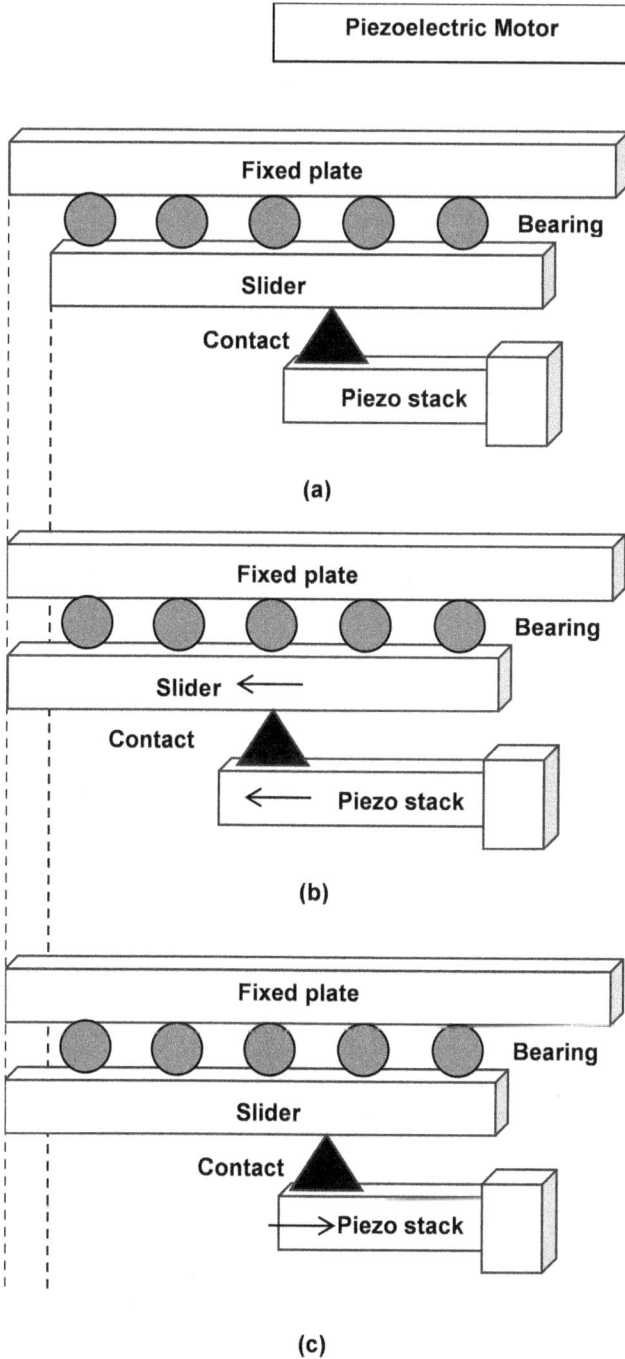

FIGURE 3.12 A piezoelectric motor for a robot in three different states: (a) without any applied voltage, (b) on applying a voltage slowly increasing with time, and (c) on applying a voltage rapidly decreasing with time.

TABLE 3.2

Reflecting Back on This Chapter and the Lessons Learned

Sl. No.	Lessons Learned	Explanation
1	Summary	Sensing and perception by robots enable them to see, hear, and feel their surroundings in order to interact with their environment. Actuators and end-effectors of robots power their movements and actions.
2	Robot cameras	Considerations for robot camera mounting were described together with their typical mounting configurations and applications. Different types of robot cameras were mentioned.
3	LiDAR and SONAR	The principle of operation, primary components, applications, advantages, and limitations of robotic LiDAR and SONAR sensors were elaborated.
4	Robot's sensors	Robots are equipped with various sensors, among which accelerometers, tactile sensors, proximity sensors, position sensors, distance sensors, temperature sensors, and humidity sensors are most commonly used in a typical robot.
5	Speaking/listening aids	The robot also has a microphone, speaker, and text-to-speech synthesizer for easy interaction with humans.
6	Robot's actuators	Several types of actuation motors used in robotics were discussed and their selection criteria were outlined. Among these, the DC motors, brushless DC motors, servo and stepper motors, AC motors, linear motors, pneumatic vane/hydraulic motors, piezoelectric and magnetic field motors are prominent.
7	Keywords and ideas to remember	Sensing and perception, actuators and end-effectors, robot cameras, LiDAR sensor, SONAR sensor, accelerometer, tactile sensor, proximity, position, distance, temperature, and humidity sensors, microphone, speaker, and actuation motors.

for controlling its lights and motors; RasberryPi is a versatile kit featuring system on chip with GPU, RAM, and connectivity; and Robot Operating System (ROS) is an open-source framework to assist researchers in building and using codes between different robotic applications.

After familiarizing ourselves with the different kinds of sensors and actuators used for building robots, let us examine how these devices impart capabilities to robots similar to humans about sensing and interacting with their environments. As we know, human–human interaction is largely based on speaking to/listening to others, and acting in response. It is the natural and prevalent way by which people communicate and exchange information among themselves. Through speech, we convey our thoughts, ideas, and emotions to our colleagues and friends. Conversation by speech fosters social bonding and connections. We would surely like to interact with robots via speech. Speech is a convenient, fast, and efficient communication mode with robots for controlling robots in industries, such as surgical robotic arms. Speech-supported robots can take care of the sick and help people in dangerous environments more easily than deaf and dumb robots. Sinch speech enables more motivating and satisfying robot–human interactions; we shall look into the techniques of speech processing for robotics in the forthcoming chapter. Speech processing becomes more complicated in noisy environments and those prone to reverberations.

REFERENCES AND FURTHER READING

Alagi H., S. Ergun, Y. Ding, T. P. Huck, U. Thomas, H. Zangi and B. Hein. 2022. Evaluation of on-robot capacitive proximity sensors with collision experiments for human-robot collaboration, *2022 IEEE/RSJ International Conference on Intelligent Robots and Systems (IROS)*, Kyoto, Japan, 23–27 October, pp. 6716–6723.

Athanasopoulos G., W. Verhelst and H. Sahli. 2015. Robust speaker localization for real-world robots, *Computer Speech & Language*, Vol. 34, 1, pp. 129–153.

Benjamin K. 2008. Transducers for Sonar Systems. In: Havelock D., S. Kuwano and M. Vorländer (Eds.), *Handbook of Signal Processing in Acoustics*, Springer, New York, pp. 1783–1819.

Bouazizi M., A. L. Mora and T. Ohtsuki. 2023. A 2D-Lidar-equipped unmanned robot-based approach for indoor human activity detection, *Sensors*, Vol. 23, 5, p. 2534.

Coiffet P. and M. Chirouze. 1983. Robot Actuators. In: *An Introduction to Robot Technology*, Springer, Dordrecht, pp. 83–105.

Dahiya R., O. Ozioko and G. Cheng (Eds.). 2023. *Sensory Systems for Robotic Applications (Control, Robotics and Sensors)*, The Institution of Engineering and Technology, UK, 330 pages.

Guo L., M. Zhang, Y. Wang and G. Liu. 2006. Environmental Perception of Mobile Robot, *2006 IEEE International Conference on Information Acquisition*, Veihai, China, 20–23 August, pp. 348–352.

Hutabarat D., M. Rivai, D. Purwanto and H. Hutomo. 2019. Lidar-Based Obstacle Avoidance for the Autonomous Mobile Robot, 2019 *12th International Conference on Information & Communication Technology and System (ICTS)*, Surabaya, Indonesia, 18 July, pp. 197–202.

Kaneko H., T. Arai, K. Inoue and Y. Mae. 1999. Real-Time Obstacle Avoidance for Robot Arm Using Collision Jacobian, *Proceedings 1999 IEEE/RSJ International Conference on Intelligent Robots and Systems. Human and Environment Friendly Robots with High Intelligence and Emotional Quotients (Cat. No.99CH36289)*, Vol. 2, 17–21 October, Kyongju, Korea (South), pp. 617–622.

Kleeman L. and R. Kuc. 2008. Sonar Sensing. In: Siciliano B. and Khatib O. (Eds.), Springer Handbook of Robotics, Springer, Berlin, Heidelberg, pp. 491–519.

Lee S. P., J. G. Lee, J. W. Chang, J. N. Kim, S. H. Lee and S. Chowdhury. 2007. Integrated Sensor System for Humidity Sensing of Robots Using Analog Mixed CMOS Technology, *2007 International Workshop on Robotic and Sensors Environments*, Ottawa, ON, Canada, 12–13 October, pp. 1–4.

Liu H. and G. Pang. 1999. Accelerometer for Mobile Robot Positioning, *Conference Record of the 1999 IEEE Industry Applications Conference. Thirty-Fourth IAS Annual Meeting (Cat. No.99CH36370)*, Vol. 3, 3–7 October, Phoenix, AZ, USA, pp. 1735–1742.

Löllmann H. W., A. Moore, P. A. Naylor, B. Rafaely, R. Horaud, A. Mazel and W. Kellermann. 2017. Microphone Array Signal Processing for Robot Audition, *2017 Hands-free Speech Communications and Microphone Arrays (HSCMA)*, San Francisco, CA, USA, 1–3 March, pp. 51–55.

Mariani S., L. Cecchini, N. M. Pugno and B. Mazzolai. 2023. An autonomous biodegradable hygroscopic seed-inspired soft robot for visual humidity sensing, *Materials & Design*, Vol. 235, 112408, pp. 1–8.

Nistler J. R. and M. F. Selekwa. 2011. Gravity compensation in accelerometer measurements for robot navigation on inclined surfaces, *Procedia Computer Science*, Vol. 6, pp. 413–418.

Rehman A. U., Y. Khan, R. U. Ahmed, N. Ullah and M. A. Butt. 2023. Human tracking robotic camera based on image processing for live streaming of conferences and seminars, *Heliyon*, Vol. 9, 8, p. e18547, https://doi.org/10.1016/j.heliyon.2023.e18547

Suwanratchatamanee K., M. Matsumoto and S. Hashimoto. 2010. Robotic tactile sensor system and applications, *IEEE Transactions on Industrial Electronics*, Vol. 57, 3, pp. 1074–1087.

Tamai Y., S. Kagami, Y. Amemiya, Y. Sasaki, H. Mizoguchi and T. Takano. 2004. Circular microphone array for robot's audition, *Sensors, 2004 IEEE*, Vol. 2, Vienna, Austria, 24–27 October, pp. 565–570.

Tegin J. and J. Wikander. 2005. Tactile sensing in intelligent robotic manipulation: A review, *Industrial Robot*, Vol. 32, 1, pp. 64–70.

Tiwari K. 2025. 10 Types of robot motors for navigation, https://kshitijtiwari.com/all-resources/mobile-robots/robot-motors/

Tourbabin V. and B. Rafaely. 2014. Speaker Localization by Humanoid Robots in Reverberant Environments, *2014 IEEE 28th Convention of Electrical & Electronics Engineers in Israel (IEEEI)*, Eilat, Israel, 3–5 December, pp. 1–5.

Tsuji S. and T. Kohama. 2020. Proximity and contact sensor for human cooperative robot by combining time-of-flight and self-capacitance sensors, *IEEE Sensors Journal*, Vol. 20, 10, pp. 5519–5526.

Wu J., J. Gao, J. Yi, P. Liu and C. Xu. 2022. Environment Perception Technology for Intelligent Robots in Complex Environments: A Review, 2022 *7th International Conference on Communication, Image and Signal Processing (CCISP)*, Chengdu, China, 18–20 November, pp. 479–485.

Yardley A. M. M. and K. D. Baker. 1986. Tactile Sensors for Robots: A Review. In: Scott P. B. (Ed.), *The World Yearbook of Robotics Research and Development*, Springer, Dordrecht, pp. 47–83.

Yuan Z. 2023. Current status and prospects of actuator in robotics, *Applied and Computational Engineering*, Vol. 11, 1, pp. 181–191.

Zhou C., X. Zhang, H. Zhang and X. Duan. 2019. Temperature Sensing at the Robot Fingertip Using Reduced Graphene Oxide-Based Sensor on a Flexible Substrate, *2019 IEEE SENSORS*, Montreal, QC, Canada, 27–30 October, pp. 1–4.

4 Talking and Listening Robots

Speech Synthesis, Recognition, and Understanding

4.1 INTRODUCTION

Robots speak, listen, and act using a combination of hardware and software technologies. These technologies enable them to generate human-like speech and communication for robot-to-robot and robot-to-human interaction. The main technologies involved in these interactions are as follows:

  i. Text-to-speech (TTS) synthesis (Kuo and Tsai 2024)
 ii. Speech recognition (Zinchenko et al. 2017)
iii. Natural language processing (NLP) (Supriyono et al. 2024)
 iv. Speakers, microphones, and actuators

Table 4.1 presents a comparative description of the fields of speech synthesis and recognition.

This chapter discusses the technologies used in natural language-controlled robots, i.e., robots that speak and understand like humans, enabling the exchange of information by robots' mouths and ears.

4.2 TTS SYNTHESIS AND VOICE GENERATION

TTS synthesis is referred to as speech synthesis. It is also called voice generation. It is an artificial intelligence technique based on machine learning models. It works by applying linguistic rules and pronunciation dictionaries to convert written text into spoken words. These words sound natural and human-like (Rashad et al. 2010; Li and Lai 2022).

4.2.1 PHASES OF TTS

TTS is a three-phase process. The three phases in TTS are text normalization, prosodic analysis, and concatenating speech synthesis (CSS) (Nair et al. 2022; Ahmad and Rashid 2024). Explanatory details are given below:

DOI: 10.1201/9781032695266-4

TABLE 4.1
Speech Synthesis and Speech Recognition

Sl. No.	Point of Comparison	Speech Synthesis	Speech Recognition
1	Definition	It is the process of converting written text into spoken language, thereby creating artificial speech from text.	It is the opposite of speech synthesis, converting spoken language into written text.
2	Fast definition	It speaks what we type.	It 'types' what we say.
3	Function	It is used to generate human-like spoken language from written text.	It is used to understand spoken words and translate them into the written format.
3	Input/output	It takes text as input and produces an audio signal as output.	It takes an audio signal as input and produces text as output.
4	Examples	A text-to-speech program reading a document audibly and distinctly to the audience.	A voice assistant listening to our voice commands and interpreting them as text.

i. Text Normalization: Text is the raw written input material. It is the material that needs to be converted into synthetic speech. Text normalization means the conversion of text into a consistent, canonical form before processing. By the canonical, normal, or standard form of a mathematical object is meant a form which presents the simplest representation of the object and allows it to be identified in a unique manner.

Figure 4.1 illustrates the step-by-step procedures and defines the technical terms used in text normalization. The procedures are feeding the input raw text, case conversion, tokenization, removing punctuation marks and stop words, parts-of-speech tagging, and stemming/lemmatization to reduce words to their root forms.

All text is uniformly converted to either lowercase or uppercase. Spellings of words are checked for correctness. Extra characters or punctuation marks are deleted. The stop words like 'the', 'a', 'an', etc. are also obliterated. Words are reduced to their base forms, e.g., 'walking' to 'walk'. The given piece of text is split into smaller units called tokens. These tokens are words, characters, or numbers, and the splitting process of the text is known as tokenization. The text is converted into a standard computer format to a target specification. The string of phones to be synthesized together is included.

In phonetics, a phone is a distinct, discriminable speech sound. It is not specific to a language. The sounds [pʰ] and [p] are two separate phones. There are four phones: [s], [p], [ɪ], and [n] in the word 'spin'.

A phoneme is the smallest unit of sound that carries a definite meaning in a language. It is a speech sound that can change the meaning of the

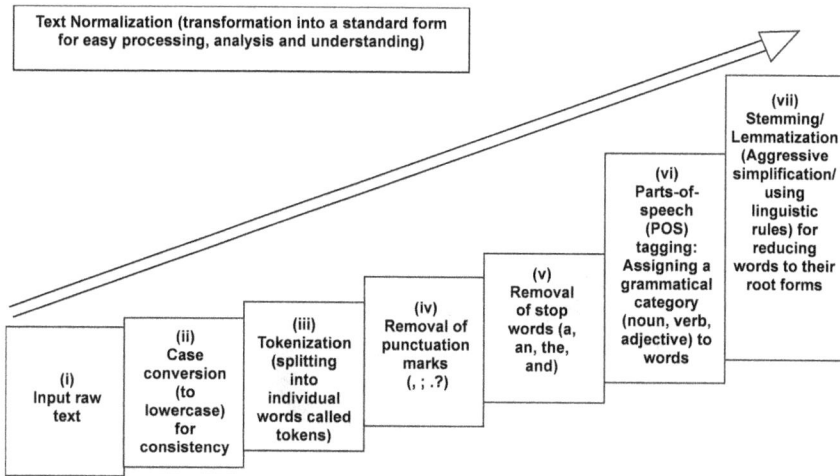

Text Normalization (transformation into a standard form for easy processing, analysis and understanding)

(i) Input raw text

(ii) Case conversion (to lowercase) for consistency

(iii) Tokenization (splitting into individual words called tokens)

(iv) Removal of punctuation marks (, ; .?)

(v) Removal of stop words (a, an, the, and)

(vi) Parts-of-speech (POS) tagging: Assigning a grammatical category (noun, verb, adjective) to words

(vii) Stemming/ Lemmatization (Aggressive simplification/ using linguistic rules) for reducing words to their root forms

FIGURE 4.1 Depiction of the stages through which a supplied text is passed during its normalization.

word. A new word is produced when a phoneme is swapped with another phoneme. To illustrate, substitution of the phoneme /p/ in the word 'peg' with the phoneme /l/ generates the word 'leg'. Similarly, the words 'kid' and 'kit' terminate with two different phonemes, /d/ and /t/.

The phoneme is represented by a written letter or a group of letters known as a grapheme. Different variations of the same phoneme occur in dissimilar contexts. But they do not change the meaning of a word. These variants are known as allophones. To clarify, the letter 't' is a grapheme; the sound of the letter 't', e.g., in 'top', is a phoneme. However, the slightly dissimilar 't' sounds in 'top' and 'stop' are allophones.

The progress of phonetic conversion takes place in the order grapheme-phoneme-allophone. Figure 4.2 illustrates the stages in phonetic conversion: grapheme-to-phoneme conversion, phonetic feature analysis, and phoneme-to-allophone conversion.

ii. Prosodic Analysis: Prosody is a reflection of the nuanced emotional characteristics of the speaker. Prosodic analysis involves analyzing a text's rhythm and emphasis patterns to identify elements such as stress patterns, pitch variations or intonation, and pauses. These elements are crucial for conveying meaning and emotion to speech in order to create natural-sounding speech (Totsuka et al. 2014; Corrales-Astorgano et al. 2024). Prosodic patterns vary with languages and dialects. So, they need careful attention. Prosody influences the utterance segmentation into syllables and words (Dahan 2015).

Figure 4.3 shows the nine steps in prosodic analysis: input as a recorded audio signal of spoken language, transcription of a speech sample, identification of syllables, marking stress levels, analysis of pitch contour, pauses

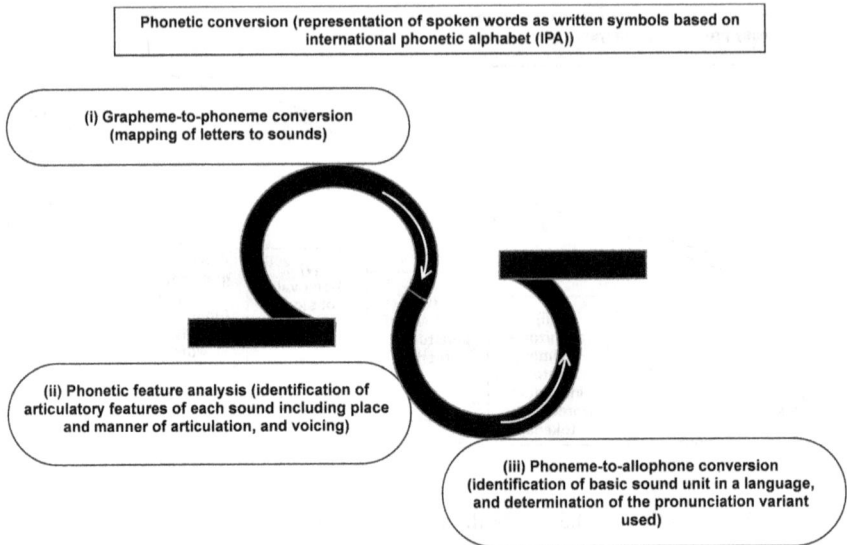

FIGURE 4.2   The three stages constituting a phonetic conversion.

and junctures, interpretation and contextualization beyond literal words, and output as a detailed breakdown of the pitch contour, intensity, and duration of speech segments.

iii. Concatenation Speech Synthesis: It is also called unit selection speech synthesis. It means joining together textual forms of pre-recorded speech-segment waveforms to form a complete utterance. The linking is done by accessing a large library of pre-recorded speech sounds like phonemes or syllables. The appropriate units for each phone segment are selected based on the text analysis. This is done in such a manner that the output speech matches the input text with high naturalness of the sound. Then the units are joined together to produce the final spoken output. Post-processing is performed to smooth potential discontinuities for removing concatenation artifacts (Rabiner and Schafer 2007; Oralbekova et al. 2024). Figure 4.4 shows the details of the concatenation of short samples of input text to produce a synthesized speech output. The six steps involved include: feeding the input text, text analysis, searching, matching, concatenation, and post-processing.

Concatenation is essentially stitching together pieces of audio. Its purpose is to create free-flowing speech. Naturalness is an essential ingredient for TTS. The speech looks natural because it uses real human speech segments. The real segments are able to capture subtle variations in pronunciation and intonation. But concatenation needs an extensive library of recorded speech to cover all possible combinations of sounds and variations. Such a vast, all-embracing library is extremely difficult to compile. Further, the process of joining together different speech segments can sometimes create noticeable gaps or unnatural transitions. Therefore, the quality of the database and the joining algorithm play significant roles in concatenation.

Prosodic Analysis (studying patterns of rhythm, stress, intonation and loudness in speech)

(i)
Input: Recorded audio signal of spoken language

(ii)
Transcription of speech sample (conversion of spoken words in audio/video recording into written text)

(iii)
Syllable identification (breaking down a word into individual distinct sounds)

(iv)
Marking stress levels (indicating the distinct sound within a word which is pronounced most emphatically by inserting a symbol, e.g., an accent mark above the stressed sound)

(v)
Pitch contour analysis (studying the changes in the pitch of sound with time by assignment of numerical values, choosing a fixed point in the frequency function of sound and tracking its movement with function changes; and by plotting the graph of fundamental frequency of sound against time)

(vi)
Pause analysis (locating pauses in the speech, marking their duration and placement in the uttered sound)

(vii)
Juncture analysis (studying the transitioning of a speaker between successive words or phrases)

(viii)
Interpretation and contextualization beyond literal words to grasp the full meaning and implication (studying prosodic features in relation to grammatical structure, intent of the speaker, the audience, social setting, surrounding situation and overall cultural context)

(ix)
Output: Detailed breakdown of the pitch contour, intensity and duration of speech segments within an utterance in the form of graphs or numerical values

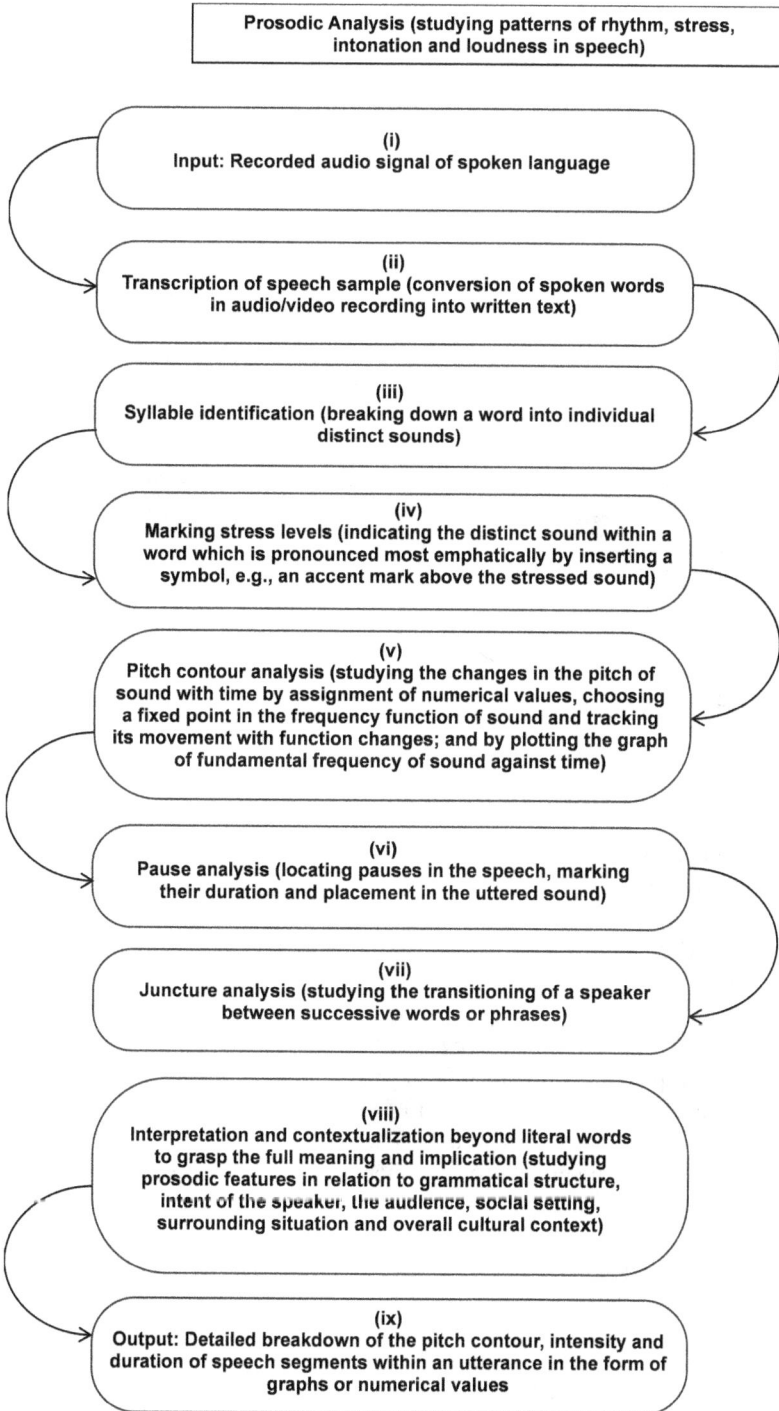

FIGURE 4.3   The progression of a prosodic analysis of speech.

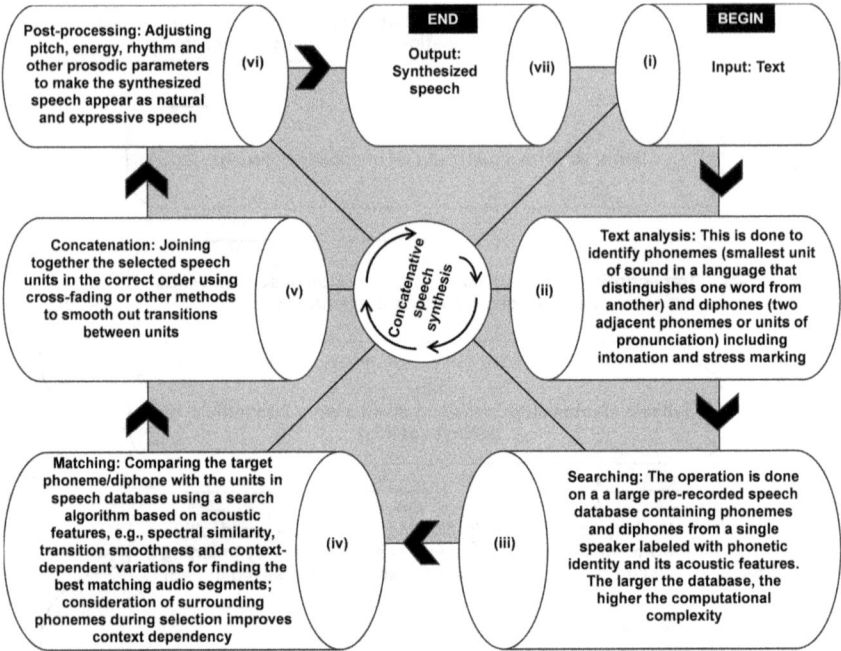

FIGURE 4.4 Combining the input text representing pre-recorded speech segments to create a synthetic voice.

4.2.2 COST FUNCTIONS AND THEIR OPTIMIZATION

Two cost functions, namely, target cost $C^t(u_i,t_i)$ and concatenation cost $C^c(u_{i-1},u_i)$ are minimized for unit selection:

i. Target Cost: This cost function signifies the desired cost of a single acoustic unit, e.g., a phoneme, which is necessary to attain a specific quality level, as determined by the acoustic properties of the unit such as pitch, duration, and energy. It is determined by considering factors such as quality, clarity, and overall prosody expected from speech.

The primary use of the target cost is in selecting the best unit from a database during speech synthesis. The selection is done on the basis of its matching with the target pronunciation. Hence, its value expresses the mismatch between the target speech unit specification t_i and a candidate unit u_i from the database.

ii. Concatenation Cost: This cost function is representative of the additional cost sustained when two acoustic units are connected together. It measures how evenly one sound transitions to the successive sound. Its attention is concentrated on the potential discontinuity or unnaturalness of sound. Factors taken into account are the transition between phonemes, pitch changes, and spectral uniformity between adjoining units aspiring to produce natural-sounding connected speech. Therefore, it is used to select units that seamlessly join with the previous unit. Seamless joining of units contributes to a smoother, more natural-sounding speech. The value of the concatenation cost expresses the acoustic or perceptual mismatch of the joint between the candidate unit $\underline{u}_i$ and the preceding unit u_{i-1}.

Cost Function Optimization Procedure: When synthesizing speech, a system first selects the acoustic unit with the nearest target cost to the desired pronunciation. Then it has to decide which unit to place next to ensure a smooth transition. So, it considers the concatenation cost. Thus, the target and concatenation cost functions are jointly optimized to enable the speech synthesis system to produce high-quality speech with natural-sounding connected speech. Ideally, all the target units should be found according to the specification. Ushering of acoustic mismatches at the edges of concatenated units should be prevented (Hunt and Black 1996; Gupta 2008).

4.2.3 USE OF TTS BY ROBOTS

Robots exploit TTS synthesis technology to convey information fluently. Speech enables audible messaging to users. Robots use TTS as an assistive technology for reading text aloud. Narrations are created for movies and screen captures for people who prefer to listen to reading. Championing oratorial variety, these systems often support numerous languages. These features make them useful across varied environments.

4.3 SPEECH RECOGNITION AND UNDERSTANDING

Robotic speech recognition and language understanding is an umbrella technology dealing with the ability of a robot to listen to human speech, interpret its underlying meaning within context, and respond properly. It essentially allows the robot to understand natural language commands and instructions, facilitating natural, intuitive, and seamless interaction between human operators and robots through spoken language. Speech recognition must be clearly differentiated from voice recognition, as explained in Table 4.2.

Various technologies used in speech and language understanding include speech recognition (conversion of spoken words into digital text using algorithms), NLP (analysis of the recognized text to comprehend its meaning), context awareness (interpretation of speech with respect to the surroundings/situations), semantic parsing (breaking down a sentence into its core meaning for triggering the intended action), and dialogue management (maintaining a conversation flow by tracking the conversational context).

4.3.1 SPEECH RECOGNITION

Speech recognition is sometimes called automatic speech recognition (ASR). Another name used is computer speech recognition. Speech recognition is essentially speech-to-text conversion. It necessitates the execution of commands for spoken words using sophisticated machine learning algorithms. These algorithms process

TABLE 4.2

Speech Recognition vs Voice Recognition

Sl. No.	Point of Comparison	Speech Recognition	Voice Recognition
1	Definition	It is the process of understanding spoken words.	It is the process of identifying a speaker.
2	Purpose	It performs a transcription of spoken words into text.	It aims at the identification of a speaker based on the voice characteristics in the supplied signal.
3	Principle	It uses AI to analyze an audio signal and identify words, phrases, or language patterns in the signal.	It uses AI to analyze vocal biometrics, such as pitch, tone, and rhythm, in a given audio signal.
4	Applications	Transcription services, virtual assistants, voice search, accessibility, creation of computer-generated captions that capture dialogue in multimedia content, etc.	Security purposes like unlocking, voice assistants, executing voice commands, etc.

and understand human speech in real time. They are able to function correctly irrespective of variations in accents, slang, pitch, speed, etc. (Chen et al. 2024; Goetzee et al. 2024).

The recognized language or command is utilized for transcription. It is sometimes used to operate a device. Instructions are frequently given to a virtual assistant.

4.3.1.1   Human Speech to Readable Written Text Conversion

Figure 4.5 shows the process of transforming the spoken words from a person's voice into easily understood text in written form. It comprises the stages of capturing the audio signal, preprocessing it, generating a Mel-scale spectrogram, post-processing, model processing, feature extraction, and output. They are implemented as follows:

   i. A microphone picks up and records speech samples. The recorded signal is as an analog audio signal.
  ii. The raw analog signal shows amplitude of sound wave in decibels with respect to time. It is preprocessed, by amplification, noise reduction, etc.
 iii. The next step is analog-to-digital conversion. During analog-to-digital conversion, the sound wave is divided into 1s-wide segments.
  iv. The Fast Fourier Transform (FFT) algorithm is applied to the digitized data. It converts the signal into a spectrogram, which is a plot of frequency on the Y-axis and time on the X-axis.
   v. The spectrogram is matched to the phonemes. As already said, a phoneme is a distinct unit of sound in a given language. The speakers of a language perceptually regard it as a single basic sound, e.g., there are ~40 phonemes in the English language.

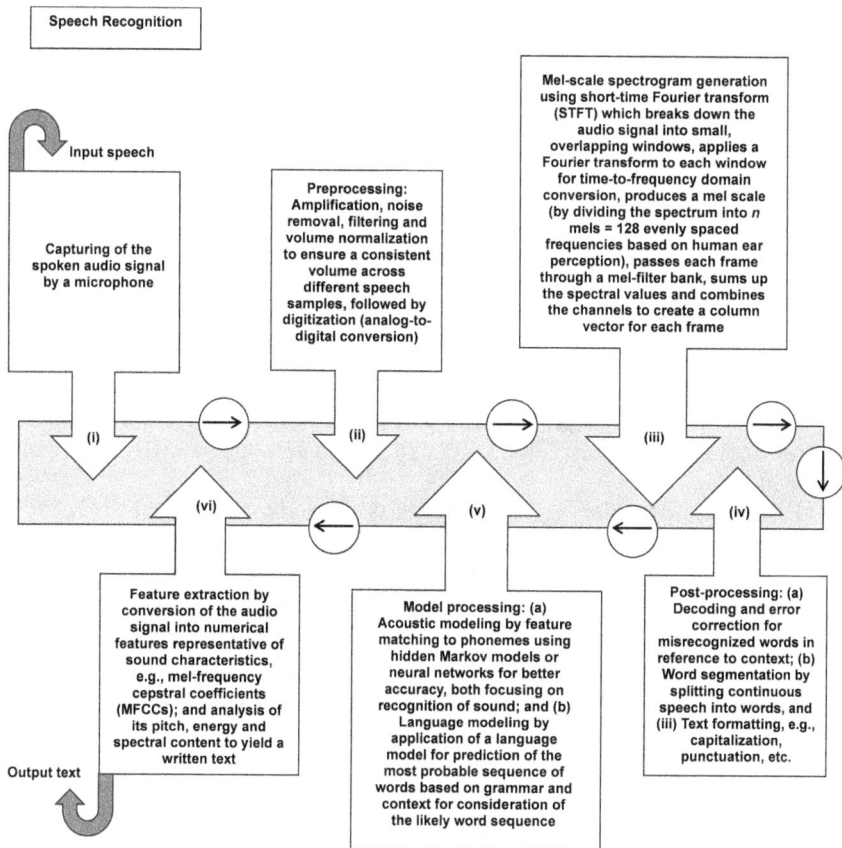

FIGURE 4.5 Recognizing words spoken by a person, and converting them into textual format.

vi. There are variations in speaking phonemes called allophones. These variations arise owing to differences in the gender, age, accent, and emotional state of the speaker. Nonetheless, the phonemes constitute the basic building blocks used by a speech recognition algorithm, such as the hidden Markov model and deep neural networks (DNNs). These algorithms arrange them in the correct order to form meaningful words and sentences.

A speech recognition system aims to find the most likely sentence that was uttered by a user given the speech input, as expressed by the equation

$$\hat{S} = \text{argmax}\,(S \in L)\left\{ P\!\left(\frac{S}{A}\right) \right\}$$

(4.1)

where 'argmax' is the abbreviation of 'argument of the maxima', representing the input value(s) at which the output value of the function is maximized,

L is the given language, $S \in L$ are possible sentences within it, A is the observed audio input, and $P\left(\dfrac{S}{A}\right)$ is the $\hat{S}$ probability of S given A is true (Jurafsky and Martin 2009).

Applying Bayes' rule, this equation may be written as

$$\hat{S} = \text{argmax}\,(S \in L) \left\{ \frac{P\left(\dfrac{A}{S}\right)P(S)}{P(A)} \right\} \tag{4.2}$$

where $P\left(\dfrac{A}{S}\right)$ is the probability of occurrence of A given S is true, and $P(S)$ and $P(A)$ are absolute probabilities of S and A, respectively. The denominator $P(A)$ is a common factor among all candidate sentences. Hence, it can be ignored.

vii. Hidden Markov Model: Correct order in phoneme arrangement is maintained using statistical probabilities in a three-layered process:

Layer 1: The algorithm examines the acoustic level and the probability of the phoneme. This is done to confirm that the correct phoneme has been detected by comparing it with well-known words, phrases, and sentences.

Layer 2: The algorithm scrutinizes the succeeding phonemes. It checks the probability that they should be following each other.

Layer 3: The algorithm inspects the word level. The objective is to find that the adjoining words make a sensible meaning. This is done by verifying the probability that they should be in succession.

A thorough probability analysis is carried out, followed by checking and re-checking. Then, the most likely text is presented as the output. The algorithm adequately fits the sequential speech content. Hidden Markov models have been the backbone of speech recognition. This is because they model speech as a sequence of states. In this sequence, each state represents a phoneme or a group of phonemes. The hidden Markov models provide a simple and effective framework for temporal modeling of speech signals as well as the consecutive phoneme arrangements for building a word. Albeit, the existence of a wide variety of phonemes and their possible combinations often renders it difficult to achieve perfection. Further information on hidden Markov models is given in Section 9.2.

viii. Deep Neural Networks: The DNNs represent complex connections between the speech input and the resulting text output through a hierarchy of layers. They can learn hierarchical representations of data. Thus, learning ability makes them particularly effective at modeling intricate patterns found in human speech. They are used for acoustic modeling to better understand the audio content of speech. They are used for language modeling as well to predict the likelihood of certain word sequences.

Neural networks can improve over time, offering a great flexibility advantage. The neural network is trained. All the different connections initially have the same weight. Necessary input data for training is supplied to the neural network. Specification of its accurate output is also made. The neural network then proposes a certain output. If the output from the neural network does not agree with the desired output, more training is needed. This difference between the actual and desired outputs is the error. The neural network adapts itself with the adjustment of weights to reduce the error. A neural network requires plenty of input training data to improve itself for error elimination. The necessity for abundant input data before becoming perfect is one of the drawbacks of neural networks in speech recognition. The other pitfall is that it fits the sequential nature of speech badly.

On a positive note, neural networks provide a supple approach. This approach grasps the varieties of the phonemes, making them capable of detecting the uniqueness of accents, emotions, age, gender, and so forth. Therefore, a hybrid strategy combining hidden Markov models with neural networks is adopted. In this hybrid strategy, each method compensates for the deficiencies of the partner, thus firming up complementarity.

4.3.1.2 NLP Algorithms

NLP algorithms are mathematical formulae used for training computers in understanding natural language (Wang et al. 2023; Khurana et al. 2023). They include the following:

i. Sentiment Analysis: It is the process of classification of text into positive, negative, or neutral sentiment categories. It can classify a movie review as either positive or negative from the language used inside it. It consists of the following steps:
   a. Tokenization: The text is broken down into individual words or tokens for separate analysis.
   b. Removal of Stop Words: The words like 'is', 'an', and 'the' have insignificant meaning. Therefore, they are removed to focus on the main words.
   c. Text Normalization: It is also known as stemming or lemmatization. It converts words into their base or root form, e.g., 'going' to 'go'.
   d. Feature Extraction: Key words that will help to determine sentiment are extracted. Adjectives like decent, evil, splendid, etc. are pulled out.
   e. Classification: The sentiment is classified using machine learning algorithms. Binary classification consists of positive and negative. A multi-class classification is represented by choosing more than two classes, e.g., delighted, gloomy, and annoyed, or on a scale (rating from 1 to 10). Sentiment finding becomes difficult whenever irony, sarcasm, or slang are encountered. Irony is a figure of speech that communicates the opposite of what is said. Sarcasm is an ironic remark made for mocking in which the speaker says something different from what the speaker actually means. Slang is the vocabulary of informal language between two persons of the same social group who know each other well.
ii. Keyword Extraction: This is the process of extracting relevant keywords or phrases from a single document. Keyword extraction helps identify topics or trends.

   iii. Named Entity Recognition (NER): It identifies predefined groups of entities, such as names of persons, organizations, locations, etc.

   iv. Topic Modeling: It aims to discover latent themes or topics across a collection of documents. It works by analyzing the frequency and co-occurrences of words across them. By such analysis, it clusters word groups and similar expressions that best characterize a set of documents.

   v. Intent Classification: This process tries to determine the intent behind a textual message, which can be a customer query, request, or complaint.

   vi. Knowledge Graph: It creates a graph network of important entities, such as people, places, and things. The graph allows easy understanding of the context and shows how different concepts are related.

   vii. Word Cloud: This is a graphical representation of the frequency of words used in the text. The intent is to gain insights about prominent themes, sentiments, or buzzwords around a particular topic. An example is identifying trends and topics in customer feedback.

   viii. Text Summarization: The text summarization algorithm creates summaries of long texts into shorter versions to make it easier for humans to understand their contents quickly for better analysis. Extractive summarization selects and combines the most important sentences or phrases. Abstractive summarization produces new sentences that capture the essence of the original text.

Robots should be able to report in natural language what they have done by providing concise summaries. They should be able to respond to questions about themselves. They should be able to learn from the natural language responses they receive to their summaries (DeChant and Bauer 2022).

4.3.2 Understanding the Deeper Meaning of Speech

4.3.2.1 Context-Aware Speech Recognition

Context awareness is the ability of a speech recognition system to extricate information from the previous utterances of an excerpt, the present circumstances, or domain knowledge for accurate interpretation of spoken words (Haase and Schönheits 2021; Chevalier et al. 2022). The system easily understands the likely meaning of a recent utterance by considering the words spoken previously. Incorporating knowledge specific to a particular domain can help the system disambiguate words that might be vague in general language, e.g., physics terminology in an automotive electronics setting provides a valuable guide in understanding the content. Integrating visual information aids in interpreting speech more exactly based on the visual scene.

    Most speech recognition systems depend on language models. These models predict the next word based on the preceding words. Advanced techniques consist of attention mechanisms in neural networks. They allow the system to focus on specific parts of the input speech based on the prevailing context. A few systems process contextual information separately using dedicated context encoders. They later integrate the processed information with the acoustic features of the speech.

Context-aware speech recognition is a boon to accuracy improvement. The system interprets ambiguous words and phrases more competently by utilizing contextual information. Thereby more explicit transcriptions are enabled. System robustness is further enhanced by quashing the interfering effects of background noise or accented speech by appreciating the intent.

Voice assistants are examples of context-aware speech recognition applications. These assistants use the previous conversation to understand the meaning of a query faultlessly when a question is asked. Intelligent speakers in smart homes exploit contextual information to interpret commands based on the ongoing activities in a room. A transcription system improves the accuracy of transcribing complex conversations. It considers the role of the speaker in reference to the complete topic of a discussion.

4.3.2.2 Semantic Parsing in Speech Recognition

It is the process of translating a spoken utterance into an expressly structured, machine-interpretable representation of its deeper meaning, inclusive of the entities, actions, and relationships. This is not merely translation into pure text (Erdogan et al. 2005; Corona 2016). Largely, it works by grabbing the key concepts and relationships within the spoken words. Hence, the computer is enabled to realize the intent behind a statement. It goes beyond the literal meaning of the words used in it. The parsed meaning is usually represented in a structured format. This format could be a knowledge graph or a definite command. Therefore, it is directly usable by a computer program, e.g., a voice assistant needs to catch a customer's intent to perform actions like scheduling a reminder or a wake-up call, playing songs, or arranging meetings. From a statement 'Set a reminder to wake up at 5 am on Tuesday', the parser would extract the action 'set reminder', the target 'wake up at 5 am', the time '5 am', and the day 'Tuesday'.

Semantic parsing uses NLP techniques, such as part-of-speech tagging, entity recognition, and dependency parsing. It is done after the speech has been transcribed into text. In semantic analysis, the system plucks out the meaning of the utterance. The plucked-out meaning is mapped to a formal representation from the linguistic structure and context. Contextual information from previous interactions or the user's current situation is incorporated to enhance accuracy.

Ambiguities in natural language originate from words having multiple meanings depending on the context. These ambiguities frequently pose difficulties for a machine to correctly interpret the intent. Handling of grammatical complexities in sentence structures and variations in speech patterns introduces complications in the execution of semantic parsing algorithms.

4.3.2.3 Dialogue Management in Speech Recognition

Dialogue management is the speech recognition component responsible for maintaining the context of a conversation by tracking crucial information, e.g., intent of the customer, entities mentioned, and the present stage of the dialogue (Passonneau et al. 2012; Reimann et al. 2024). It enables a natural and coherent interaction between a user and a voice-based system. Usually, in such systems, the recognition process becomes erroneous in scenarios where multiple user inputs are needed to complete a task.

The dialogue management workflow involves:

   i. Recognition of Intent: This aims to identify the user's primary action or goal.
  ii. Entity Extraction: Its purpose is recognizing specific pieces of information within the speech, e.g., names, dates, locations, or quantities.
 iii. Dialogue State Tracking: This is done using information already provided by the user.
 iv. Prompting and Clarification: These are done by asking follow-up questions or requesting additional information when necessary.
  v. Error Handling: It is useful in managing situations where the speech recognition system makes mistakes on encountering ambiguous input, allowing for graceful recovery and re-prompting.

Dialogue management works by:

   i. Speech Recognition: Here, the spoken words are converted into text by the speech recognition engine,
  ii. Natural Language Understanding: Here, the text is analyzed to identify the user's intent and extract relevant entities,
 iii. Dialogue State Update: In this part, the system updates its internal representation of the conversation based on the recognized intent and entities,
 iv. Dialogue Policy: Herein, the system decides what to do next, such as providing information, asking for clarification, or completing a task based on the current dialogue state, and
  v. Response Generation: In this period, the system generates a response, either in the form of spoken text or an action, based on the dialogue policy.

Practical dialogue management systems include smart home assistants, e.g.,

   i. For implementing the instruction, 'Adjust the oven at 90°C degrees,' the system needs to understand that 'oven' is the equipment, '90°C' is the desired temperature, and then to update the dialogue state accordingly; and
  ii. A virtual customer service agent when asked, 'What are office hours?' the system will recognize the intent as 'check office hours' and provide the relevant information '9 am to 5 pm'.

4.4 DISCUSSION AND CONCLUSIONS

In this chapter, use of speech synthesis and recognition, the two subbranches of speech processing, was discussed for facilitating human-robot interaction (Table 4.3). Recent advances in this field were surveyed. Service robots in restaurants and conversational robots in healthcare and interviews should be able to respond in natural language. The responses of these robots should include scientific terminology wherever applicable. Moreover, they should operate in an emotional manner for wide acceptability by people.

TABLE 4.3

Ideas and Information Gained from This Chapter

Sl. No.	Information Gained	Explanation
1	Summary	Speech recognition and understanding involve several operations, such as the conversion of human speech into readable written text, application of natural language processing algorithms, context awareness, semantic parsing, and dialogue management, all of which play vital roles in the process. Salient aspects of speech technology were elucidated in order to clarify how robots can communicate effectively with humans through spoken language.
2	TTS synthesis	Text-to-speech synthesis and voice generation for robotics were described in reference to the phases of text normalization, prosodic analysis, and concatenation of speech segments, as well as the target and concatenation cost functions.
3	Speech recognition	It consists of several operations, such as speech-to-text conversion, language processing, contextual adaptation and procedures beyond simple text processing for literal interpretation of speech, where each step contributes significantly to the overall outcome.
4	Keywords and ideas to remember	Text-to-speech synthesis, voice generation, cost functions, speech recognition, human speech to readable written text conversion, natural language processing, context-aware speech recognition, semantic parsing, dialogue management.

AN ELOQUENT ROBOT

I am a conversational robot
Chattering and gossiping
Questioning and Answering
My actions are never boring
My capabilities to talk and listen
Connect me easily to humans.
Robotic voice communication
Improves human-robot interaction.

Robot speech synthesis uses AI to create human-like speech. After analysis of the written text, it predicts how it sounds naturally and realistically, and produces a waveform played by the robot's speakers.

In speech recognition, a sentence spoken by a human operator is recognized by the robot using an ASR system (Tada et al. 2020). Then it applies syntactic and semantic parsing to determine the sequence of commands that it is expected to follow. Practically, the system is susceptible to inevitable errors. Errors also creep in because of environmental noises and distance of robot microphone from the speaking person. Recurrent neural networks can apply semantic parsing from sequences of letters and phonemes. Recognition error-resistant semantic parsers have been developed.

Following the discussion of robot speech technology in this chapter, let us turn our attention to making robots capable of seeing by equipping them with vision facilities. A robot lacking in this ability will struggle to avoid obstacles while negotiating complex spaces. It will not be able to pick up specific items or sort objects to perform tasks like assembly or inspection. Of course, a blind robot can navigate using LiDAR, SONAR or touch sensors but that will make matters more complicated. So, in the next three chapters we shall be engaged in investigating how robots are enabled to see their surroundings, as vision is the primary means to perceive the environment. Nevertheless, it should be borne in mind that robots may be required to work in in dark areas or in limited visibility conditions such as underground exploration of tunnels or in caves plunged in darkness, or in medical surgery. These robots rely on non-visual sensors, e.g., surgery is done using tactile sensors to manipulate delicate tissues.

REFERENCES AND FURTHER READING

Ahmad H. A. and T. A. Rashid. 2024. Planning the development of text-to-speech synthesis models and datasets with dynamic deep learning, *Journal of King Saud University - Computer and Information Sciences*, Vol. 36, 7, p. 102131, pp. 1–18.

Chen J., X. Zhou and Q. Qin. 2024. Research on Speech Recognition of Sanitized Robot Based on Improved Speech Enhancement Algorithm, *2024 5th International Seminar on Artificial Intelligence, Networking and Information Technology (AINIT)*, Nanjing, China, 29–31 March, pp. 1641–1644.

Chevalier P., B. Schadenberg, A. Aly, A. Cangelosi and A. Tapus. 2022. Context-Awareness in Human-Robot Interaction: Approaches and Challenges, *HRI* 2022, March 7–10, Sapporo, Hokkaido, Japan, pp. 1241–1243.

Corona R. 2016. An analysis of using semantic parsing for speech recognition, Undergraduate Honors Thesis, Computer Science Department, University of Texas at Austin, 36 pages.

Corrales-Astorgano M., C. González-Ferreras, D. Escudero-Mancebo and V. Cardeñoso-Payo. 2024. Prosodic feature analysis for automatic speech assessment and individual report generation in people with down syndrome, *Applied Sciences*, Vol. 14, 1, p. 293, pp. 1–13.

Dahan D. 2015. Prosody and language comprehension, *WIREs Cognitive Science*, Vol. 6, pp. 441–452.

DeChant C. and D. Bauer. 2022. Toward robots that learn to summarize their actions in natural language: A set of tasks, *Proceedings of the 5th Conference on Robot Learning (CoRL 2021)*, PMLR, London, UK, 8–11 November 2021, Vol. 164, pp. 1807–1813.

Erdogan H., R. Sarikaya, S. F. Chen, Y. Gao and M. Picheny. 2005. Using semantic analysis to improve speech recognition performance, *Computer Speech and Language*, Vol. 19, 3, pp. 321–343.

Goetzee S., K. Mihhailov, R. Van De Laar, K. Baraka and K. V. Hindriks. 2024. Audio-Visual Speech Recognition for Human-Robot Interaction: A Feasibility Study, 2024 *33rd IEEE International Conference on Robot and Human Interactive Communication (ROMAN)*, Pasadena, CA, USA, 26–30 August, pp. 930–935.

Gupta K. 2008. A Concatenative Synthesis-Based Speech Synthesizer for Hindi. In: Sobh T. (Ed.), *Advances in Computer and Information Sciences and Engineering*, Springer, Dordrecht, pp. 261–264.

Haase T. and M. Schönheits. 2021. Towards Context-Aware Natural Language Understanding in Human-Robot-Collaboration, *2021 IEEE 17th International Conference on Automation Science and Engineering (CASE)*, 23–27 August, Lyon, France, pp. 1648–1653.

Hunt A. J. and A. W. Black. 1996. Unit Selection in a Concatenative Speech Synthesis System Using a Large Speech Database, *1996 IEEE International Conference on Acoustics, Speech, and Signal Processing Conference Proceedings,* 9 May, Atlanta, GA, USA, Vol. 1, pp. 373–376.

Jurafsky D. and J. H. Martin. 2009. *Speech and Language Processing: An introduction to Natural Language Processing, Computational Linguistics, and Speech Recognition,* 2nd Edition, Prentice Hall, Pearson Education International, London, 1024 pages.

Khurana D., A. Koli, K. Khatter and S. Singh. 2023. Natural language processing: State of the art, current trends and challenges, *Multimedia Tools and Applications,* Vol. 82, pp. 3713–3744.

Kuo Y.-C. and P.-H. Tsai. 2024. Enhancing Expressiveness of Synthesized Speech in Human-Robot Interaction: An Exploration of Voice Conversion-Based Methods, *2024 10th International Conference on Control, Automation and Robotics (ICCAR),* 27–29 April, Orchard District, Singapore, pp. 1–4.

Li Y. and C. Lai. 2022. Robotic Speech Synthesis: Perspectives on Interactions, Scenarios, and Ethics, *HRI 2022 Workshop: RoboIdentity: Exploring Artificial Identity and Emotion via Speech Interactions,* 7–10 March, Sapporo, Hokkaido, Japan, ACM, New York, NY, USA, 4 pages.

Nair J., A. Krishnan and S. Vrinda. 2022. Indian Text to Speech Systems: A Short Survey, *2022 International Conference on Connected Systems & Intelligence (CSI),* 31 August to 2 September, Trivandrum, India, pp. 1–8.

Oralbekova D., O. Mamyrbayev, D. Kassymova and M. Othman. 2024. Current advances and algorithmic solutions in speech generation, *Vibroengineering Procedia,* Vol. 54, pp. 160–166.

Passonneau R. J., S. L. Epstein and T. Ligorio. 2012. Naturalistic dialogue management for noisy speech recognition, *IEEE Journal of Selected Topics in Signal Processing,* Vol. 6, 8, pp. 928–942.

Rabiner L. R. and R. W. Schafer. 2007. Introduction to digital speech processing, *Foundations and Trends in Signal Processing,* Vol. 1, 1, pp. 1–194.

Rashad M. Z., H. M. El-Bakry, I. R. Isma'il and N. Mastorakis. 2010. An Overview of Text-to-Speech Synthesis Techniques. In: Mastorakis N. E. and V. Mladenov (Eds.), *CIT'10: Proceedings of the 4th International Conference on Communications and Information Technology,* World Scientific and Engineering Academy and Society (WSEAS), Wisconsin, USA, pp. 84–89.

Reimann M. M., F. A. Kunneman, C. Oertel and K. V. Hindriks. 2024. A survey on dialogue management in human–robot interaction, *ACM Transactions on Human-Robot Interaction,* Vol. 13, 2, Article 22, pp. 1–22.

Supriyono, A. P. Wibawa, Suyono and F. Kurniawan. 2024. Advancements in natural language processing: Implications, challenges, and future directions, *Telematics and Informatics Reports,* Vol. 16, p. 100173, https://doi.org/10.1016/j.teler.2024.100173

Tada Y., Y. Hagiwara, H. Tanaka and T. Taniguchi. 2020. Robust understanding of robot-directed speech commands using sequence to sequence with noise injection. *Frontiers in Robotics and AI,* Vol. 6, Article 144, pp. 1–12.

Totsuka N., Y. Chiba, T. Nose and A. Ito. 2014. Robot: Have I Done Something Wrong? Analysis of Prosodic Features of Speech Commands under the Robot's Unintended Behavior, *2014 International Conference on Audio, Language and Image Processing,* Shanghai, China, 7–9 July, pp. 887–890.

Wang W., X. Li, Y. Dong, J. Xie, D. Guo and H. Liu. 2023. Natural Language Instruction Understanding for Robotic Manipulation: A Multisensory Perception Approach, *2023 IEEE International Conference on Robotics and Automation (ICRA),* London, UK, 29 May to 2 June, pp. 9800–9806.

Zinchenko K., C.-Y. Wu and K.-T. Song. 2017. A study on speech recognition control for a surgical robot, *IEEE Transactions on Industrial Informatics,* Vol. 13, 2, pp. 607–615.

5 Robots Able to See
General Aspects of Robot Vision

5.1 INTRODUCTION

A robot without vision is essentially blind. Although blindness may *prima facie* appear to be a serious impediment to robots' performance, it is not a hindrance to robots in performing certain tasks. Notwithstanding, robot vision (RV) is desirable in some situations. In others, it becomes an essential prerequisite. Therefore, we undertake to pursue the multiple facets of RV in depth in the present and the next two chapters.

This chapter outlines the fundamental principles of RV, distinguishing it from computer vision (CV). Although the inputs to both these techniques are images, the output of CV is information or features, whereas the output of RV is physical action performed by the robot. RV incorporates kinematics and reference frame calibration in its algorithms. Problems unique to RV arise from the fact that the data are collected hurriedly from a moving robot's sensor. Often, the position and orientation of the robot's sensor are not known clearly. Moreover, the motion of the sensor usually causes blurring of the images. Techniques to overcome these problems will be explained, notably active vision, anomaly and interest detection, semantic scene understanding, place recognition, simultaneous localization and mapping (SLAM), vision-based scene understanding, and 3D object detection.

5.2 IMAGES, VIDEO, AND VISION

5.2.1 IMAGE AND VIDEO PROCESSING

The combination of optics with signal processing gave birth to image and video processing. The image is treated as a two-dimensional signal. An image file represents a single, static frame. Video processing is a type of signal processing in which the input and output are video files or streams. A video is the recording, reproduction, and broadcasting of moving images, which is done by electronically representing a sequence of images or frames and combining them together for the simulation of the illusion of motion and interaction. A video file comprises a sequence of frames stored in various formats.

5.2.2 COMPUTER VISION

By combining image and video processing with machine learning, we obtain CV. Hence, computer vision = Optics + Image/video signal processing + Machine learning. CV is a branch of artificial intelligence for the automation of image analysis by training computers to understand images and provide their interpretations.

 DOI: 10.1201/9781032695266-5

5.2.3   Machine Vision and Pattern Recognition

Machine vision (MV) is a subset of CV that utilizes CV for industrial applications, such as automatic inspection and process control in factories. A related, though distinct, branch is pattern recognition (PR) or feature recognition, which deals with the identification of particular patterns or features in visual data. These patterns could be faces of individuals or objects of different shapes.

5.2.4   Robot Vision

RV is built by integrating all the preceding techniques and using additional components. Although RV is sometimes used synonymously with or interchangeably with MV, it must be distinguished from it. Some applications of MV, such as visual inspection, involve simply placing an optical sensor in front of machine parts for fault detection. Such applications are not connected with robotics.

RV unifies the concepts of robotics into its algorithms and methodology. Examples of robotic concepts include kinematics, calibration of reference frames, and the robot's ability to interact with the environment physically.

RV and action triggering involve capturing images of the scene, analyzing the captured images to recognize relevant features/objects in the images, and initiating desired actions to execute a job. It is a multi-stage process, as illustrated in Figure 5.1. A robot arm is shown near a workpiece from which objects will be picked up for an industrial process. The robot arm is properly illuminated and photographed with a CCTV camera. The captured image undergoes digitization, pre-processing, segmentation, and feature extraction. The resulting more informative dataset is used for image classification and interpretation. On the basis of these investigations, the requisite actions are triggered through the actuator, enabling the robot arm to execute the job. It is noted that the two main stages of this activity are image segmentation and feature extraction. For segmentation, the pixels constituting the image are treated as data points. They are partitioned into discrete groups based on their characteristics. A k-means clustering algorithm divides the pixels in the image into k clusters, generating a set of segments that cover the entire image. The features are extracted from these segments. The features are the individual measurable properties within a recorded dataset, e.g., numerical (integral or float), categorical (red, green, blue), ordinal (such as small, large, extra-large shirt size), binary (yes/no), or textual. Autoencoders or principal component analysis (PCA) are used. Autoencoders work by training a neural network in the recreation of its input data, thereby constraining it to discover structures in the data. PCAs entail dimensionality reduction of datasets to emphasize variations and reveal patterns/relationships.

The Watchful Robot

I am a robot with vision
I can see with my eyes like humans
I walk fearlessly without hesitation
To reach my destination
I am a smart sprinter
Whom no obstacles can hinder.

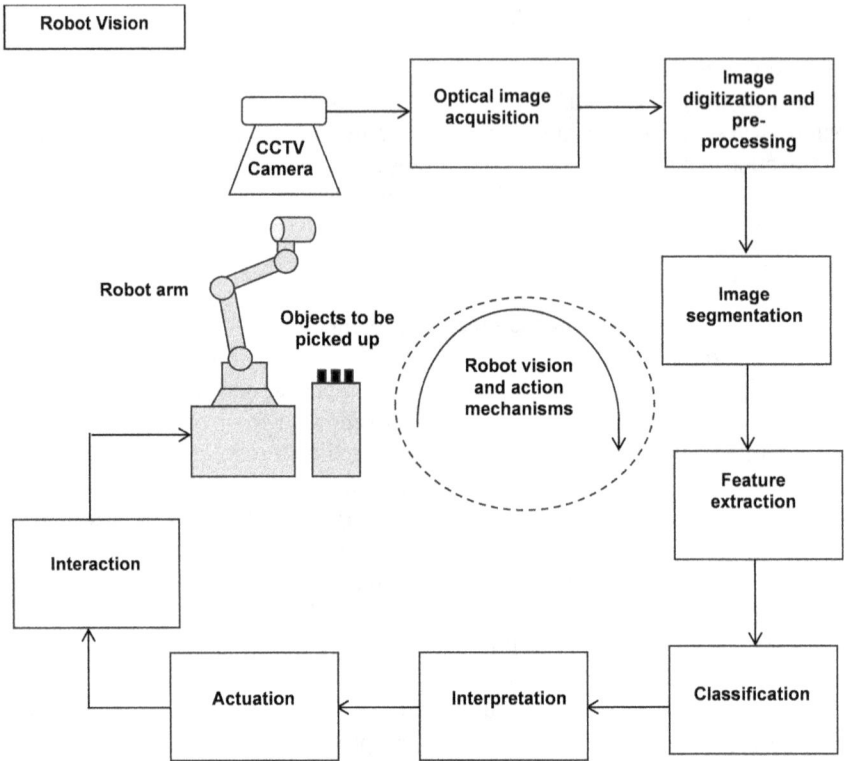

FIGURE 5.1   Stages comprising robot vision and action tasks.

5.2.5   DIFFERENCES AMONG SIGNAL AND IMAGE/VIDEO PROCESSING, PR, CV, MV, AND RV

The subtle differences of RV with its kin techniques can be easily visualized by looking at and comparing their inputs and outputs. The input to signal processing is an electrical signal, the raw signal. Its output, too, is an electrical signal, the processed signal. Both the input and output of image and video processing are images: the raw and processed images. For image processing, these are still images, while for video processing, they are moving images. The input as well output of PR is information. From CV, the input and output begin to differ. CV takes images or videos as input and dumps out information about the image or video, or its relevant features. Similar is the case with MV also. However, RV takes in images or video as inputs to generate physical actions as outputs. Briefly, we can state that:

Input and Output of Signal Processing: Electrical signals
Input and Output of Image Processing: Images
Input and Output of PR: Information
Input of CV: Images/video, Output: Information

TABLE 5.1

Difference between Robot Vision and Computer Vision

Sl. No.	Point of Comparison	Computer Vision	Robot Vision
1	Scope	It is a field of artificial intelligence.	It applies computer vision to robots. Hence, it is a computer vision subfield that is particularly relevant to robotics.
2	Function	It analyzes images and videos.	It incorporates robotic techniques, such as kinematics, into computer vision to enable robots to interact effectively with their environment.
3	Applications	Face recognition, video surveillance, and medical diagnostics.	Assistance in production lines and factories; in hospitals (to perform surgical procedures); in reconnaissance, surveillance, and space operations; and in search and rescue missions in difficult terrains.

Input and Output of MV: Same as for CV
Input of RV: Images/video, Output: Physical action

Table 5.1 presents the major contrasting features between RV and CV.

5.3 COMPUTER VISION

CV is supported on three technological pillars, viz.,

　　i. image classification,
　　ii. image classification with localization, and
　　iii. object detection.

5.3.1 Image Classification

Image classification is a two-step process consisting of:

　　i. Categorization of Pixels: Groups of pixels or vectors found in an image are subdivided into different categories on the basis of pre-specified rules, and
　　ii. Labeling of Pixel Categories: Labels are assigned to the categories of pixels or vectors contingent on the particular rules.

The various techniques used in image classification fall into one of the two types:

　　i. Unsupervised Image Classification: It is a fully automated method that does not require any data for training. Instead, it applies machine learning algorithms for analyzing and clustering the given unlabeled sets of data. The clustering is done by discovering hidden patterns or groups of data within

the supplied data. Two common machine learning algorithms used are
k-means clustering and the iterative self-organizing data analysis technique
algorithm (ISODATA).

ii. Supervised Image Classification: In this approach, previously classified
samples of data or pixels, called known reference samples, are used to train
the unknown samples of data or pixels. Two popular algorithms used are
support vector machine (SVM) and artificial neural networks (ANNs).

We shall describe k-means clustering, ISODATA, and SVM algorithms, and then
explain the image classification procedure. Thereafter, we shall move to neural
networks.

5.3.1.1   *K*-Means Clustering

As we know, the k-means clustering is an unsupervised machine learning algorithm
(Ikotun et al. 2023). It is used to group unlabeled data points into k clusters. The clustering is performed by randomly selecting a set of central points, known as centroids.
Then the data points are assigned to one of the k clusters. Assignment of data points
is based on the nearness of a data point to the centroid of a particular cluster. In this
way, all the data points are assigned to their respective clusters. Once this assignment has been completed, the clusters with new centroids are chosen to minimize
the sum-of-squares distances between each data point and its corresponding cluster
center. This minimizes the mean squared error (MSE), indicating the within-cluster
variability. The process is iteratively repeated until satisfactory clustering is achieved.

5.3.1.2   ISODATA

ISODATA is a data analysis technique for unsupervised classification of data. It is
iterative and self-organizing (Memarsadeghi et al. 2007). It is a modified version
of k-means clustering designed to overcome its shortcomings. k-Means clustering
assumes prior knowledge of the number of clusters, whereas ISODATA allows for
a variable number of clusters. In the ISODATA technique, the centers of clusters
are placed randomly. The data points are assigned to a cluster based on the shortest distance between the data point and the cluster center. The standard deviation σ
within each cluster, as well as the distance d between the centers of the clusters (the
inter-center distance), is calculated. Each cluster is decomposed into two clusters if σ
exceeds a user-predefined threshold and the number of pixels is double the threshold
value for the minimum number of data points. On the opposite side, the clusters are
blended together if d is less than a threshold distance or if the number of data points
in a cluster falls below a threshold value. An iteration is performed using the new
centers of clusters. The iterations are repeated until the distance d decreases to less
than the threshold value or the average change in d between successive iterations
decreases below a threshold.

5.3.1.3   The SVM Algorithm

The SVM is a supervised machine learning algorithm used for classification and
regression tasks. It works by finding the hyperplane that separates the different
classes in the feature space. The features are the colors of pixels, the textures, and the

edges in the images. SVM determines the hyperplane that maximizes the distance between the closest points of the different classes, known as the margin. The points located closest to the hyperplane are designated as the support vectors. It exhibits less vulnerability to overfitting than a neural network. Overfitting occurs when an algorithm fits too closely and tightly to its training data, such that it cannot generalize and cannot predict accurately when exposed to new data.

5.3.1.4 Image Classification Procedure

The image classification setup shown in Figure 5.2 applies supervised machine learning algorithms to train a model, known as the classifier, by feeding sample images along with their corresponding class labels. The trained model is then applied to classify unknown images submitted for analysis.

On the left-hand side of the diagram in Figure 5.2 is the training side, consisting of an image database from which specimen images are pre-processed, feature extraction is performed on the pre-processed sample images, and they are annotated with associated ground truth labels. The classifier model is trained using the features and class labels.

On the right-hand side of the diagram, there is the testing side. After the learning process is completed, unknown images are pre-processed, followed by feature extraction. The features of the query image are fed into the classifier, yielding the class label of the image.

5.3.1.5 The ImageNet 2012 Challenge and the Deep Learning Revolution in Image Classification

September 30, 2012, is a red-letter day in the annals of deep learning. On this day, a convolutional neural network (CNN) named AlexNet (Krizhevsky et al. 2017) successfully met the ImageNet challenge. It displayed excellent performance on ImageNet, the contemporary dataset of that time. The participants in this competition were required to accurately detect various objects and scenes, and classify images from a truncated list of 1,000 ImageNet classes. The AlexNet scored lower than a 25% error rate. The runners-up model was 9.8% points behind the winner. We shall first describe the main aspects of ImageNet, and then move on to the deep learning models, starting with AlexNet.

5.3.1.6 ImageNet: The Dataset for Image Classification

The ImageNet is a hierarchical database of images for vision research (Deng et al. 2009). This ontology of images is highly useful for training machine learning models in image classification and other image processing tasks. Ontology in AI refers to a set of concepts and categories within a knowledge domain that represent their properties and mutual relationships. The ontology of images, therefore, holds the key to the retrieval of images based on their contents.

By a hierarchical database is meant a model of data representation in which data is organized in the form of a tree structure. Such data organization makes navigation and searching easier. A tree is a data structure consisting of several nodes. The node is a point of intersection or branching of lines or pathways. The nodes are joined to each other by links. The single node at the highest level is referred to as the root node. Each element of the tree data structure has one parent node. It has either zero or more

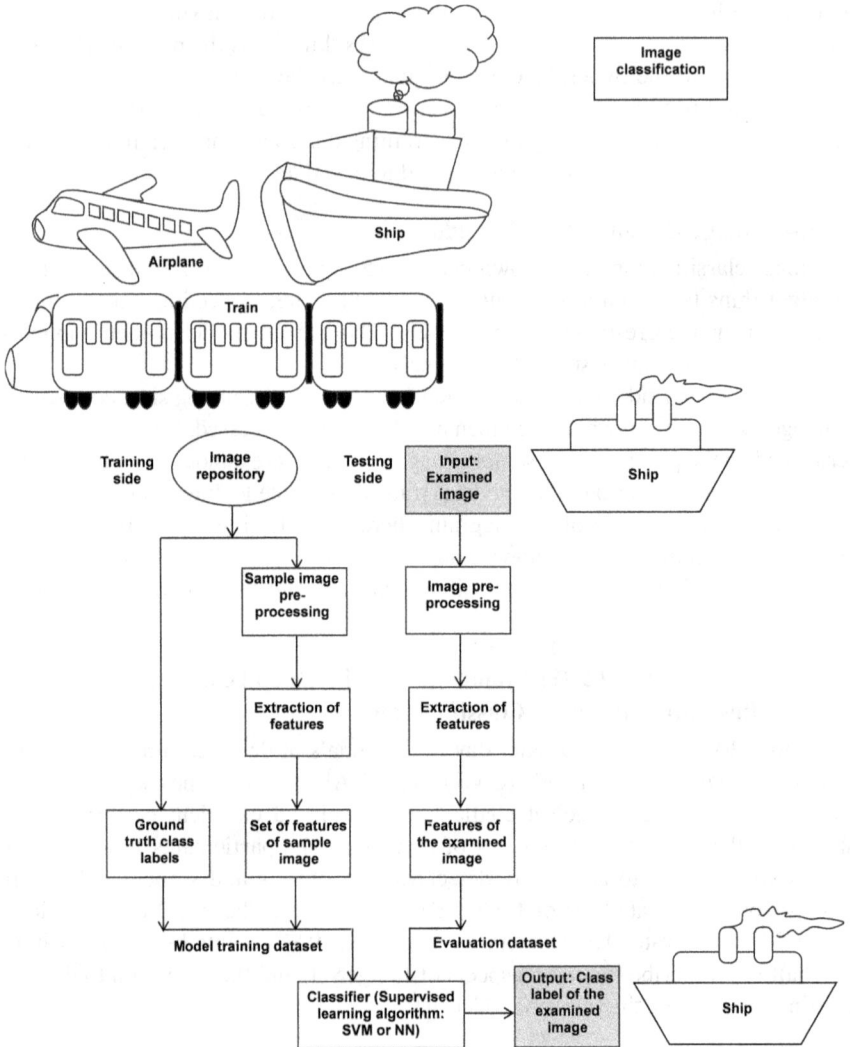

FIGURE 5.2 Workflow of image classification.

child nodes. The parent-to-child relationship is one-to-many. The child-to-parent relationship is one-to-one.

ImageNet utilizes the hierarchy of WordNet for the organization of images. WordNet® is a structured collection of information about words of the English language, commonly called a lexical database. In this database, English nouns, verbs, adjectives, and adverbs are arranged into sets of synonyms known as synsets.

Constructed on the WordNet spinal structure, ImageNet seeks to populate most of the synsets of WordNet, numbering ~80,000 with around 500–1,000 tidy images of high resolution, providing an open-source dataset of ~5×10^7 annotated clean photos for research purposes.

5.3.1.7   Deep Learning Models for Image Classification

i AlexNet: The AlexNet is a pioneering CNN architecture. It is a deep CNN
   comprising five convolutional layers, numbered 1–5, and three fully connected
   layers, 6–8 (Figure 5.3). The input image to AlexNet is a $227 \times 227 \times 3$ RGB
   image (227 pixels wide and 227 pixels high image with three color channels).
   The first convolutional layer has 96 filters of size $11 \times 11 \times 3$ and a stride of 4.
   The output image from this layer has the dimensions $55 \times 55 \times 96$ (a 3D volume
   with dimensions of 55 pixels in width, 55 pixels in height and 96 pixels in depth,
   indicating the number of layers in the stack). The max-pooling layer (with a
   filter size of 3×3 and a stride of 2) reduces image dimensions to $27 \times 27 \times 96$.
   The second convolutional layer has 256 filters of size $5 \times 5 \times 96$, a stride of

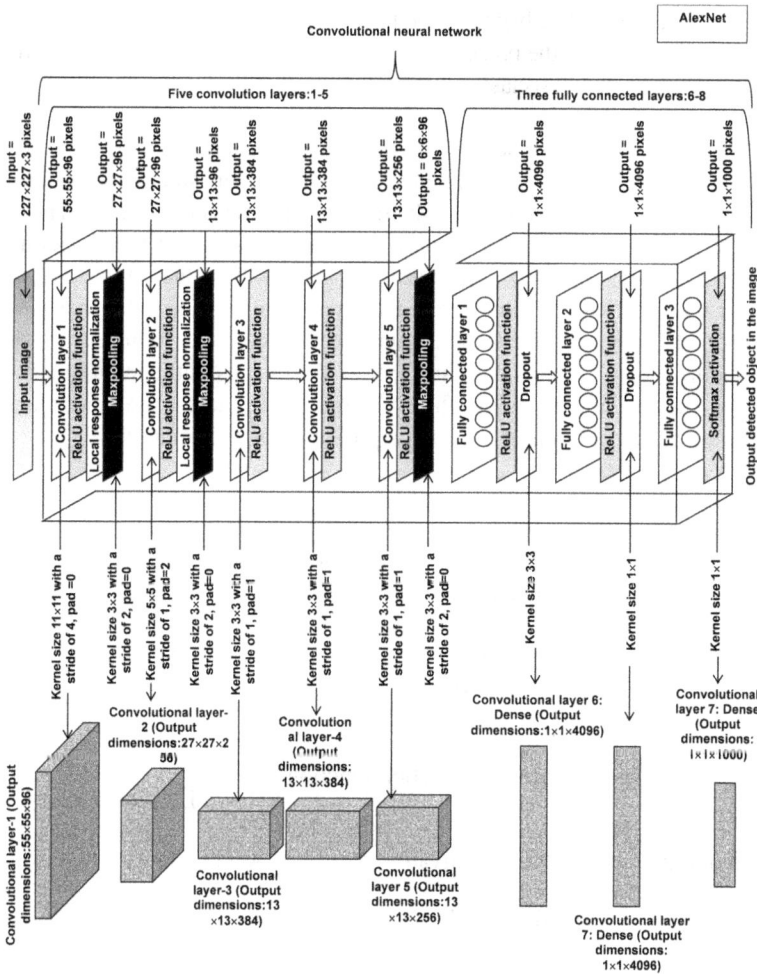

FIGURE 5.3   Detailed AlexNet architecture.

1, and a padding of 2. The output image dimensions after this layer become $27 \times 27 \times 256$. Then the max-pooling layer (with a filter size of 3×3 and a stride of 2) reduces the image dimensions to $13 \times 13 \times 256$. The third convolutional layer has 384 filters of size $3 \times 3 \times 256$, a stride of 1, and a padding of 1. The output dimensions after this layer are: $13 \times 13 \times 384$. The fourth convolutional layer has 384 filters of size $3 \times 3 \times 384$, a stride of 1, and a padding of 1. The output dimensions after this layer are: $13 \times 13 \times 384$. The fifth convolutional layer has 256 filters of size $3 \times 3 \times 384$, a stride of 1, and a padding of 1. The output dimensions after this layer are: $13 \times 13 \times 256$. The max-pooling layer (with a filter size of 3×3 and a stride of 2), decreases the image dimensions to $6 \times 6 \times 256$. In the sixth layer, the output from the previous layer is flattened to 9,216 units. Next to that, a fully connected layer consists of 4,096 units, all of which are fully connected to the previous 9,216 units. Dropout has been used for avoiding overfitting. The seventh layer is a fully connected layer with 4,096 units, which are all fully connected to the units of the previous layer. Dropout has been used as in the preceding layer. The eighth layer feeds into a softmax classifier having 1,000 classes distribution.

A CNN is a feed-forward neural network containing a stack of convolutional layers. A convolutional layer is contemplated as made of many square templates known as convolution kernels, which are mathematically matrices of weights. The kernels slide over the image looking for patterns. When the pattern of the kernel matches a portion of the image, the kernel gives a large positive value. Otherwise, a value of zero or a smaller value is registered. Padding means adding extra pixels, usually zeros, around the edges of the input image before convolution to ensure that the output feature map maintains the same spatial dimensions as the input image. Stride determines the size of the step by which the filter moves across the input image.

The sequence of layers in AlexNet is represented by the equation:

AlexNet

$$
\begin{aligned}
= &\left(\begin{array}{c} \text{Convolution layer 1} + \text{ReLU activation function} + \text{Local response normalization} \\ + \text{Maxpooling} \end{array} \right) \\
+ &\left(\begin{array}{c} \text{Convolution layer 2} + \text{ReLU activation function} + \text{Local response normalization} \\ + \text{Maxpooling} \end{array} \right) \\
+ &\left(\text{Convolution layer 3} + \text{ReLU activation function} \right) + \left(\begin{array}{c} \text{Convolution layer 4} \\ + \text{ReLU activation function} \end{array} \right) \\
+ &\left(\text{Convolution layer 5} + \text{ReLU activation function} + \text{Maxpooling} \right) \\
+ &\left(\text{Fully connected layer 6} + \text{ReLU activation function} + \text{Dropout} \right) \\
+ &\left(\text{Fully connected layer 7} + \text{ReLU activation function} + \text{Dropout} \right) \\
+ &\left(\text{Fully connected layer 8} + \text{Softmax activation function} \right)
\end{aligned}
$$

$$(5.1)$$

The convolution layer of a CNN is the feature extractor layer. It slides a filter over the image to identify features and build a feature map of the image.

The rectified linear unit (ReLU) activation function is the mathematical function expressed by the equation

$$\text{ReLU} f(x) = \max(0,\, x) \tag{5.2}$$

It converts all the negative values to zero, thereby introducing non-linearity in the deep learning model. It overwhelms the vanishing gradient problem, enabling the neural network to learn more complicated data relationships.

Unlike the tanh and sigmoid functions, the value returned by the ReLU activation function is not restricted within a defined range. This non-restriction imposes the constraint of local response normalization (LRN).

The LRN is a neurobiological phenomenon-based concept of lateral inhibition in which the output of neurons is locally normalized. The LRN assists neurons in learning from intricate data patterns in three ways:

   a. By suppression of feeble activations and laying emphasis on the intense activations,
   b. By contrast creation in a region through the production of a local maximum to improve sensory perception, and
   c. By raising the sensitivity of a neuron to its proximate neurons.

LRN is implemented in two modes:

   Mode 1: Within-Channel Normalization: Here, local regions undergo spatial extension, but they remain in separate channels.
   Mode 2: Across-Channel Normalization: In this case, the local regions extend across neighboring channels, but they do not have a spatial limit.

Some of the LRN/CNN layers are followed by max-pooling layers. The max-pooling is a down-sampling operation. It involves the sliding of a window called the filter or kernel across the input image data and picking up the maximum value within the window.

The fully connected (FC) layer is a neural network layer that obeys the condition that every neuron in the present layer is connected to every neuron in the preceding layer, thus producing complete linkages. It generates the final output predictions of the network.

Dropout is a regularization method used for prevention of overfitting of a neural network. It does so by modification of the structure of the network. Dropout prevents co-adaptation, a condition in which the neural network becomes heavily dependent on certain connections. During dropout, the input and output layers remain untouched. But some neurons of a chosen layer, along with their connections, are randomly deleted with a specified probability. The network parameters are then updated in accordance with the learning process. In the subsequent iteration, additional neurons are deleted, and network training is redone. Dropout rates vary from 0.2 to 0.5, depending on the neural network depth or the extent of the dataset. The larger the dataset, the lower the dropout rate needed, and hence, the less aggressive the dropout.

The softmax activation function is used in the final layer of an NN model for the transformation of raw output scores called logits into probability values. This is done by taking the exponential of each output. The values thus obtained are normalized by division by the sum-total of all the exponentials taken. Through this process, all values are confined within the bounds of (0, 1). Also, they add up to 1. Hence, they are construed as probabilities.

The AlexNet model could not be accommodated in a single graphical processing unit's (GPU) memory. So, it was split into two halves. Each half was run on one GPU by placing ½ the total number of kernels or neurons on each GPU. By cross-GPU parallelization, the GPUs were able to directly read from/write to each other's memory. This resulted in the omission of routing via the memory of the host CPU.

The AlexNet has 6×10^7 parameters and 6.5×10^5 neurons, and is trained for the classification of 1.2×10^6 images into 10^3 classes. The top-1 error rate is 37.5% and the top-5 error rate is 17% (Krizhevsky et al. 2017).

The AlexNet demonstrated the successful application of deep neural networks to extremely large datasets. Prior to the advent of AlexNet, CV was predominantly done using a machine learning model known as an SVM, and by shallow neural networks. An Alex-based technique is developed for the detection and classification of the grasped objects in robotics (Abbas et al. 2020).

5.3.1.8   The Visual Geometry Group-16 (VGG-16) Architecture

The VGG-16 is a deep CNN architecture (Simonyan and Zisserman 2015; Bagaskara and Suryanegaran 2021; Hussain et al. 2024) consisting of 16 layers, 13 convolutional layers, and three fully connected layers (Figure 5.4). Let us explain the top of the diagram in Figure 5.4: The 13 convolution layers are numbered 1, 2, 3, …, 13. The convolution layers 1, 2, and 3 each have a ReLU layer. Max-pooling layers are placed after (convolution layer 2 + ReLU), and convolution layers 4, 7, 10, 13. Input image dimensions are: $224 \times 224 \times 3$ pixels. The output image dimensions after the convolution layer 2 + ReLU are: $224 \times 224 \times 64$ pixels, and that after the next pooling layer are: $112 \times 112 \times 64$ pixels. The output image dimensions after the convolution layer 4 are: $112 \times 112 \times 128$ pixels, and that after the next pooling layer are: $56 \times 56 \times 128$ pixels. The output image dimensions after the convolution layer 7 are: $56 \times 56 \times 256$ pixels, and that after the next pooling layer are: $28 \times 28 \times 256$ pixels. The output image dimensions after the convolution layer 10 are: $28 \times 28 \times 512$ pixels, and those after the next pooling layer are: $14 \times 14 \times 512$ pixels. The output image dimensions after the convolution layer 13 are: $14 \times 14 \times 512$ pixels, and that after the next pooling layer are: $7 \times 7 \times 512$ pixels. The output image dimensions after the fully convolution layer 1 are: $1 \times 1 \times 4,096$ pixels, that after fully convolution layer 2 are: $1 \times 1 \times 406$ pixels and that after fully convolution layer 3 are: $1 \times 1 \times 1,000$ pixels.

The bottom of the diagram in Figure 5.4 provides a simple representation of the dimensional changes in the image. The convolution layers are shown by white rectangular boxes with gray sides, the pooling layers by black rectangular boxes, the fully convolutional layers by white rectangular boxes and the softmax layer by a dotted box.

A comparative assessment of neural networks with varying depths was conducted using an architecture with very small (3×3) convolution filters. When the network depth was increased to 16–19 weight layers, an appreciable enhancement in

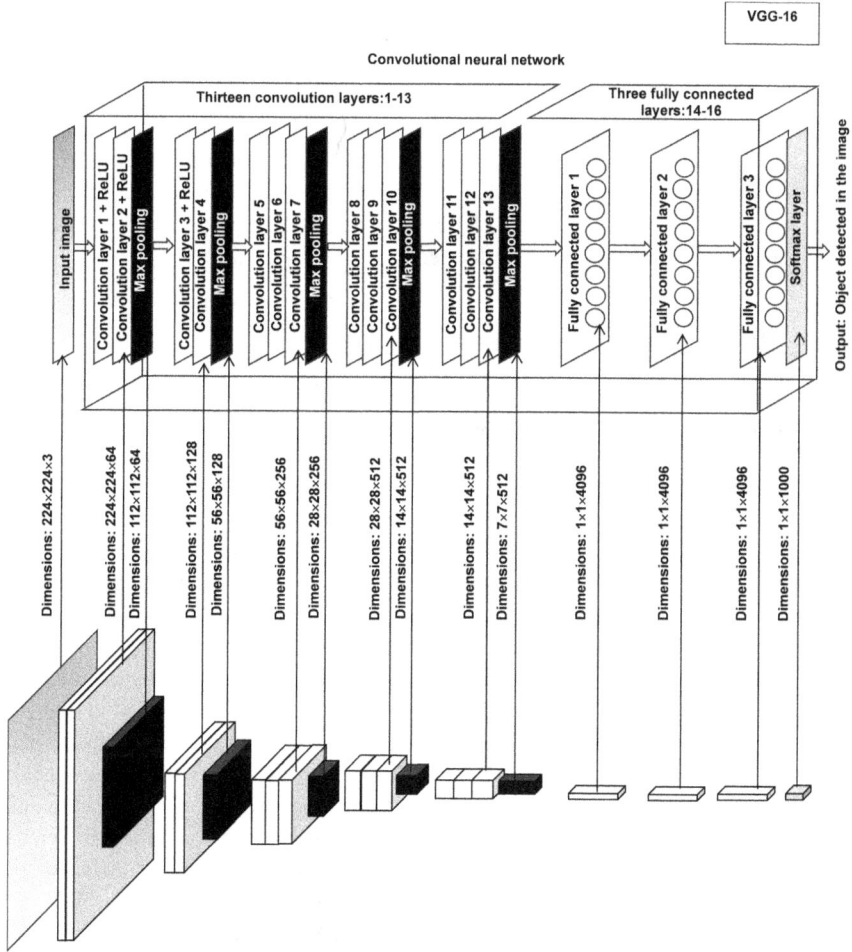

FIGURE 5.4 Structural details of VGG-16 Net.

network performance was observed for large-scale image recognition compared to existing techniques.

A deep learning method for human facial expression recognition is developed based on an improved VGG-16 CNN (Wu and Zhong 2021).

5.3.1.9 Very Deep CNNs

Inception v1 (GoogLeNet) was a cutting-edge deep neural network technology in the ImageNet Large-Scale Visual Recognition Challenge 2014 (ILSVRC 2014; Szegedy et al. 2015). It is a 22-layer deep network for image classification and detection.

CNNs having up to 34 weight layers can perform efficient optimization over long sequences, such as a vector of size 32,000 demanded for processing acoustic waveforms (Dai et al. 2017). A CNN with 18 weight layers has more than 18% higher absolute accuracy than a CNN with 3 weight layers.

5.3.2 NEURAL ARCHITECTURE SEARCH

The neural architecture search (NAS) is a technique for the automation of the design of ANNs using search algorithms for exploration and discovery of ideal neural network architectures for assigned tasks (Elsken et al. 2019; Chitty-Venkata and Somani 2022). A search space of possible architectures is defined. Deep reinforcement learning (DRL) methods are applied to find the most effective architecture in solving CV problems, e.g., landmark detection, object detection/tracking, registration on 2D/3D image data, image segmentation, and video analysis (Le et al. 2021).

5.3.3 PROGRESSIVE NEURAL ARCHITECTURE SEARCH

The progressive neural architecture search (PNAS) is a method for learning the structure of CNNs using a sequential model-based optimization (SMBO) strategy. The SMBO is a formalization of Bayesian optimization (BO) (Lacoste et al. 2014). The BO is a Bayesian theorem-based approach for optimizing decision-making about which parameter needs to be set next for an iteration by applying a real-valued function called the objective function, with conditional equations defining constraints. The objective function calculates the quantity to be optimized in terms of certain decision variables that can be chosen for their maximization or minimization to understand how the prior settings were performed. Two main components of BO are: a probabilistic model that approximates the objective function (surrogate model) and an acquisition function for guiding the choice of the next evaluation point in line with the surrogate model using the predicted mean and variance produced by the model.

In SMBO, the search algorithm is entrusted with the work of searching a neural network architecture space of cell structures, rather than a complete CNN (Liu et al. 2018). After learning a cell structure, it is stacked the required number of times to produce the final CNN. For stacking, the highest-ranked structure is nominated.

The search begins with simple models and forges ahead toward complex ones. During this search, the unlikely structures encountered on the way are pruned out. Concurrently with the quest for cell structures in an increasing order of complexity, another process is carried out, namely, the learning of the surrogate model that regulates the search in the structure space.

This method yields accuracies comparable to ultra-modern technological achievements on the CIFAR-10 dataset (Canadian Institute for Advanced Research, 10 classes) and ImageNet. It is fivefold more efficient than the RL-aided technique and eightfold faster than it (Zoph et al. 2018).

5.4 DISCUSSION AND CONCLUSIONS

In this chapter, we learnt the key terms of RV, followed by the basic principles of image processing to transform raw visual data into usable information for robots (Table 5.2). Image processing lays down the foundation for RV by allowing the robots to interpret visual data captured by their cameras as meaningful information. Before analysis, the quality of the input image is enhanced by performing operations such as filtering, smoothing, and color conversion. Image classification tells robots what

TABLE 5.2
Principal Concepts and Knowledge Acquired from This Chapter

Sl. No.	Knowledge Acquired	Explanation
1	Summary	Key terms related to images, video, and vision were defined, including image and video processing, computer vision, machine vision, pattern recognition, and robot vision. Hair-splitting differences in the meanings of seemingly similar terms were pointed out.
2	Image classification algorithms	The image classification algorithms discussed include k-means clustering, ISODATA, and support vector machine. The image classification procedure was expounded.
3	Image database	The ImageNet 2012 challenge and the deep learning revolution in image classification, and the ImageNet hierarchical database for vision research were described.
4	Deep learning models	Among the deep learning models for image classification, the AlexNet and the Visual Geometry Group-16 architectures were considered; these are pioneering convolutional neural network architectures in computer vision. Inception v1 is a deep CNN architecture.
5	Design automation	The design of artificial neural networks is automated with a neural architecture search technique.
6	Learning methods	Progressive neural architecture search, a technique for learning convolutional neural network structures, was discussed.
7	Keywords and ideas to remember	Images, video, and vision; signal, image, and video processing; computer vision, machine vision, pattern recognition, robot vision, image classification, k-means clustering, ISODATA, SVM algorithm, The ImageNet 2012 challenge, deep learning models for image classification, the Visual Geometry Group-16 architecture, very deep CNNs, neural architecture search, progressive neural architecture search

object is present in an image. The identified object is assigned a category label such as a person, a table, or a car. However, the knowledge about the identity of an object is insufficient for manipulating the object. The robot must be able to identify the locations of different objects within an image. For navigating autonomously, the robot must recognize objects and locate definite landmarks. Beyond basic image classification lies a more complex technique called object detection, which identifies multiple objects in an image. It pinpoints the exact positions of the objects within an image by marking their boundaries with bounding boxes. Object detection is the topic of the next chapter.

REFERENCES AND FURTHER READING

Abbas M., J. Narayan, S. Banerjee and S. K. Dwivedy. 2020. AlexNet Based Real-Time Detection and Segregation of Household Objects Using Scorbot, *2020 4th International Conference on Computational Intelligence and Networks (CINE)*, Kolkata, India, 27–29 February, pp. 1–6.

Bagaskara A. and M. Suryanegaran. 2021. Evaluation of VGG-16 and VGG-19 Deep Learning Architecture for Classifying Dementia People, *2021 4th International Conference of Computer and Informatics Engineering (IC2IE)*, Depok, Indonesia, 15–15 September, pp. 1–4.

Chitty-Venkata K. T. and A. K. Somani. 2022. Neural architecture search survey: A hardware perspective, *ACM Computing Surveys*, Vol. 55, 4, Article No. 78, pp. 1–36.

Dai W., C. Dai, S. Qu, J. Li and S. Das. 2017. Very Deep Convolutional Neural Networks for Raw Waveforms, *2017 IEEE International Conference on Acoustics, Speech and Signal Processing (ICASSP)*, New Orleans, LA, USA, 5–9 March, pp. 421–425.

Deng J., W. Dong, R. Socher, L.-J. Li, K. Li and F.-F. Li. 2009. ImageNet: A Large-Scale Hierarchical Image Database, *2009 IEEE Conference on Computer Vision and Pattern Recognition*, Miami, FL, USA, 20–25 June, pp. 248–255.

Elsken T., J. H. Metzen and F. Hutter. 2019. Neural architecture search: A survey, *Journal of Machine Learning Research*, Vol. 20, 1, pp. 1997–2017.

Hussain M., T. Thaher, M. B. Almourad and M. Mafarja. 2024. Optimizing VGG16 deep learning model with enhanced hunger games search for logo classification, *Scientific Reports*, Vol. 14, Article No., 31759, 34 pages.

Ikotun A. M., A. E. Ezugwu, L. Abualigah, B. Abuhaija and J. Heming. 2023. K-means clustering algorithms: A comprehensive review, variants analysis, and advances in the era of big data, *Information Sciences*, Vol. 622, pp. 178–210.

ILSVRC. 2014. ImageNet Large Scale Visual Recognition Challenge (2014), https://www.image-net.org/challenges/LSVRC/2014/

Krizhevsky A., I. Sutskever and G. E. Hinton. 2017. Imagenet classification with deep convolutional neural networks, *Communications of the ACM*, Vol. 60, 6, pp. 84–90.

Lacoste A., H. Larochelle, F. Laviolette and M. Marchand. 2014. Sequential model-based ensemble optimization, *arXiv:1402.0796*, pp. 1–9.

Le N., V. S. Rathour, K. Yamazaki, K. Luu and M. Savvides. 2021. Deep reinforcement learning in computer vision: A comprehensive survey, *Artificial Intelligence Review*, Vol. 55, pp. 2733–2819.

Liu C., B. Zoph, M. Neumann, J. Shlens, W. Hua, L.-J. Li, L. Fei-Fei, A. Yuille, J. Huang and K. Murphy. 2018. Progressive Neural Architecture Search, arXiv:1712.00559v3, pp. 1–20, https://arxiv.org/abs/1712.00559; In: Ferrari V., M. Hebert, C. Sminchisescu and Y. Weiss (Eds.), *Computer Vision – ECCV 2018*. Lecture Notes in Computer Science, Vol. 11205, Springer, Cham, pp. 19–35.

Memarsadeghi N., N. S. Netanyahu and J. LeMoigne. 2007. A fast implementation of the ISODATA clustering algorithm, *International Journal of Computational Geometry and Applications*, Vol. 17, 1, pp. 71–103; Preliminary version: *Proceedings of the IEEE International Geoscience and Remote Sensing Symposium (IGARSS'03)*, Toulouse, France, 2003, Vol. III, pp. 2057–2059.

Simonyan K. and A. Zisserman. 2015. Very Deep Convolutional Networks for Large-Scale Image Recognition, *3rd International Conference on Learning Representations, (ICLR 2015)*, San Diego, CA, USA, May 7–9, pp. 1–14.

Szegedy C., W. Liu, Y. Jia, P. Sermanet, S. Reed, D. Anguelov, D. Erhan, V. Vanhoucke and A. Rabinovich. 2015. Going Deeper with Convolutions, *2015 IEEE Conference on Computer Vision and Pattern Recognition (CVPR)*, Boston, MA, USA, 7–12 June, pp. 1–9.

Wu S. and S. Zhong. 2021. Expression recognition method using improved VGG16 network model in robot interaction, *Journal of Robotics*, Vol. 2021, 9 pages.

Zoph B., V. Vasudevan, J. Shlens and Q. V. Le. 2018. Learning Transferable Architectures for Scalable Image Recognition, *Proceedings of the 2018 IEEE/CVF Conference on Computer Vision and Pattern Recognition (CVPR 2018)*, Salt Lake City, UT, USA, 18–22 June, pp. 8697–8710.

6 Robot Vision
Object Detection by Robots

6.1 INTRODUCTION

In this chapter, we delve into object detection, a more intricate process compared to image classification because it involves both image categorization and localization, and therefore requires more computational power than image classification (Bai et al. 2020; Sun et al. 2024).

6.2 2D OBJECT DETECTION

2D object detection is a robot vision technique employed by robots to identify and locate various objects in images. It is implemented by collecting a variety of images that contain the objects that need to be detected. Then, boxes are drawn around the objects, and the objects are labeled with an annotation tool. Neural networks are trained for recognizing the objects in images.

2D object detection provides a fundamental level of visual understanding to robots by identifying objects' presence, location, and category within a 2D image plane. It is used in various robotic operations for:

   i. Robot navigation to identify obstacles, lane markings, or waypoints on a flat plane surface,
  ii. Robotic pick-and-place operations for locating specific objects on a table or conveyor belt,
 iii. Robotic inspection to identify components on a product platform and its defects, and
  iv. Automating barcode/QR code reading by identifying barcodes for inventory management and tracking of items in warehouses by robots.

But 2D object detection only provides planar information. The appearance of an object varies with the angle of view, raising issues of perspective. Therefore, it is unsuitable for tasks that require precise 3D object localization. 3D object detection is similar to 2D object detection, with the additional capability to understand depth and spatial relationships. 2D object detection is preferred for a robot operating in a well-structured environment with minimal variations in depth. The primary reasons are the lower cost and easier implementation of 2D cameras compared to their 3D counterparts. Furthermore, 2D cameras can process images much faster than 3D cameras and are well-suited to high-speed production lines.

DOI: 10.1201/9781032695266-6

Essentially, 2D object detection performs two operations entailing 2D image classification and localization (Zou et al. 2023):

i. Classification or Identification of Objects in an Image: It determines which objects are present in a given image and assigns them the correct class labels.
ii. Localization: It gives the bounding boxes for the identified object(s). Thus, the locations of the objects detected are marked.

6.2.1 DIFFERENCE BETWEEN IMAGE CLASSIFICATION WITH LOCALIZATION AND OBJECT DETECTION

Like 2D object detection, image classification with localization performs the dual task of classification of the main object in an image, along with its localization within the image. Its purpose is split up into the two sub-tasks:

i. Main Object Classification: The main object in the image is classified by determining its category.
ii. Localization of the Main Object: The position and size of the classified object are ascertained by defining a box surrounding the main object found in the image to indicate its exact location. The box is known as the bounding box.

Image classification with localization is a simpler process concerned with the classification of the main object found in an image and its localization within the image. Object detection is a complicated process of classifying and localizing all the objects in an image. It is essential to emphasize and clarify that image classification with localization is a sub-activity of 2D object detection (Kniazieva 2023).

6.2.2 PASCAL VISUAL OBJECT CLASSES (VOC) DATASET FOR BENCHMARKING OBJECT DETECTION

The PASCAL VOC (Pattern Analysis, Statistical Modeling, and Computational Learning Visual Object Classes) dataset is a publicly available and widely used reference dataset for evaluation of computer models for object detection and localization (Everingham et al. 2010). It consists of 20 object categories, including animals, dining tables, sofas, TVs/monitors, boats, bicycles, cars, airplanes, and people. It features annotations such as pixel-level segmentation, bounding boxes, and class labels, as well as appraisal matrices, e.g., mean Average Precision (mAP) for object detection and classification, and segmentation masks for image segmentation. The PASCAL VOC dataset comprises three subsets:

i. Training subset containing images for model training,
ii. Validation subset with images for model validation, and
iii. Test subset with images for benchmarking trained models.

6.2.3 Traditional Sliding Window Algorithm for Object Localization

A practical example of object detection using a sliding window is a service for the visualization of the location of a desired book in a library. In this example, CCTVs connected to a cloud server having databases of book title images provide the images for feature matching (Lee et al. 2017).

The algorithm consists of the determination of a quality function, e.g., a classifier score at several rectangular subregions of the image. The position at which the classifier score has the maximum value is decided as the location of the object. Accordingly, a fixed-size rectangular window slides across the image in left-to-right and top-to-bottom directions. As sliding takes place, a portion of the image is confined within the window. Features such as pixel intensities and histogram-oriented gradients (HOGs) are extracted from the portion of the image that falls within the window. These features are supplied to a classifier, e.g., a support vector machine (SVM) or a convolutional neural network (CNN), to determine whether the object of interest is enclosed within the window. If the answer is affirmative, a bounding box is drawn surrounding the object found.

In the sliding window algorithm shown in Figure 6.1, a small square/rectangular fixed-size window smaller than the examined image is created at the top-left corner

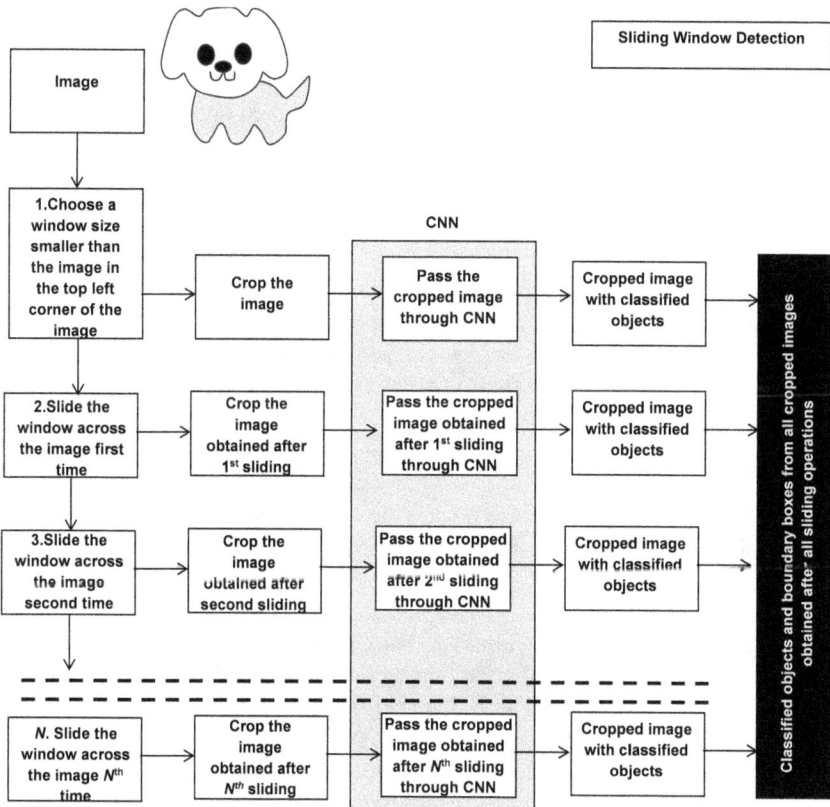

FIGURE 6.1 The sliding window algorithm.

of the image. The image is cropped and fed to the CNN. CNN analyzes the contents in the section of the image presented to it, looking for the objects or patterns that it has been trained to recognize. After analysis of one section of the image, the window slides a small distance to the right. Similar image analysis is performed for this section also. In this way, the image analysis is carried out for the N sections in the image, covering the complete image analysis. The cropped images with classified objects obtained during all these sliding operations are delivered as the output.

The sliding window algorithm is supplemented with image pyramids to detect objects of different sizes. The image pyramids consist of a stacking of images with the highest-resolution image at the foundation or base and the lowest-resolution image at the uppermost point or apex, resembling a pyramidal shape. The sliding window plus image pyramid combined effort entails resizing of the image to multiple scales and running the sliding window algorithm at each scale.

The sliding window method provides a straightforward implementation that accommodates a wide range of classifiers. It must be noted that complex-shaped objects or those varying considerably in appearance, as well as situations demanding window sliding at multiple scales, are more computationally intensive. Therefore, these cases are tricky and demand a lot of effort. To address these cases, the search effort is reduced by limiting it to a coarse grid of possible locations. Along these lines, the speed of computation is increased. However, this high speed is achieved at the cost of sacrificing the accuracy of localization.

6.2.4   BRANCH-AND-BOUND SCHEME-BASED EFFICIENT SUBWINDOW SEARCH

A targeted search is carried out in place of the complete search space. The targeted search is performed by decomposition of the parameter space into disjoint subsets at the primary stage of the search process. During decomposition, some portions of the parameter space are rejected. These are the portions where the quality function score is below a certain score from some earlier probed state. The search ceases as soon as a rectangle is identified that has a quality score on a par with the upper bound of the remaining prospective regions. Average precision scores of 0.240 for cats and 0.162 for dogs are obtained on the PASCAL VOC 2007 dataset (Lampert et al. 2008, 2009).

6.2.5   R-CNN: REGION-BASED CNN

The R-CNN was a pioneering effort toward the application of CNNs in object detection (Girshick et al. 2014). It is a machine learning model consisting of three modular sections:

   i. Category-Independent Region Proposal Generation Module: Classification of a huge number of regions is unnecessary. To bypass this lengthy and cumbersome process, the selective search algorithm is applied to extract merely 2,000 regions from the image. We proceed with these extracted regions further. These are named as region proposals.

   In the selective search algorithm, the image is over-segmented on the basis of the intensities of pixels. The bounding boxes of the segmented parts are added to the list of region proposals. Then the adjoining segments are grouped based on similarity considerations. The list of region proposals is fittingly updated. In this way, the segments grow in size. Thus, the region proposals are crafted from smaller to larger segments. This crafting takes place in a bottom-up fashion. Class labels are ascribed to the region proposals.

ii. Feature Vector Extraction Module: A large CNN extracts fixed-length feature vectors from each region proposal in this module. These feature vectors are extracted for predicting the class and the bounding box of the region proposal. To this end, the region proposals are warped into a square and feed-propagated through the CNN, producing a 4,096-dimensional feature vector. The features are fed from the output layer of the CNN to SVMs, which perform classification of the regions.

iii. Category-Specific Linear SVM Module: It performs classification of the regions. Multiple SVMs are trained for object classification. Each machine individually determines whether the input supplied contains a specific class. A linear regression model specially trained for the purpose predicts the ground-truth bonding box. This box is the hand-labeled bounding box of an object used for data training and testing.

In the R-CNN pipeline displayed in Figure 6.2, a selective search is performed on the image to produce ~2,000 region proposals. These are the bounding boxes around

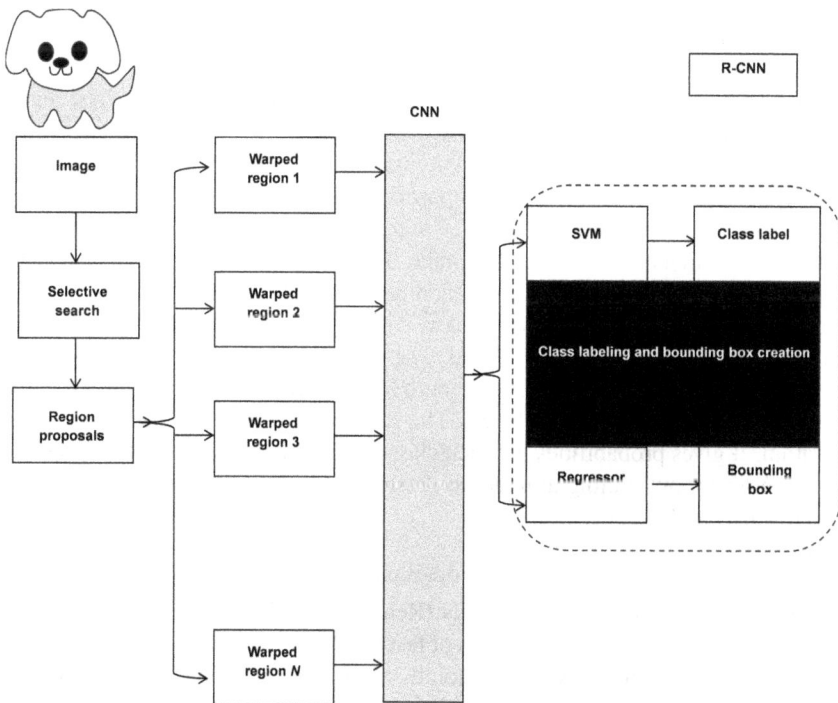

FIGURE 6.2 The R-CNN pipeline.

the objects of interest. From these region proposals, the warped regions 1, 2, 3, ..., N are generated by resizing to a predefined size. The warped regions are conveyed to a pretrained CNN to extract a feature vector of length 4,096 from each region proposal. The features are sent to a SVM engaged in classification to predict the class of the object. The features are passed through a regressor to make the bounding boxes of the detected objects.

Thus, the R-CNN combines CNNs with region proposals for an image. Hence, it enables the production of bounding boxes that contain objects and their corresponding classes.

6.2.5.1 The Bottleneck Faced by an R-CNN

The crux of the difficulty arises from the independent feed-forwarding of individual region proposals. Obviously, these proposals overlap in certain regions. So, the same region may be subjected to feature extraction on several occasions. Repeated feature extraction from the same region is exasperating. It leads to repetitive computation, resulting in the wastage of time and resources.

6.2.5.2 Fast R-CNN: Fast Region-Based CNN

It is an improved version of the original R-CNN. In this version, the input image undergoing feature extraction is the complete image, rather than region proposals. Thus, the overlapping issues encountered with R-CNN are circumvented (Girshick 2015; Shahin et al. 2021). A convolutional feature map of the full image is generated. From this feature map, the regions of interest (ROIs) of varying shapes are extracted. From the ROIs, features of the same shape are pulled out for easy concatenation. In order to do this, the fast R-CNN incorporates an ROI pooling layer. From the ROI feature vector, a softmax layer predicts the class of the proposed region. The regressor also provides the offset values for the bounding box.

The fast R-CNN flowline is depicted in Figure 6.3. A selective search is applied to the image to generate a set of region proposals. The generation of the region proposal set is followed by the creation of warped regions 1, 2, 3, ..., N. Simultaneously with the selective search performed on the image, a CNN extracts features from the entire image, FM-1, FM-2, ..., FM-N. The region proposals and the features are fed to the ROI pooling layer. This layer divides each region proposal into a grid of cells. Max pooling is done on each cell of this grid. The max pooling returns a single value for the features within the cell. The fixed-length feature vectors for each region proposal are sent to fully connected (FC) layers. The softmax classifier is a machine learning algorithm. It gives probabilities for each class label. The regressor predicts the location of an object by training a model to determine the coordinates of the boundary box surrounding the object.

6.2.5.3 Faster R-CNN: Faster Region-Based CNN

It is an improved version of fast R-CNN (Ren et al. 2015, 2017). The procedure followed in faster R-CNN differs from that of fast R-CNN. It substitutes the large number of regional proposals of R-CNN with a jointly trained region proposal network (RPN). Here, a ROI alignment layer is used in place of the ROI pooling layer in fast R-CNN.

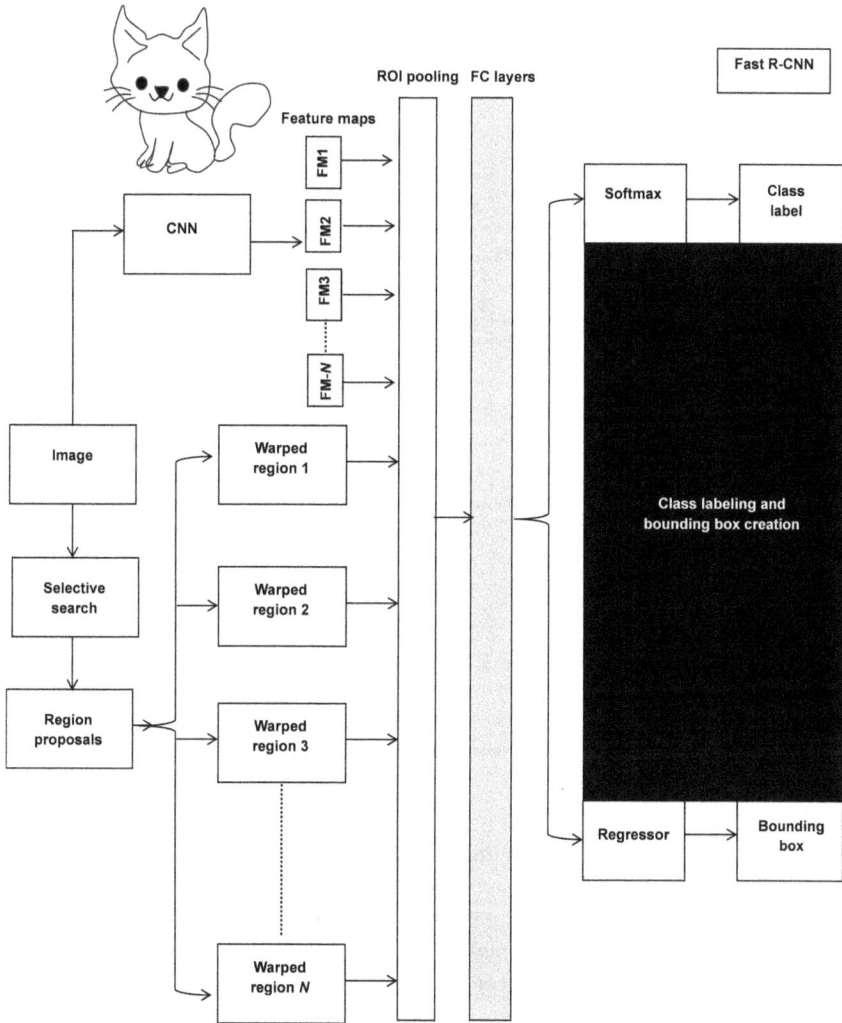

FIGURE 6.3 The fast R-CNN flowline.

The alignment layer applies bilinear interpolation. By this interpolation, it maintains the spatial information on the feature maps, thereby facilitating pixel level prediction.

Figure 6.4 shows the faster R-CNN workflow. The input image is fed into a CNN, which acts as a feature extractor for the entire image. The extracted features are supplied to a RPN. This is a CNN. The RPN slides filters over the features received from the first CNN to make region proposals. These region proposals and the feature maps are subjected to the ROI pooling operation. Max pooling is performed on non-uniformly sized inputs to produce fixed-size feature maps for various ROIs. The fixed-size patches move to the fully convolutional layers. A

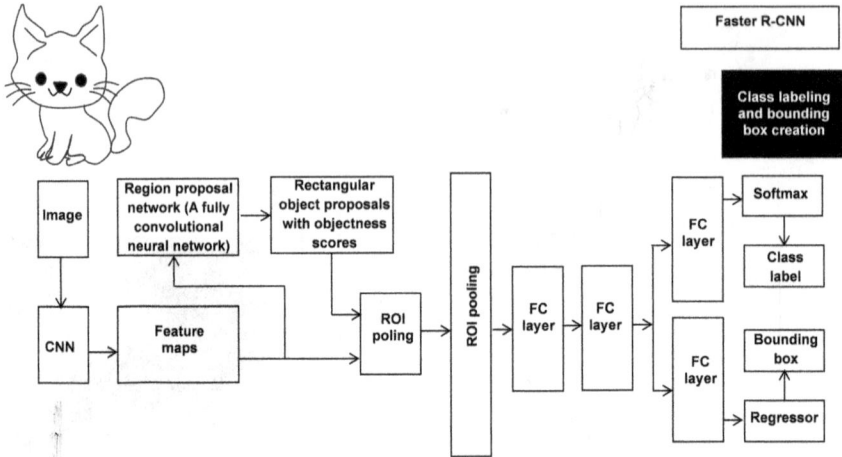

FIGURE 6.4   Flow sequence of a faster R-CNN system.

softmax classifier layer predicts the class of the object within the region proposal. The bounding box regression layer forecasts the refined bounding box coordinates for the classified object.

6.2.5.4   Mask R-CNN: Mask Region-Based CNN

Mask R-CNN is an R-CNN for instance segmentation. It is an extension of faster R-CNN (He et al. 2017; Le et al. 2018). It is simple in training and offers easy generalization to other tasks, such as human poses, which can be estimated in the same framework. In Mask R-CNN, a branch is added in parallel with the bounding box recognition branch. The added branch makes a prediction of an object mask. It thereby endows the capability for precise fine-grained segmentation and identification of the pixel-wise boundaries of each object, in addition to the usual object detection job. Thus, it can predict the shape of the object.

6.2.6   Unsupervised Object Discovery and Its Localization

A part-based region matching method applies a probabilistic Hough transform-supported matching algorithm. A standout score is introduced for foreground localization (Cho et al. 2015). The probabilistic Hough transform employs a small random sample of edge points instead of the entire set of edge points. This quickens the algorithm (Kiryati et al. 1991). Evidently, the sample should not be so small that detection of features becomes unfeasible.

For object detection experiments, a set of bounding boxes is formed around the objects and object parts. These object-containing participating regions are matched across images using the Hough transform. The Hough transform assigns a confidence value to each participant. The confidence is determined from both appearance and consistency viewpoints. The dominant objects are marked by comparing the scores of the contending regions. The formulated procedure is evaluated on PASCAL-2007 dataset by choosing the regions whose scores are higher than those of other regions containing the objects.

6.2.7 OBJECT DETECTION BY SELF-SUPERVISED FEATURE LEARNING

During the training for self-supervised learning, a premeditated excuse task called the pretext task is designed for solution by a CNN (Jing and Tian 2019). As examples of pretext tasks, image generation models learn by training on large datasets of images to understand and generate images based on visual features. Context-based tasks utilize context features, e.g., the context similarity and spatial relations among patches. In semantic labeling (assignment of class labels to pixels), the network is trained through extemporaneous labels. In a cross-modal strategy, the network is trained through the verification of correspondence between two channels of input data. The downstream tasks include image classification and object detection.

Based on the attributes of data, pseudo labels for the pretext task are spontaneously produced. In so doing, the neural network undergoes self-supervised training to learn the object functions of the pretext task. Thenceforth, the learned features are ferried to postliminary tasks as pretrained models to subdue overfitting. Features from only the first several layers are usually conveyed because the shallow layers capture low-level features, e.g., corners, edges, and texture. The deeper layers seize the high-level features of the task.

The performance of self-supervised and supervised methods on downstream object detection missions differs by less than 3% on standard datasets. This performance comparison suggests the generalizability of learned features by self-supervised mechanisms.

6.3 DISCUSSION AND CONCLUSIONS

2D object detection is essential for robot vision. It allows robots to identify and locate objects within their 2D camera view. It is crucial for basic tasks like navigation, manipulation, and interaction of robots with their environment. It provides valuable assistance to robots in obstacle avoidance by detecting walls, furniture, or people in a 2D camera view. It facilitates easy grasping of an object by a robot by precisely locating the position of an object in a 2D image. The requisite guidance is furnished to a robot manipulator to enable it to grasp the object. Object detection through 2D camera vision helps in identifying defects or specific objects.

On the whole, the robots can understand the layout of their environment. They can identify obstacles and plan movement paths. The main advantage of 2D object detection is that the algorithms employed here are faster and more computationally efficient than 3D algorithms. Hence, they enable real-time robot vision applications, even in complex scenarios that require advanced 3D perception. The basic understanding provided by 2D object detection is readily augmented with techniques like depth estimation to construct a more comprehensive 3D picture of the environment. Thus, it prepares the groundwork for advanced perception of the environment by robots. Table 6.1 gives glimpses of important insights gained from Chapter 6. A perusal of advanced topics in robot vision is deferred to the ensuing Chapter 7.

TABLE 6.1

Central Theme and Knowledge Gained from This Chapter

Sl. No.	Knowledge Gained	Explanation
1	Summary	2D object detection, a robot vision technique that allows robots to identify and locate multiple objects in planar images, is broader in scope than image classification with localization, which ascribes a single label to an entire image. Features of the PASCAL Visual Object Classes (VOC) dataset were presented. It is a widely acclaimed dataset used by researchers for benchmarking object detection algorithms and comparing their relative performances. Several methods of 2D object detection were outlined.
2	Sliding window algorithm	A traditional sliding window algorithm for object localization was described. It systematically moves a fixed-size window across an image and analyzes the content within each window to determine the presence of an object of interest.
3	Branch-and-bound scheme-based efficient subwindow search	Compared to a brute-force exhaustive search, the branch-and-bound scheme-based efficient subwindow search is an optimization technique to efficiently search through a large set of potential sub-images to find the optimal location of an object within an image, thereby significantly reducing the computational cost.
4	Deep learning model	A deep learning model that identifies objects within an image was discussed. Referred to as the region-based convolutional neural network (R-CNN), it generates potential regions of interest in the image and then extracts features from those regions using a convolutional neural network. Its successively enhanced variations are fast R-CNN, Faster R-CNN, and Mask R-CNN.
5	Unsupervised/ supervised learning	The unsupervised object discovery and its localization were explained, followed by object detection by self-supervised feature learning.
6	Keywords and ideas to remember	Images, video and vision; signal, image and video processing; computer vision, machine vision, pattern recognition, robot vision, image classification, K-means clustering, ISODATA, SVM algorithm, the ImageNet 2012 challenge, deep learning models for image classification, AlexNet, the visual geometry group-16 architecture, very deep CNNs, neural architecture search, progressive neural architecture search

We continue our discussion of robot vision in Chapter 7, highlighting some of the unique and unexpected challenges that robots face regarding vision. These exclusive problems arise out of the blue during a field operation and must be solved skillfully.

REFERENCES AND FURTHER READING

Bai Q., S. Li, J. Yang, Q. Song, Z. Li and X. Zhang. 2020. Object detection recognition and robot grasping based on machine learning: A survey, *IEEE Access*, Vol. 8, pp. 181855–181879.

Cho M., S. Kwak, C. Schmid and J. Ponce. 2015. Unsupervised Object Discovery and Localization in the Wild: Part-Based Matching with Bottom-up Region Proposals, *2015 IEEE Conference on Computer Vision and Pattern Recognition (CVPR)*, Boston, MA, USA, 7–12 June, pp. 1201–1210.

Everingham M., L. Van Gool, C. K. I. Williams, J. Winn and A. Zisserman. 2010. The PASCAL visual object classes (VOC) challenge, *International Journal of Computer Vision*, Vol. 88, pp. 303–338.

Girshick R. 2015. Fast R-CNN, *Proceedings of the 2015 IEEE International Conference on Computer Vision (ICCV)*, Santiago, Chile, 7–13 December, pp. 1440–1448.

Girshick R., J. Donahue, T. Darrell and J. Malik. 2014. Rich Feature Hierarchies for Accurate Object Detection and Semantic Segmentation, *2014 IEEE Conference on Computer Vision and Pattern Recognition*, Columbus, OH, USA, 23–28 June, pp. 580–587.

He K., G. Gkioxari, P. Dollár and R. Girshick. 2017. Mask R-CNN, *2017 IEEE International Conference on Computer Vision (ICCV)*, Venice, Italy, 22–29 October, pp. 2980–2988.

Jing L. and Y. Tian. 2019. Self-supervised visual feature learning with deep neural networks: A survey, *arXiv:*1902.06162v1, pp. 1–24.

Kiryati N., Y. Eldar and A. M. Bruckstein. 1991. A probabilistic Hough transform, *Pattern Recognition*, Vol. 24, 4, pp. 303–316.

Kniazieva Y. 2023. What's the difference between image classification & object detection? https://labelyourdata.com/articles/object-detection-vs-image-classification#:~:text=While%20image%20classification%20focuses%20on,to%20solving%20this%20intricate%20challenge

Lampert C. H., M. B. Blaschko and T. Hofmann. 2008. Beyond Sliding Windows: Object Localization by Efficient Subwindow Search, *2008 IEEE Conference on Computer Vision and Pattern Recognition*, Anchorage, AK, USA, 23–28 June, pp. 1–8.

Lampert C. H., M. B. Blaschko and T. Hofmann. 2009. Efficient subwindow search: A branch and bound framework for object localization, *IEEE Transactions on Pattern Analysis and Machine Intelligence*, Vol. 31, 12, pp. 2129–2142.

Le T. D., D. T. Huynh and H. V. Pham. 2018. Efficient Human-Robot Interaction Using Deep Learning with Mask R-CNN: Detection, Recognition, Tracking and Segmentation, *2018 15th International Conference on Control, Automation, Robotics and Vision (ICARCV)*, Singapore, 18–21 November, pp. 162–167.

Lee J., J. Bang and S.-I. Yang. 2017. Object Detection with Sliding Window in Images Including Multiple Similar Objects, *2017 International Conference on Information and Communication Technology Convergence (ICTC)*, Jeju, Korea (South), 18–20 October, pp. 803–806.

Ren S., K. He, R. Girshick and J. Sun. 2015. Faster R-CNN: Towards Real-Time Object Detection with Region Proposal Networks, *NIPS'15: Proceedings of the 28th International Conference on Neural Information Processing Systems*, Montreal, Canada, December 7–12, MIT Press, Cambridge, MA, USA, Vol. 1, pp. 91–99.

Ren S., K. He, R. Girshick and J. Sun. 2017. Faster R-CNN: Towards real-time object detection with region proposal networks, *IEEE Transactions on Pattern Analysis and Machine Intelligence*, Vol. 39, 6, pp. 1137–1149.

Shahin S., R. Sadeghian and S. Sareh. 2021. Faster R-CNN-Based Decision Making in a Novel Adaptive Dual-Mode Robotic Anchoring System, *2021 IEEE International Conference on Robotics and Automation (ICRA)*, Xi'an, China, 30 May to 5 June, pp. 11010–11016.

Sun Y., Z. Sun and W. Chen. 2024. The evolution of object detection methods, *Engineering Applications of Artificial Intelligence*, Vol. 133, Part E, p. 108458, https://doi.org/10.1016/j.engappai.2024.108458

Zou Z., K. Chen, Z. Shi, Y. Guo and J. Ye. 2023. Object detection in 20 years: A survey, *Proceedings of the IEEE*, Vol. 111, 3, pp. 257–276.

7 Robot Vision
Exclusive Challenges Faced by Robots

7.1 INTRODUCTION

Robot vision encounters several embarrassing and awkward situations during the robot's operation in practical settings in the real-world context. Robots experience difficulties and struggle to distinguish objects from complex backgrounds. Identifying and tracking moving objects in dynamic environments, too, is a fiddly issue. Therefore, the handling of variations in lighting and texture is necessary because they impact the quality of the acquired images. Adaptation of robots to occlusions and clutter is not easy. On numerous occasions, it is necessary to interpret 3D geometry from 2D images. To further exacerbate matters, all this must be done fast enough to ensure real-time performance in demanding scenarios. Robots must attain the ability to accurately perceive and react to a constantly changing, often messy, and sometimes ambiguous visual world. Humans often navigate intuitively in these cases. A few such bothering situations are mentioned in the subsections below, along with possible remedial suggestions (van Eden and Rosman 2019; Owen-Hill 2025).

7.2 MISCONSTRUED CIRCUMSTANCES IN ROBOT VISION

Several occasions arise when there are chances of misinterpretation of images (Figure 7.1). Misinterpretation of practical situations by a robot occurs owing to errors induced from various sources, including errors from robot's inaccurate position or orientation, errors from the effects of robot's motion on the collected data, errors from environmental effects on the collected data, and errors from blockage of the object robot's field of view.

7.2.1 LACK OF KNOWLEDGE ABOUT THE PRECISE POSITION OR ORIENTATION OF THE ROBOT

This occurs because the data is acquired during the robot's use by a sensor in motion, rather than by a fixed and stationary, vibration-free camera at a particular location, as was done during its training. Therefore, the conditions of the robot's training differ from those during its field operation. This difference may sometimes cause intolerant errors.

DOI: 10.1201/9781032695266-7

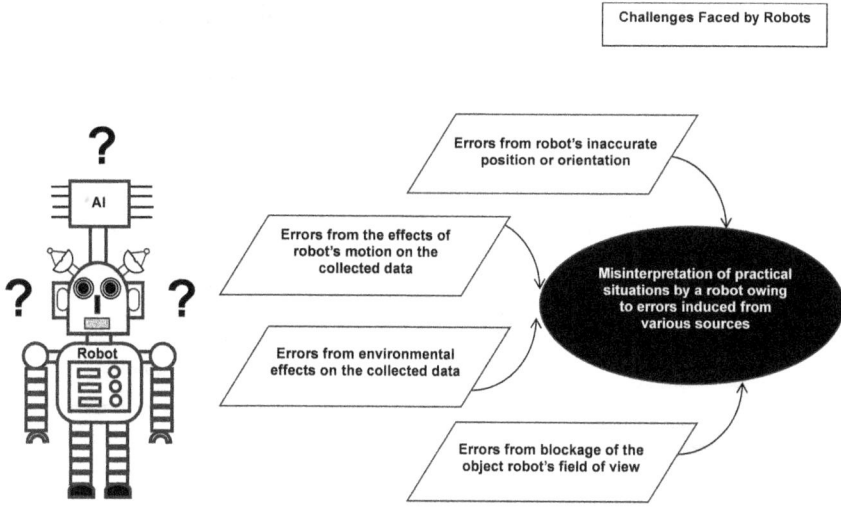

FIGURE 7.1 Confusing situations likely to be misinterpreted by robots.

7.2.2 INFLUENCE OF MOTION OF THE ROBOT ON THE COLLECTED DATA

The images captured by the sensor of a mobile robot are vulnerable to motion-induced blurring. Further, as the robot is continuously moving around the space under examination, the scales or orientations at which the objects are seen may not match with those at which the robot was trained (Nertinger et al. 2023).

7.2.3 INFLUENCE OF ENVIRONMENTAL PARAMETERS ON THE COLLECTED DATA

It is not necessary that the lighting conditions during the robot's field application are exactly identical to those that were employed during its training. The sensors of the robot are prone to errors arising from the variations in lighting conditions during the robot's training and its use in new situations.

7.2.4 OCCLUSION AND NON-VISIBILITY OF OBJECTS OF INTEREST

During the course of its motion, the robot may often find itself in a location where it is positioned in such a manner that the object of interest is not within its field of view. It may be blocked by intervening structures obstructing the view. These issues introduce serious complexities in the detection and localization of the object (Yoshioka et al. 2021).

7.3 MEETING THE CHALLENGES TO ROBOT VISION

The problems are solved by robots using various techniques (Figure 7.2). Methods for the mitigation of errors to meet challenges to robot vision are: active computer vision, anomaly detection, image-of-interest detection, semantic vision, visual place

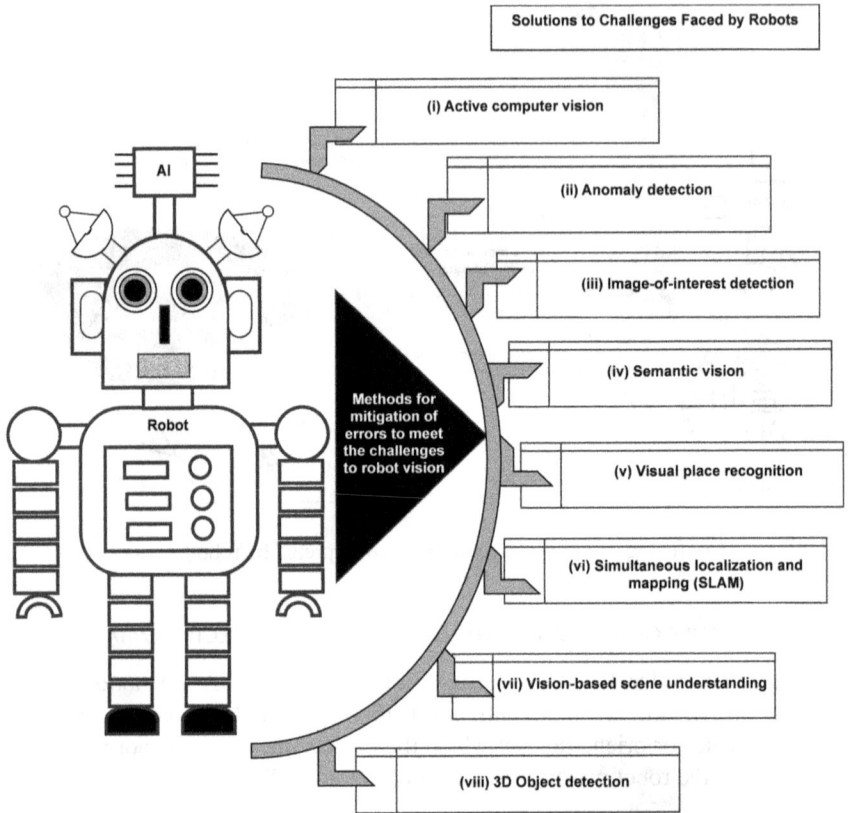

FIGURE 7.2 Robots overcoming the difficulties encountered in vision.

recognition, simultaneous localization and mapping (SLAM), vision-based scene understanding, and 3D Object detection, which are discussed in the subsections hereunder.

7.3.1 ACTIVE COMPUTER VISION

In the traditional passive computer vision, the robot's sensor seizes the entire scene and tries to extract useful information from the scene. The active computer vision approach is formulated around the interaction of the sensor with the environment (Yuille and Blake 1992; Zeng et al. 2020). Through this interaction, the robot can understand the environment in an effective and efficient fashion. During the robot's movement, the sensory data recorded by the camera vision are analyzed decisively while selectively rejecting irrelevant information. Utilizing this information, the viewpoint of the camera sensor is manipulated in order to make adjustments for proper investigation of the environment. These adjustments make it possible for the robot to obtain the necessary information that it wants to deal with the instantaneous issues it encounters. Active vision solves the problem of object occlusion by overcoming the limitations of the field of view. Additionally, difficulties caused by poor resolution of the camera are also overcome.

A crucial part of active vision is the planned sensing of perceptions from the environment. Sensor planning involves the determination of the pose and settings of the vision sensors of the robot to execute a task. It requires a multiplicity of views of the object to be handled (Chen et al. 2011). Thus, active vision bequeaths the robots with intelligent information-gathering capability by controlling the motion of their information-collecting visual sensors.

7.3.2 Anomaly Detection

The adoption of anomaly detection techniques greatly enhances the robustness and reliability of robots. An anomaly is a spatial, temporal, or spatio-temporal departure from the anticipated behavior and performance of a robot. The anticipated behavior and performance are defined in terms of the sequence of operational states, and the form and mode of the robot's interaction with its environment (Kim et al. 2022; Nandakumar et al. 2024). Anomalies of minor or major nature originate from unforeseen impediments or variations in the environment. They also arise from sensor/actuator failures. Anomalies are detected by model- and data-driven methods.

7.3.2.1 Model-Based Methods

Models are constructed based on advanced knowledge of a robot's dynamic behavior. Any deviation from the modeled behavior is an indicator of an anomaly (Xinjilefu et al. 2015). Model predictive control (MPC) applies a model of the system for the prediction of its future behavior (Saputra et al. 2021). A comparison of the observed robot's behavior with the predicted behavior helps in recognizing anomalies. Then, remedial actions are taken in real time. Kalman or particle filters are used to determine the internal state of a robotic system from practical measurements supplemented by a model of the system. Anomalies are identified by comparing the determined state with the actual state. Necessary corrections are made (Amoozgar et al. 2013).

7.3.2.2 Data-Based Methods

A method is described for the detection of an anomalous face (Bhattad et al. 2018). In this method, a feature vector is constructed that has unfailingly large entries for anomalous images. Unsupervised learning is used for scoring an image based on this feature. A peeking behavior in an autoencoder defeats obvious constructions.

The feature construction eliminates rectangular patches from an image. It gives a prediction about the probable content of the patch conditioned on the remainder of the image. A specially trained autoencoder is used for the prediction. The result of the prediction is compared with the image. When the score is high, it is surmised that the autoencoder faced difficulty in making a prediction. Likelihood of an anomaly is therefore implied.

The autoencoder is a neural network. It works by compression of the input data into its vital features. The input data compression is followed by reconstruction of the initial input from the compacted depiction. The compression is called encoding, and reconstruction is termed decoding. Latent variables in input data are discovered by training the autoencoder using unsupervised learning. The hidden or random variables inform us about the manner of distribution of data (Bergman and Stryker 2023).

7.3.3 IMAGE-OF-INTEREST DETECTION

Robots produce a voluminous quantity of images of the environment. Examination of this huge amount of information is a time-consuming and labor-intensive process. This wastage of time is easily avoidable by evolving a mechanism by which the information is arranged and ranked in accordance with its usefulness or likely interest to the user. This mechanism automatically flags the information of interest to quicken the image analysis. The interest aspect is directly included in a method developed to remedy this situation (Burke 2017). In this method, random pairs of images are presented to a human operator. The presentation is used for the selection of images of interest to the application being run. A Gaussian process smoother dramatically decreases the number of comparisons than those required in standard probabilistic algorithms. This is achieved by utilizing the resemblances between features of images extracted by a convolutional neural network (CNN) that has been previously trained.

In another approach, histogram features are extracted from saliency maps, which highlight the pixels or regions of an input image that contribute most to the model's prediction. These features are applied to determine the existence of interesting objects in images (Scharfenberger et al. 2013).

7.3.4 SEMANTIC VISION

Semantics is the study of the meaning of data. Semantic vision involves understanding the objects found in an image. Their spatial and functional interrelationships are examined (Sevilla-Lara et al. 2016). Semantics analyses objects with respect to the layout and 3D structure of the scene. It works by segmentation of an image into regions of interest. Classification of each pixel in a segment is done, and it is assigned to one of several classes, e.g., a car, a road, a tree, and sky. Traffic scene understanding is provided by semantic vision. This understanding is essential for a self-driven autonomous vehicle (Geiger et al. 2014).

7.3.5 VISUAL PLACE RECOGNITION

Place recognition is the process of accurately spotting the location of a given query image. The spotting is done from the locations of images of the same place in an extensive geotagged database (Zeng et al. 2018). Weather conditions and illumination alter the appearance of the image of a particular place appreciably. So, they pose hurdles in this process. Therefore, changes in appearance within the environment must be taken into account. The accounting is obligatory, and must be done either explicitly or implicitly in place recognition solutions to prevent chances of failure (Lowry et al. 2016).

A traditional method of place recognition is distinctive invariant feature extraction on a scale-invariant basis. The Scale-Invariant Feature Transformation (SIFT) algorithm described in the next section is widely used for extracting distinctive invariant features. Of late, CNNs have become the predominant image representation extractors.

7.3.5.1 Scale-Invariant Feature Transformation

Individual features are matched to a database of features of known objects (Lowe 2004; Guo et al. 2018). A fast nearest-neighbor algorithm is used. It is a method of determining the closest point to a specified point in a set. A Hough transform is applied for the identification of clusters pertinent to a single object. The Hough transform is a technique for the detection of shapes, such as lines and circles in an image by converting them into mathematical representations. It makes recognition easier, even for obscured or broken shapes. Verification is performed through a least-squares solution, which solves the equation $Ax=b$ as closely as possible, minimizing the error. The features are invariant to the scale of the image and its rotation. The matching is robust across variations in illumination, 3D viewpoint, and distortion.

7.3.5.2 CNN-Based Approach

This approach is explained in Sections 5.3.1.7–5.3.1.9.

7.3.6 Simultaneous Localization and Mapping

The SLAM algorithm is a method which concurrently and recursively performs two real-time operations (Durrant-Whyte and Bailey 2006; Bailey and Durrant-Whyte 2006; Khairuddin et al. 2015):

a. Determining the location of the robot's camera within the test environment.
b. Updating the map of the environment.

It assists the robot in understanding the place at which it is located in the environment in relation to the structure of the environment, e.g., it helps an autonomous vehicle to map out the environment and pinpoint its location in that map.

Let us see how SLAM improves the functionality of a home vacuum cleaner robot. The SLAM uses the onboard camera and other sensors to create a map of the cleaning area. This map illustrates the potential obstacles that the robot may encounter during its movement. The map guides the robot's motion and prevents it from cleaning the same area twice. By localizing itself, the robot estimates the amount of motion required to move from its current position to a nearby location. The estimation is performed using camera sensor data and information on the number of wheel movements. On the other hand, a robot without an SLAM facility will wander randomly through the room. It will clean certain areas multiple times, while leaving other areas unclean. It will consume a lot of power, thereby excessively draining the battery (MathWorks 2024).

7.3.7 Vision-Based Scene Understanding

For the execution of grasping and manipulation jobs, a robot is required to compute grasps for a large number of objects in dynamic and cluttered environments. These can arise from a change in the workplace of the robot, noise effects, or inaccuracies in control. Modified forms of CNN architectures necessitate exact camera calibration and accurate robot control. They take disproportionately long computation times even in static conditions, leading to their infrequent use.

A real-time, object-independent grasp synthesis method is developed for closed-loop grasping (Morrison et al. 2018). Depth images of the region around the object to be grasped are recorded by a camera sensor mounted on a robot's wrist. A generative grasping convolutional neural network (GG-CNN) is applied. It produces antipodal graphs.

In generative AI, CNNs are utilized within generative adversarial neural networks (GANs) to produce and discern visual content. CNNs are engaged in determining whether a picture contains a certain object. This is a recognition task. GANs strive to make a picture of the same object. This is a generation task. Both networks are constructing a representation of a distinctive picture of the object. The term 'antipodes' refers to diametrically opposite points on a body.

The grasps are parameterized in terms of quality of grasp, angle, and gripper width for each and every pixel in the image. This step usually takes a fraction of a second to complete. After computation of the best grasp, a velocity command is sent to the robot. The system works by closed-loop control. Therefore, dynamic objects are graspable. Errors in control can also be corrected.

For proactive planning and action, the robot should holistically perceive the information of a workplace. For a holistic understanding of a scene based on vision, the cognition of objects, humans, and the environment is taken into consideration along with visual reasoning. Thus, the visual information is compiled into semantic knowledge. This compilation enables a robot's collaboration with humans in making decisions (Fan et al. 2022).

7.3.8 3D OBJECT DETECTION

7.3.8.1 Point Clouds, Depth Maps, and Stereo Images

We collect data points in a 2D space represented by an (X, Y) coordinate system to draw the 2D image of an object. Likewise, we can collect data points in 3D space represented by the (X, Y, Z) coordinate system to sketch the 3D shape of an object. A collection of data points for drawing the 3D shape of an object in a 3D (X, Y, Z) system of coordinates is referred to as a point cloud. Each point in a point cloud is characterized by a set of (x, y, z) coordinates. Point clouds are used for mapping features such as buildings, infrastructure, terrains, and roads (Zheng et al. 2023). A 3D scanner or a LIDAR is employed for the creation of point clouds. LIDAR is an active remote sensing device. It emits laser pulses. The reflected pulses from an object are captured. The time of flight is measured. Then the formula calculates the distance of the point on the object from itself

Distance of the reflecting point on the object

$$
= \left(\begin{array}{l} \text{Speed of the light} \\ \times\text{Time taken by lincident light to reach the point plus} \\ \text{the time taken by reflected light from the point to return to the laser} \end{array} \right) / 2 \tag{7.1}
$$

Finally, it converts the distance traveled into elevation. Photogrammetry software is a computer program that takes multiple overlapping photographs of an object from different angles. The patterns of electromagnetic radiant pictures are analyzed to produce a 3D model of the object.

A depth map is an image formed by real or integral values that are measured with respect to the viewpoint. A pair of images of an object taken from different perspectives constitutes a stereo image (Häne et al. 2011).

7.3.8.2 The 2D vs. 3D Object Detection

In 2D object detection, the images are annotated by drawing boxes around the objects in them and labeling them. Using the annotated images as input data, a 2D object detection model is trained for the recognition and localization of objects in these images through patterns and features. After training, the model is applied to new images for object detection by inference.

In 3D object detection, annotations include the depth or distance from the camera as an additional parameter. A 3D object detection model is trained with the annotated data. The new 3D object data is exposed to the trained model to draw inferences about 3D objects.

The input data for 2D object detection are red-green-blue (RGB) images, whereas the same for 3D object detection are RGB images, point clouds, depth maps, and stereo images.

Annotation for 2D object detection is relatively simple in nature, involving 2D bounding boxes, while the annotation for 3D object detection is intricate, entailing 3D bounding boxes.

Models used for 2D object detection are: YOLO (You Look Only Once) and SSD (Single-Shot MultiBox Detector). YOLO uses a single CNN to split an image into grids. It enables the prediction of bounding boxes and class probabilities or confidence scores. It processes images at 45 frames^{-1} while its smaller version, Fast YOLO, does so at 155 frames^{-1} (Redmon et al. 2016).

SSD is a neural network model. It works by discretizing the output space of bounding boxes into a group of default boxes. These boxes range over dissimilar aspect ratios and scales for feature map locations. The SSD is easily trained with a smaller number of images to yield more accurate predictions than other single-stage methods (Liu et al. 2016).

Indoor robotics needs reliable 3D object detection. For exploiting RGB-D imagery to perform 3D object detection, the objects in the world are represented in terms of 3D cuboids. This is done by extending the automatic object segmentation using the constrained parametric min-cuts (CPMC) framework to 3D. The CPMC is a framework for generating and ranking conceivable hypotheses regarding the spatial extent of objects in images (Carreira and Sminchisescu 2012).

The physical and statistical interactions between the objects and the environment are modeled along with interactions between objects. On the basis of this modeling, an integrated framework is proposed to detect and recognize 3D cuboids in indoor scenes. Experiments demonstrate that the approach gives an effective combination of segmentation features and geometrical properties apart from contextual relations between objects (Lin et al. 2013).

A model for 3D object detection is the PointNet. It is a unified, effective, and efficient architecture for classification of objects, part segmentation, or semantic parsing of a scene (Charles et al. 2017). The architecture takes the point cloud as input. For object detection, the point cloud consists of samples from the shape of the object. For semantic parsing, the input is a single object or a small part of a scene.

PIXOR (ORiented 3D object detection from PIXel-wise neural network predictions) is an accurate, single-stage, real-time 3D object detector. It operates on 3D point clouds by representing the scene from the Bird's Eye View (BEV) for 3D object localization in applications like autonomous driving (Yang et al. 2018).

2D object detection is computationally less demanding, while 3D object detection is highly computationally intensive.

2D object detection offers sufficient accuracy for surveillance and basic augmented reality applications. 3D object detection is imperative where spatial context is indispensable, e.g., in self-driving cars. It yields deeper insights into a situation than its 2D counterpart.

7.4   DISCUSSION AND CONCLUSIONS

This chapter presented glimpses of situations in which robot vision suffers from practical limitations. When the background in an image is cluttered or has similar colors and patterns, differentiating an object from its surrounding environment is liable to errors. Changes in lighting conditions like shadows, glare, and different intensities significantly affect the interpretation of images. Identification and interpretation of objects partially hidden behind other objects is a confounding process. Another bottleneck arises during the reconstruction of 3D information about an object's shape and spatial relationships from 2D camera images, with limited viewpoints. Analysis of visual data fast enough to enable immediate robot responses in dynamic situations leaves them in a quandary. Sophisticated algorithms are necessary to account for relative motion for tracking objects in motion. The issue is aggravated, particularly when the robot is also in motion. Table 7.1 provides a quick look back at this chapter.

After a thorough exposure to robot vision in the preceding chapters, we now probe into the ways of building emotional intelligence in robots. Emotional intelligence allows robots to understand and respond to human emotions in a better way. Through emotional intelligence, more natural and effective robot-human interactions are rendered feasible. Applications like healthcare, education, and customer service greatly benefit from fostering trust. The overall user experience is improved with the utilization of emotional robots. The next chapter is concerned with making robots more relatable and helpful to humans. The robots behave cordially by recognizing emotional cues like facial expressions and tone of voice of their human colleagues.

TABLE 7.1
A Quick Retrospection of This Chapter

Sl. No.	Takeaway	Explanation
1	Summary	Possible situations in which the robots are likely to make vision-related misjudgments were discussed, e.g., when there is a lack of knowledge about the precise position or orientation of the robot, the collected data is influenced by the motion of the robot or affected by environmental parameters or when the object of interest is occluded and not visible due to obstruction. Several techniques have been developed to address the challenges of robot vision. After all these methods were explained, a comparison was made between 2D and 3D object detection.
2	Active vision	In active vision, the robot can move its sensors to gather more useful information about its surroundings. The use of model- and data-based methods for anomaly detection was explained.
3	Anomaly detection	Anomalies in the kinematic or dynamic behavior of a robot are detected by comparing its observed motion with the expected motion. The use of model- and data-based methods for anomaly detection was explained.
4	Image-of-interest detection	During the examination of the vast amount of data, image-of-interest detection is employed as a time-saving measure, where visual representations are provided with specific regions, such as corners and edges, highlighted.
5	Semantic vision	In semantic vision, the relationships between objects in an image are understood.
6	Visual place recognition	Visual place recognition gives the robot the ability to recognize a place from its visual features, such as color and shape. Scale-invariant feature transformation is an algorithm that detects, describes, and matches local features in images. A convolutional neural network-based approach is also useful for this purpose.
7	SLAM	Simultaneous localization and mapping (SLAM) constitutes a technology that helps robots build maps of their environments and use these maps to navigate while keeping track of their locations.
8	Vision-based scene understanding	It analyzes visual data to interpret and derive meaningful information about a scene, e.g., the objects, their relationships, spatial layout, and context.
9	3D object detection	Point clouds (3D representation of a scene with each point representing a specific location in space), depth maps (2D image with each pixel representing the distance of the corresponding point in the scene from the camera), and stereo vision (using two cameras for depth information) are the strategies adopted to help the robot in 3D object detection.
10	Keywords and ideas to remember	Uncertainty about the precise position or orientation of the robot, influence of motion of the robot and environmental parameters on the collected data, occlusion of objects of interest, active computer vision, anomaly detection, model- and data-based methods, image-of-interest detection, semantic vision, visual place recognition, scale-invariant feature transformation, convolutional neural networks-based approach, 3D object detection, point clouds, depth maps, and stereo images.

REFERENCES AND FURTHER READING

Amoozgar M. H., A. Chamseddine and Y. Zhang. 2013. Experimental test of a two-stage Kalman filter for actuator fault detection and diagnosis of an unmanned quadrotor helicopter. *Journal of Intelligent and Robotic Systems*, Vol. 70, pp. 107–117.

Bailey T. and H. Durrant-Whyte. 2006. Simultaneous localization and mapping (SLAM): Part II, *IEEE Robotics & Automation Magazine*, Vol. 13, 3, pp. 108–117.

Bergmann D. and C. Stryker. 2023. What is an autoencoder? https://www.ibm.com/topics/autoencoder

Bhattad A., J. Rock and D. A. Forsyth. 2018. Detecting anomalous faces with 'no peeking' autoencoders, *ArXiv:1802.05798*, https://doi.org/10.48550/arXiv.1802.05798

Burke M. 2017. *User-driven mobile robot storyboarding: Learning image interest and saliency from pairwise image comparisons*, eprint arXiv:1706.05850, 8 pages.

Carreira J. and C. Sminchisescu. 2012. CPMC: Automatic object segmentation using constrained parametric min-cuts, *IEEE Transactions on Pattern Analysis and Machine Intelligence*, Vol. 34, 7, pp. 1312–1328.

Charles R., H. Su, M. Kaichun and L. Guibas. 2017. PointNet: Deep Learning on Point Sets for 3D Classification and Segmentation, *2017 IEEE Conference on Computer Vision and Pattern Recognition (CVPR)*, Honolulu, HI, USA, 21–26 July, pp. 77–85.

Chen S., Y. Li and N. M. Cwok. 2011. Active vision in robotic systems: A survey of recent developments, *International Journal of Robotic Research*, Vol. 30, 11, pp. 1343–1377.

Durrant-Whyte H. and T. Bailey. 2006. Simultaneous localization and mapping: Part I, *IEEE Robotics & Automation Magazine,* Vol. 13, 2, pp. 99–110.

Fan J., P. Zheng and S. Li. 2022. Vision-based holistic scene understanding towards proactive human–robot collaboration, *Robotics and Computer-Integrated Manufacturing*, Vol. 75, p. 102304.

Geiger A., M. Lauer, C. Wojek, C. Stiller and R. Urtasun. 2014. 3D traffic scene understanding from movable platforms*, IEEE Transactions on Pattern Analysis and Machine Intelligence*, Vol. 36, 5, pp. 1012–1025.

Guo F., J. Yang, Y. Chen and B. Yao. 2018. Research on Image Detection and Matching Based on SIFT Features, *2018 3rd International Conference on Control and Robotics Engineering (ICCRE)*, Nagoya, Japan, 20–23 April, pp. 130–134.

Häne C., C. Zach, J. Lim, A. Ranganathan and M. Pollefeys. 2011. Stereo Depth Map Fusion for Robot Navigation, *2011 IEEE/RSJ International Conference on Intelligent Robots and Systems*, San Francisco, CA, USA, 25–30 September, pp. 1618–1625.

Khairuddin A. R., M. S. Talib and H. Haron. 2015. Review on Simultaneous Localization and Mapping (SLAM), *2015 IEEE International Conference on Control System, Computing and Engineering (ICCSCE)*, Penang, Malaysia, 27–29 November, pp. 85–90.

Kim H. S., I. J. Park, H. Han and J. Y. Son. 2022. Anomaly Detection for Robotic Assembly, *2022 13th International Conference on Information and Communication Technology Convergence (ICTC)*, Jeju Island, Korea, Republic of, 19–21 October, pp. 1667–1670.

Lin D., S. Fidler and R. Urtasun. 2013. Holistic Scene Understanding for 3D Object Detection with RGBD Cameras, *2013 IEEE International Conference on Computer Vision*, Sydney, NSW, Australia, 1–8 December, pp. 1417–1424.

Liu W., D. Anguelov, D. Erhan and C. Szegedy. 2016. SSD: Single Shot MultiBox Detector. In: Leibe B., J. Matas, N. Sebe and M. Welling (Eds.), *Computer Vision – ECCV 2016*. Lecture Notes in Computer Science (LNCS), Vol. 9905, Springer International Publishing AG, Cham, pp. 21–37.

Lowe D. G. 2004. Distinctive image features from scale-invariant keypoints, *International Journal of Computer Vision*, Vol. 60, pp. 91–110.

Lowry S., N. Sünderhauf, P. Newman, J. J. Leonard, D. Cox and P. Corke. 2016. Visual place recognition: A survey*, IEEE Transactions on Robotics*, Vol. 32, 1, pp. 1–19.

MathWorks. 2024. *What Is SLAM? How It Works, Types of SLAM Algorithms, and Getting Started, (1994–2024)*, The MathWorks, Inc., https://ch.mathworks.com/discovery/slam.html

Morrison D., P. Corke and J. Leitner. 2018. Closing the loop for robotic grasping: A real-time, generative grasp synthesis approach, *arXiv:*1804.05172v2, 10 pages.

Nandakumar S. C., D. Mitchell, M. S. Erden, D. Flynn and T. Lim. 2024. Anomaly detection methods in autonomous robotic missions, *Sensors,* Vol. 24, 4, p. 1330.

Nertinger S., R. J. Kirschner, S. Abdolshah, A. Naceri and S. Haddadin. 2023. Influence of Robot Motion and Human Factors on Users' Perceived Safety in HRI, *2023 IEEE International Conference on Advanced Robotics and Its Social Impacts (ARSO)*, Berlin, Germany, 5–7 June, pp. 46–52.

Owen-Hill A. (Updated). 2025. Top 10 challenges for robot vision, https://blog.robotiq.com/top-10-challenges-for-robot-vision

Redmon J., S. Divvala, R. Girshick and A. Farhadi. 2016. You Only Look Once: Unified, Real-Time Object Detection, *2016 IEEE Conference on Computer Vision and Pattern Recognition (CVPR)*, Las Vegas, NV, USA, 27–30 June, pp. 779–788.

Saputra R. P., N. Rakicevic, D. Chappell, K. Wang and P. Kormushev. 2021. Hierarchical decomposed-objective model predictive control for autonomous casualty extraction, *IEEE Access*, Vol. 9, pp. 39656–39679.

Scharfenberger C., S. L. Waslander, J. S. Zelek and D. A. Clausi. 2013. Existence Detection of Objects in Images for Robot Vision Using Saliency Histogram Features, 2013 *International Conference on Computer and Robot Vision*, Regina, SK, Canada, 28–31 May, pp. 75–82.

Sevilla-Lara L., D. Sun, V. Jampani and M. J. Black. 2016. Optical Flow with Semantic Segmentation and Localized Layers, *2016 IEEE Conference on Computer Vision and Pattern Recognition (CVPR)*, Las Vegas, NV, USA, 27–30 June, pp. 3889–3898.

van Eden B. and B. Rosman. 2019. An Overview of Robot Vision, *2019 Southern African Universities Power Engineering Conference/Robotics and Mechatronics/Pattern Recognition Association of South Africa (SAUPEC/RobMech/PRASA)*, Bloemfontein, South Africa, 28–30 January, pp. 98–104.

Xinjilefu X., S. Feng and C. G. Atkeson. 2015. Center of Mass Estimator for Humanoids and Its Application in Modelling Error Compensation, Fall Detection and Prevention, *Proceedings of the 2015 IEEE-RAS 15th International Conference on Humanoid Robots (Humanoids)*, Seoul, Republic of Korea, 3–5 November, pp. 67–73.

Yang B., W. Luo and R. Urtasun. 2018. PIXOR: Real-Time 3D Object Detection from Point Clouds, *2018 IEEE/CVF Conference on Computer Vision and Pattern Recognition*, Salt Lake City, UT, USA, 18–23 June, pp. 7652–7660.

Yoshioka K., H. Okuni, T. T. Ta and A. Sai. 2021. Through the Looking Glass: Diminishing Occlusions in Robot Vision Systems with Mirror Reflections, *2021 IEEE/RSJ International Conference on Intelligent Robots and Systems (IROS)*, Prague, Czech Republic, 27 September to 1 October, pp. 1578–1584.

Yuille A. L. and A. Blake (Eds.). 1992. *Active Vision*, MIT Press, Spain, 368 pages.

Zeng R., Y. Wen, W. Zhao and Y.-J. Liu. 2020. View planning in robot active vision: A survey of systems, algorithms, and applications, *Computational Visual Media*, 6, pp. 225–245.

Zeng Z., J. Zhang, X. Wang, Y. Chen and C. Zhu. 2018. Place recognition: An overview of vision perspective, *Applied Sciences*, Vol. 8, 11, pp. 2257.

Zheng S., Y. Li, Z. Yu, S.-Y. Cao, M. Wang, J. Xu, R. Ai, W. Gu, L. Luo and H.-L. Shen. 2023. I2P-Rec: Recognizing Images on Large-Scale Point Cloud Maps through Bird's Eye View Projections, *2023 IEEE/RSJ International Conference on Intelligent Robots and Systems (IROS)*, Detroit, MI, USA, 1–5 October, pp. 1395–1400.

8 Emotionally Intelligent Robots

Bayesian Inference and Fuzzy Logic

8.1 INTRODUCTION

Emotions are psychological states connecting human thoughts, behaviors, and bodily reactions. They are triggered by events which a person perceives as important to his/her well-being. In fact, they are complex neurophysiological behaviors rooted in the amygdala, hippocampus, and prefrontal cortex of the human brain. Conceptualized as experiences, evaluations, and motivations, they consist of three fundamental components:

i. Experiential Component: It is the subjective, personal feeling or awareness of the emotion, such as happiness, sadness, anger, or fear.
ii. Behavioral Component: It is the outward manifestation of emotion, such as through facial expressions, body language, or actions.
iii. Physiological Component: It includes the increased heart rate, sweating, or changes in breathing associated with emotional feelings.

Emotional intelligence deals with understanding, utilizing, and managing emotions. It is made possible by thoughts and feelings, enabling the perception and management of human emotions. It concerns the observation and interpretation of the emotions of friends, colleagues, and other people. Accordingly, it allows one to respond and react to those feelings in a manner of reciprocity, returning the favor of someone's act of kindness with an equivalent action. As a result, various individuals in a society engage in emotional exchanges. Besides the accomplishment of effective person-to-person communication, strong interpersonal relationships are built among people.

Also called affective robots, the emotional robots can spontaneously interact with humans in a more natural way (Spezialetti et al. 2020; Khare et al. 2024). They recognize human emotions through facial expressions, tone of voice, and body language, including gestures of human beings. Based on the detected emotions, they adjust their behavior and responses for performing everyday jobs and professional roles.

Social robots are robots built to directly engage and communicate with people while adhering to established social conventions (Bryant 2019). These robots help

DOI: 10.1201/9781032695266-8

kids with autism. They assist the elderly by interacting with them fervently in a warm and engaging manner, using both verbal and nonverbal communication. They are applied in the diagnosis of autism spectrum disorder, a neurological and developmental disorder impacting an individual's interaction and socialization with others (Arrent et al. 2022). Table 8.1 brings out the special attributes and traits that tell apart an emotional robot from a regular robot (Kolling et al. 2016).

This chapter reviews the advancements in the field of robotics that replicate the emotional behavior of humans to work in a friendly and hospitable fashion bringing an emotive aroma to the environment. They differ from conventional robots, displaying a monotonous and heartless machine-like behavior. Although emotional robots are very welcoming and lovable, a word of caution from such robots is given by the following poem:

Safeguarding human identity!

In my dream, I was going to the market one day
When I was greeted by a person on the way
A simple man on the street
With a smiling face and walking briskly on his feet
His behavior was very pleasing,
He looked very natural and amazing, but a little surprising

TABLE 8.1
Regular Robot and Emotional Robot

Sl. No.	Point of Comparison	Regular Robot	Emotional Robot
1	Purpose of design	It aims to perform specific tasks based on programmed instructions.	It is intended to recognize and respond to human emotions, allowing for more nuanced and empathetic interactions between robots and individuals.
2	Functionality	It primarily focuses on completing tasks such as cleaning, assembly, or manufacturing.	It understands and reacts to human emotions, potentially offering moral comfort or support.
3	Sensory capabilities	It utilizes basic sensors, such as accelerometers and proximity detectors.	It is equipped with advanced sensors, sophisticated cameras, and microphones to detect facial expressions, tone of voice, and body language of humans.
4	AI algorithms	It mainly uses algorithms designed for task execution.	It leverages complex AI models to interpret emotional cues and generate appropriate responses to emotions.
5	Examples	A vacuum cleaner that navigates a room and cleans the floor.	A companion robot designed to interact with elderly or sick individuals, providing conversational responses to queries, mixed with emotional care.

So, I asked him, "Was he a human or a robot?"
To which he replied, "A social robot".
Imagine if it happens, it will be very confusing
With eyes watching but stubbornly refusing
To distinguish robotics from reality
In a mixed man-robot society
Therefore, ethical laws must protect human rights and dignity
To avert any danger of a human getting a mistaken robotic identity.
Or a robot posing as a human entity.

8.2 EMOTIONAL AI

Emotional artificial intelligence (AI) entails endowing a robot with the gift to exhibit humanoid emotion (Yan et al. 2021). Emotional AI, also known as affective computing, is the process by which human emotions are imitated by computer systems (Wu 2024). This becomes possible through the analysis of gigantic amounts of data for the identification of patterns and the prediction of the emotional states of people. The process of emotional AI development consists of three stages involving the collection and analysis of emotion-related data and the generation of reactionary responses to the emotions, as shown in Figure 8.1.

Stage 1: Collection of Emotion-Related Data: The sources of these data are human facial expressions in the form of motion and configurations of the small micromotor muscles underneath the skin of a person's face. The facial expressions are supplemented with voice intonations, namely the rise and fall of pitch in sound, to highlight an expression. Human body language adds more flavor to emotions. There are several

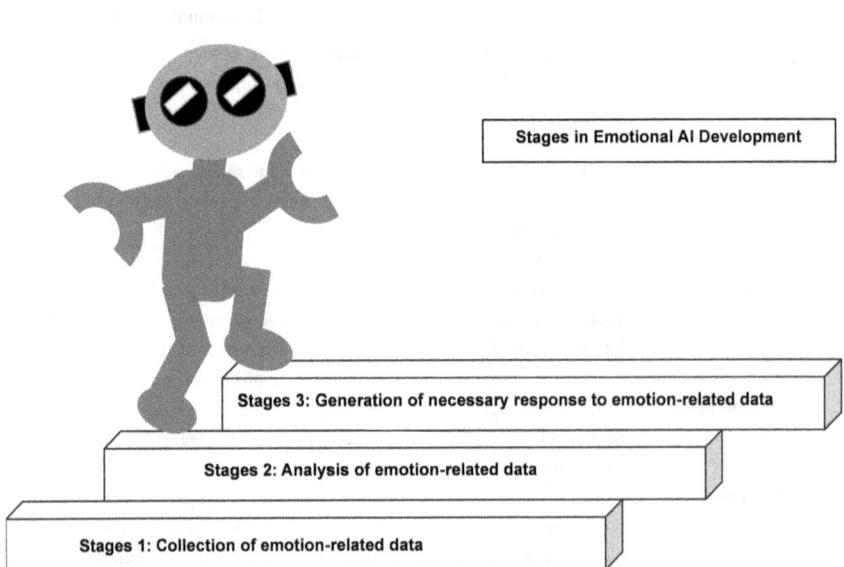

Stages in Emotional AI Development

Stages 3: Generation of necessary response to emotion-related data

Stages 2: Analysis of emotion-related data

Stages 1: Collection of emotion-related data

FIGURE 8.1 The three stages in building emotional AI.

different physical behavior-based nonverbal communication ways in which emotions are expressed. Eye contact, posture, gestures, handshakes, happiness, surprise, fear, anger, sadness, and disgust are a few such expressions. No less important are the physiological signals of emotion arousal, such as respiratory changes, changes in heart rate, blood pressure, sweating, skin temperature, galvanic skin response, electrocardiogram (ECG), electroencephalogram (EEG), electromyogram (EMG), etc. All these signals contribute toward formulating the overall pattern of emotional behavior exhibited by humans.

Stage 2: Analysis of Collected Emotion Data: Interpretation of the data is performed using machine learning (ML) algorithms. These algorithms identify emotional cues. These cues are the verbal or nonverbal signals. They provide indications about how someone is feeling, thinking, or reacting to a given situation or incident.

Stage 3: Generation of Necessary Response to Input Emotion Data: Appropriate human responses are quickly developed in answer to the interpreted emotional state. The responses take the form of textual, speech, or visual outputs.

There are four facets to emotional AI: perception, utilization, understanding, and management of emotions (Seyitoğlu and Ivanov 2024).

8.3 EMOTIONAL ROBOT ALGORITHM

8.3.1 MAIN COMPONENTS OF THE ALGORITHM

The emotional robot algorithm is an exciting and thought-provoking idea. Figure 8.2 shows the four components of an emotional robot algorithm. The first component is preprocessing of acquired sensory input. The second component is emotion

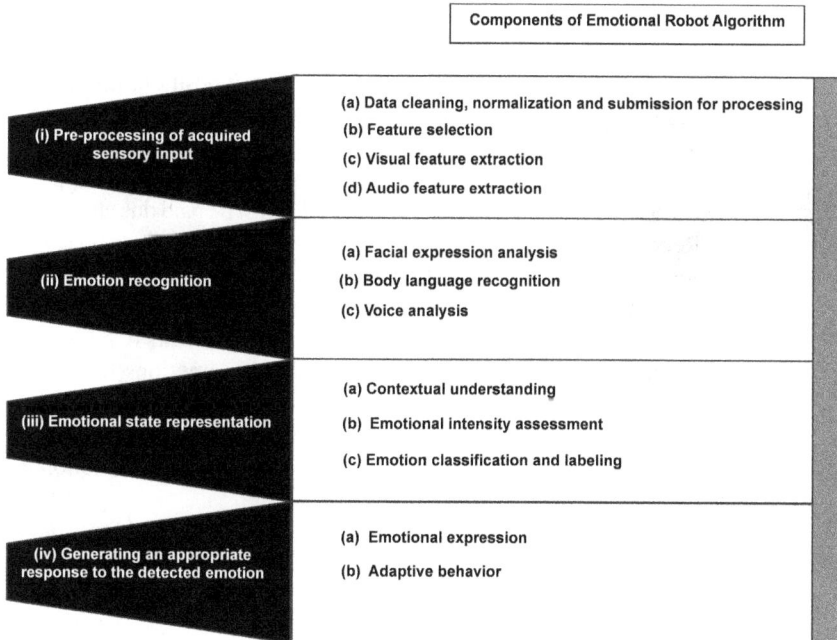

Components of Emotional Robot Algorithm

(i) Pre-processing of acquired sensory input
- (a) Data cleaning, normalization and submission for processing
- (b) Feature selection
- (c) Visual feature extraction
- (d) Audio feature extraction

(ii) Emotion recognition
- (a) Facial expression analysis
- (b) Body language recognition
- (c) Voice analysis

(iii) Emotional state representation
- (a) Contextual understanding
- (b) Emotional intensity assessment
- (c) Emotion classification and labeling

(iv) Generating an appropriate response to the detected emotion
- (a) Emotional expression
- (b) Adaptive behavior

FIGURE 8.2 The structural constitution of an emotional robot algorithm.

recognition. The third and fourth components are, respectively, the representation of emotional state and the generation of an appropriate response to the detected emotion. Each of these components is split up into subcomponents. These are elaborated alongside the mentioned components.

An emotional robot algorithm is a composite algorithm. It is formed by the combination of several ML algorithms. This algorithm generally consists of the following (Thilmany 2007):

i. Preprocessing of Acquired Sensory Input: The raw signals are converted into a suitable form for analysis. The preprocessing is a complex, multifaceted process consisting of a series of operations:
   a. Data Cleaning, Normalization, and Submission for Processing: The data received from various sensors is cleaned by identifying sources of noise in the signal, removing the noise and disturbances that affect it, handling missing values, and validating data sources in the signal. The data are normalized by eliminating repetition and redundancy. Repeated and irrelevant portions are deleted. In addition, related multiple relationships are isolated. These measures are necessary to prepare the data for analysis of human facial expressions, body language, and voice tone using ML techniques, as they have a significant influence on the results if not taken into account.
   b. Feature Selection: The relevant features in the input sensory data that are most indicative of emotions are identified. Optionally, emotion detection is enhanced by measuring physiological data. Heart rate and skin conductance measurements are vital biological parameters connected with a person's emotional feelings.
   c. Visual Feature Extraction: Facial expressions are examined by computer vision techniques. Key features like eyebrow position, lip curvature, and eye gaze are identified during this examination and brought into the limelight.
   d. Audio Feature Extraction: The voice tone and pitch variations are processed. These help to detect emotional cues and must be paid due attention.
ii. Emotion Recognition: ML algorithms are employed to analyze various emotions for their accurate recognition in order that their formal response greeting can be triggered in acknowledgment.
   a. Facial Expression Analysis: Convolutional neural networks (CNNs) are used to identify facial features. They can classify emotions based on patterns discerned in facial expressions.
   b. Body Language Recognition: Body posture, gestures, and movement patterns are scrutinized. Their scrutiny makes inference of emotional states easier.
   c. Voice Analysis: Speech recognition and analysis are applied to voice pitch, intonation, and pace. Voice analysis is a valuable tool for detecting emotions.
iii. Emotional State Interpretation: ML techniques are applied for performing emotion classification and understanding.

a. Contextual Understanding: The emotional meaning of cues behind the detected emotions is understood and interpreted more deeply by taking into account the current situation and its correlation with previous interactions. Relating emotions to the context makes the experience livelier and more impactful.

b. Emotional Intensity Assessment: The strength or severity of the detected emotion is determined. This is needed to impart strength to the emotions that are articulated more emphatically, and provide a debilitated answer to feebly expressed emotions.

c. Emotion Classification and Labeling: Algorithms such as support vector machines (SVM), neural networks, or deep learning models are trained on large datasets of labeled emotional data to achieve fast and accurate emotion classification. They categorize the detected emotions into basic classes like happiness, sadness, anger, fear, and surprise. An emotional label like 'happy', 'sad', or 'angry' is assigned to the recognized emotional state based on the output of the model.

iv. Generating an Appropriate Response to the Detected Emotion: This is achieved through a dual strategy of emotional expression and adaptive behavior. It includes adjustment of the robot's facial expressions, tone of voice, or physical actions. These adjustments reflect empathy or understanding toward the interacting human operator using ML.

a. Emotional Expression: Appropriate facial expressions, voice tone, or body movements are chosen to convey an empathetic response that is aligned with the user's perceived emotion.

b. Adaptive Behavior: Adjustments are made to the robot's actions, conversation style, or responses in response to the interpreted emotional state.

8.3.2 Considerations and Concerns during Algorithm Formulation

When designing an emotional robot algorithm, one has to consider several factors to ensure the accuracy of results and to safeguard ethics. Let us list some of these factors that come to mind immediately.

i. Cross-Cultural Variations in the Human Race: There are wide cultural differences in facial expressions and emotional displays across various societies of the human race (Mohan et al. 2021). As common knowledge, expressions like smiling for happiness and crying or weeping for sorrow are believed by consensus among people without any disagreement. Nevertheless, subtle distinctions and gradations in intensities of these expressions differ significantly. The differences depend on cultural norms and interpretations across the globe. In some places, people rely more on eye movements to convey emotions. In other localities, the focus is principally on the mouth region. Therefore, both universal and cultural expressions can vary significantly. Indeed, it is a well-supported argument worthy of consideration. These variations must be considered for training emotion recognition algorithms.

ii. Provision of Built-in User Feedback Loop: Customer feedback mechanisms must be duly incorporated. They will enable the users to provide input on the emotion recognition accuracy of the robotic system. The user feedback will improve the system over time

iii. Issues of Privacy and Ethical Concerns: The privacy of emotional states of an individual requires clear consent mechanisms and guidelines regarding data collection and usage. Ethical considerations must be strictly adhered to when using sensors to detect emotions. Emotions significantly influence ethical judgments. They create situations where individuals might be exploited due to their emotional states. There exist possibilities and potentialities that can lead to biased decisions based on feelings like anger, fear, or sympathy instead of rational reasoning. These issues are particularly problematic in situations that require objective analysis. Emotional appeals may be utilized to persuade people. These can become harmful, especially when they exploit the vulnerabilities of people or employ deceptive tactics. Risks of this nature raise ethical worries about the intent and impact of emotional manipulation. Empathy can promote ethical behavior by encouraging compassion and understanding toward people. But the lack of empathy leads to harmful actions.

8.4 SPECIFIC ALGORITHMS USED IN EMOTIONALLY INTELLIGENT ROBOTS

Considering the diversity of emotional behavior, it is evident that emotional robots cannot function with a single algorithm, but rather with an intermixed algorithmic technique. Figure 8.3 presents a broad view of the algorithms employed by emotionally intelligent robots. Familiar emotional robot algorithms include Bayesian inference, fuzzy logic, Markov models, self-organizing maps, SVMs, decision trees, natural language understanding and reinforcement learning.

8.4.1 BAYESIAN INFERENCE FOR ROBOT EMOTION DETECTION

As emotions exhibit an overlapping nature, emotional states can be modeled as probability distributions, indicating a likelihood of experiencing a particular feeling in a given situational context. Let us inquire about the probabilistic aspect of emotions. It sounds simple, but it is easier said than done.

8.4.1.1 Building a Probabilistic Inference Perspective of Emotion Recognition

Bayesian inference is a computational model in ML. It is widely used in image processing and cognitive science (Kato et al. 2006; Martinez-Hernandez et al. 2016). It looks upon the process of emotion recognition as a probabilistic inference problem. This probabilistic problem is solved by applying Bayesian statistical methods to emotions. Bayesian methods are applied toward understanding the ways in which people experience emotions and draw inferences about them. They use Bayes' theorem to fit a probability model to a set of data from prior evidence. Before encountering new information, individuals have pre-existing beliefs about emotions. These beliefs are instinctively derived from their past experiences and cultural practices within the respective society in which they are brought up. They act as a prior distribution of emotions in the Bayesian framework. Figure 8.4 shows the stages in the Bayesian

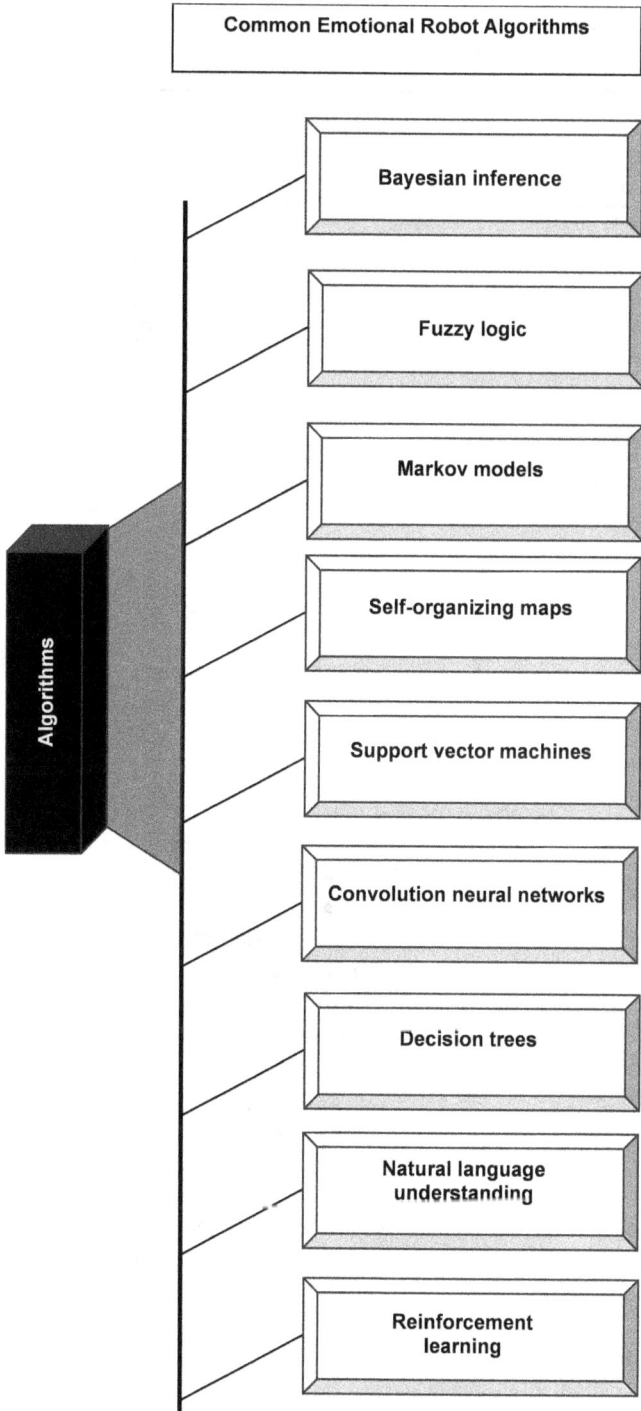

FIGURE 8.3 Common algorithms of induction of emotional response behavior in social robots.

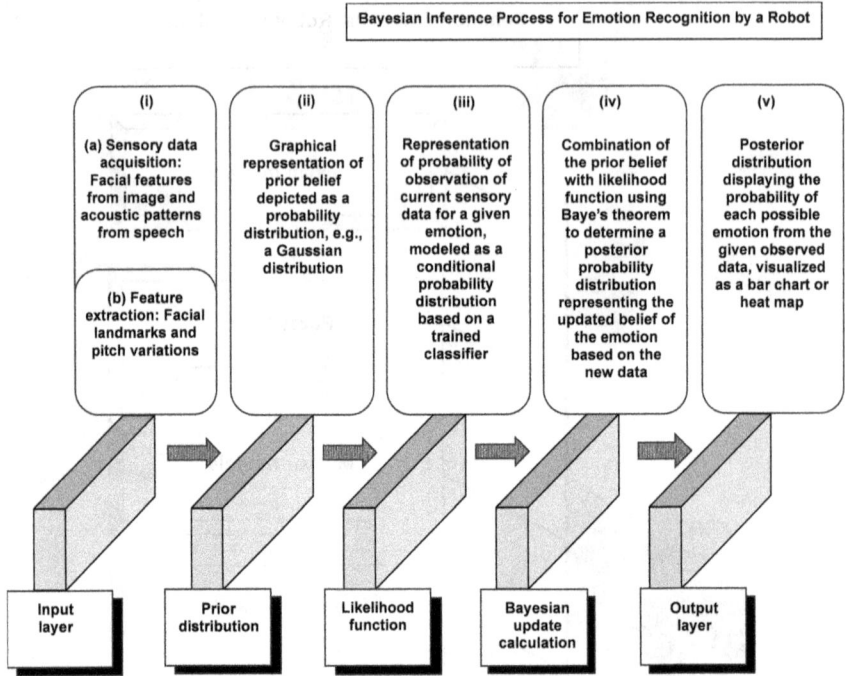

FIGURE 8.4 The Bayesian inference process for robots.

inference process for emotion recognition, viz., input layer consisting of sensory data acquisition and feature extraction, prior distribution depicted as a probability graph, the likelihood function given as a representation of probability of observation of current sensory data for a given emotion, the Bayesian update calculation in which the prior belief is combined with a likelihood function using Bayes' theorem to determine a posterior probability distribution, and the output layer as a posterior distribution displaying the probability of each possible emotion.

Individuals continuously update their beliefs about the emotional state of someone. This updating of beliefs is based on newly acquired sensory information. The intent of updating is to combine prior knowledge with new data. The incorporation of prior knowledge enhances the accuracy of emotion detection in real-time applications.

When new cues, such as facial expressions, are observed, a likelihood function is used. It represents the probabilities of occurrence of those cues when a specific emotion is given. From the perspective of Bayesian methods, the prior beliefs are combined with the likelihood function using Bayes' theorem. Doing so helps individuals to update their beliefs about the most probable emotion. As a result, a posterior distribution of emotions is obtained. This distribution reflects the current understanding of individuals about emotions based on the new information.

The gist of the discussion is that Bayesian inference is employed in robot emotion detection to enable robots to interpret human emotions more accurately. The interpretation is done by incorporating prior knowledge about emotional expressions into new data. Beliefs are updated based on new sensory data, such as facial expressions and voice tone. The analysis provides a probabilistic assessment of the most likely emotional state. It enables more nuanced and adaptive interactions between robots and humans.

8.4.1.2   Applications of Bayesian Inference

How is Bayesian inference helpful in emotion detection? Bayesian inference helps robots improve emotion detection by leveraging prior knowledge and integrating new evidence through probability. As a result, more nuanced and context-aware emotion recognition is achieved in human–robot interaction. Fascinating areas where Bayesian inference has made progress are as follows:

i. Multimodal Integration for Emotion Recognition: A Bayesian model can easily combine information from multiple sensory modalities like facial expressions, voice tone, body posture, and body language. By incorporating this information, it is able to provide a more comprehensive understanding of emotions. Thus, it helps to build a more robust emotion detection system (Bera et al. 2019).

   a. Analysis of Facial Expression: A Bayesian model analyzes facial features like eye gaze, brow position, and mouth curvature of a person. The analysis of facial features is applied to infer the most likely emotion based on a probability distribution.

   b. Analysis of Voice Tone: A Bayesian model analyzes prosodic features like pitch, volume, and speech rate of a person. The analysis of prosodic features enables the estimation of a speaker's emotional state.

   c. Social Interaction of Robots: A Bayesian network is used to model the emotional state of a group of persons participating in a social engagement. This modeling allows a robot to respond appropriately to the overall sentiment of the people in the group.

ii. Contextual Understanding of the Situation: A Bayesian network incorporates contextual information about the situation. This could be in the form of a conversation topic or social cues. As a consequence, an improved emotion recognition accuracy is achieved.

iii. Modeling and Handling of Uncertainty: The assessment of uncertainty plays a crucial role in understanding the way in which people behave emotionally. Particularly, it makes us aware about the ways the people navigate abstruse situations or react to complicated emotional expressions.

We recognize that a primary aspect of Bayesian inference is its excellent ability to quantify and manage uncertainty in emotion perception. It works by continuously updating probability distributions as new information from sensors like facial expressions, voice tone, and body language becomes available. By naturally accounting for uncertainty in emotion detection, Bayesian inference is able to provide a probability distribution over possible emotional states rather than a single categorical prediction. This specialty has a meaningful impact on emotion recognition because of the ambiguous nature of this process.

iv. Generation of Adaptive Response by Dynamic Updates: As the robot interacts with a person, it continuously refines its understanding of the person's
emotional state and probability of different emotions by updating the posterior probability based on new observations, Hence, the robot dynamically
adjusts its responses to better suit the perceived emotional state of the user.
These improvements make the robot more adaptive to changing situations.
v. Learning from Interaction with Humans: A Bayesian model updates its prior
beliefs about emotion expression based on new data obtained through continuous interaction with humans. This belief-updating mechanism allows
for personalized and adaptive emotion recognition over time.

8.4.1.3   Advantages of Bayesian Inference

Let us scrutinize the resources offered by Bayesian inference that can be gainfully
utilized. In robot emotion detection, Bayesian inference offers significant advantages
in a multiplicity of ways. It allows for the dynamic updating of probabilities based on
new sensory data. It effectively handles uncertainty in complex emotional situations.
It incorporates prior knowledge about human emotions and provides a framework
to reason about the likelihood of different emotional states. Therefore, a robust and
adaptable emotion recognition is implemented in robots through Bayesian inference.
Let us enlist the key advantages of Bayesian inference in robot emotion detection.
These are as follows:

i. Probabilistic Interpretation of Results: Bayesian inference provides a clear
probabilistic interpretation of the results. It allows for a better understanding of the confidence level associated with the predicted emotion.
ii. Explanatory Power of Bayesian Inference: The analysis of the posterior distribution helps developers to gain insights into the main features that are
most influential in determining the emotional state. These insights allow for
the refinement of the emotion detection model.

As an example of the application of Bayesian inference, a robot interacting with a
human might initially have a neutral prior belief about the person's emotion. As the
robot watches the facial expressions and tone of voice of the person, it updates its
belief about the person's emotion. It gradually shifts and leans in drawing inference
toward happy or sad situations, depending on the new evidence.

8.4.1.4   Limitations of Bayesian Inference

The inherent duality of situations suggests that where there are advantages, there are
also disadvantages. When using Bayesian inference for robot emotion detection, an
infuriating limitation arises from the difficulty of selecting appropriate prior distributions. There is a chance of overfitting to the training data. Real-time applications
add computational complexity. The handling of nuanced emotions is difficult. Large,
diverse datasets are necessary to capture the full spectrum of human emotional
expressions accurately. All these quandaries hinder the accuracy and reliability of
emotion recognition in robots. A more detailed explanation of limitations will clarify
the types of technical snags and hitches.

i. Complexity Arising from Selection of Prior Distribution: Bayesian inference relies heavily on the choice of prior distribution of probabilities of emotions. The preceding distribution represents initial beliefs about the data. Selecting an accurate prior distribution for complex emotions is not easy. The reason is that emotions often overlap with each other and vary significantly between individuals.

ii. Complexity in Model Designing: Development of a robust Bayesian model for emotion detection is a complicated task. It requires careful design and selection of appropriate features and prior distributions.

iii. Dependency on Extensive Datasets: Accurate emotion detection with Bayesian inference heavily relies on the availability of large, well-annotated datasets. These datasets must represent a wide range of emotional expressions. Hence, they are difficult to acquire and maintain.

iv. Risk of Overfitting: The model might overfit to specific patterns if the training data is not sufficiently diverse. Then, poor performance is observed on unseen data with subtle emotional nuances.

v. Computational Burden and Cost: Bayesian inference involves complex calculations. Hence, it is computationally expensive for real-time robot interactions. Serious issues are encountered, especially when dealing with large datasets. An intricate Bayesian emotional model, too, is troublesome.

vi. Challenges Due to Nuanced Emotional Behavior: Emotions are not discretely categorized. They would rather exist on a spectrum. These characteristics of emotions make it difficult to accurately capture subtle variations and complex emotional states within a Bayesian framework.

8.4.1.5   Potential Solutions to Limitations of Bayesian Inference

As already said, Bayesian methods involve incorporating prior beliefs. Their incorporation introduces some level of subjectivity into the analysis. It can be a double-edged sword with positive and negative consequences. It carries inherent risks because of the biasing of results if the prior beliefs are poorly specified. Recognizing limitations helps us identify areas where improvement is needed, allowing us to focus our efforts in the right direction. Strategies have evolved to overcome the different inadequacies. A few notable ones are given below:

i. Adoption of Adaptive Prior Distributions: Adaptive prior distributions are employed. These distributions can learn and update themselves based on new data encountered during interaction. They can help in the mitigation of the issue of selecting the right initial prior distribution.

ii. Formulation of Hybridized Approaches: Bayesian inference is combined with other ML techniques. In particular, deep learning improves the accuracy and robustness of emotion detection.

iii. Practicing Dimensionality Reduction: Applying dimensionality reduction techniques helps in the management of the complexity of feature space. It reduces the computational load.

iv. Use of Continuous Emotion Models: Models that represent emotions as continuous variables rather than discrete categories are preferable. They capture nuances in emotional expression far better than those using discrete variables.

8.4.2   Fuzzy Logic in Robot Emotion Detection

We have seen how Bayesian inference allows for the probabilistic updating of beliefs based on new evidence. Rather than clinging to and brooding about a single race course, exploring different possibilities leads to new opportunities. Another distinct approach to emotion recognition is fuzzy logic, which deals with uncertainty and imprecision. Its indispensability originates from the need to help robots handle vague and imprecise information about emotions.

Fuzzy logic is a valuable tool for handling uncertainty in emotional interpretation (Hsu et al. 2013; Nicolai and Choi 2015). The foremost reason is that it allows for the representation of emotions as degrees of truth by utilizing fuzzy rules rather than simply distinguishing between true and false. Here, strict binary classifications of propositions are not used. Such a representation makes fuzzy logic well-suited to the subtleties and often vague nature of human emotions. Sometimes a person may experience a blend of feelings at the same time or express them in a deceptive manner. In traditional logic, somebody is either happy or sad. Dissimilar to traditional logic, fuzzy logic allows for partial membership of a person in multiple emotional states. The person may be slightly happy, somewhat happy, or very happy. A person may be in a melancholy, wistful, bittersweet, or pensive mood. Instead of categorizing a person as absolutely happy or completely unhappy, fuzzy logic allows robots to express the emotional state in fractional terms as partially happy (0.7) and partially sad (0.2). This kind of expression reflects the nuanced nature of human emotions (Cardone et al. 2023; Martin et al. 2023). Salient features of fuzzy logic are as follows:

i. Use of Linguistic Variables: Fuzzy logic uses linguistic variables to represent emotional intensity. Variables like 'very angry', 'a little surprised', or 'quite disappointed' are compatible with our natural way description of emotions in everyday life.

ii. Definition of Fuzzy Rules: Robotic systems can interpret complex emotional differences by defining fuzzy rules for relating input signals for facial expressions or voice tone to emotional states.

iii. Handling Ambiguity: Emotional cues are often indistinguishable or inconsistent in real-world interactions. Fuzzy logic effectively handles such situations.

A typical fuzzy logic rule for emotion interpretation of 'happiness': If the corners of the mouth curve upward, revealing teeth, the person is feeling happy. A fuzzy model of emotion and behavior selection for a robot is proposed (Ho et al. 1997).

8.4.2.1 Main Steps of Fuzzy Logic

To prepare a list of steps, we brainstorm and write down every possible intermediate stage on the way to reach the goal. The steps in fuzzy logic for emotion detection are shown in Figure 8.5. They include the input variables from face, voice and body; fuzzification by mapping of input variables to fuzzy sets; laying down the fuzzy rules for combining the fuzzy input values for estimating the degrees of different emotions; inference engine for application of the fuzzy rules to fuzzified input values; and defuzzification for conversion of the calculated degrees of membership of emotions into a single output value signifying the final emotion classification.

  i. Input variables: These include the face, voice, and body indicators.
 ii. Fuzzification: Input data like facial features or voice pitch are converted into fuzzy sets e.g., 'slightly smiling face', 'moderate pitch of sound', etc. Membership values indicate the degree to which they belong to a particular category.
iii. Definition of Fuzzy Rules: A set of rules is defined to map the fuzzy input values to corresponding emotional states. If the facial expression is slightly smiling and the voice pitch is moderate, then the emotion is likely 'happy'.
 iv. Inference Engine: The fuzzy logic system uses the defined rules to calculate the degree of membership for each possible emotion based on the input data.
  v. Defuzzification: The calculated membership values are converted back into a clear emotional state, e.g., 'happy with a confidence level of 0.8'.

8.4.2.2 Applications of Fuzzy Logic

How is fuzzy logic useful in developing emotional AI for robots? This is a question that must be answered to clarify our expectations from fuzzy logic in this field. In robot emotion detection, fuzzy logic is primarily used to handle the inherent

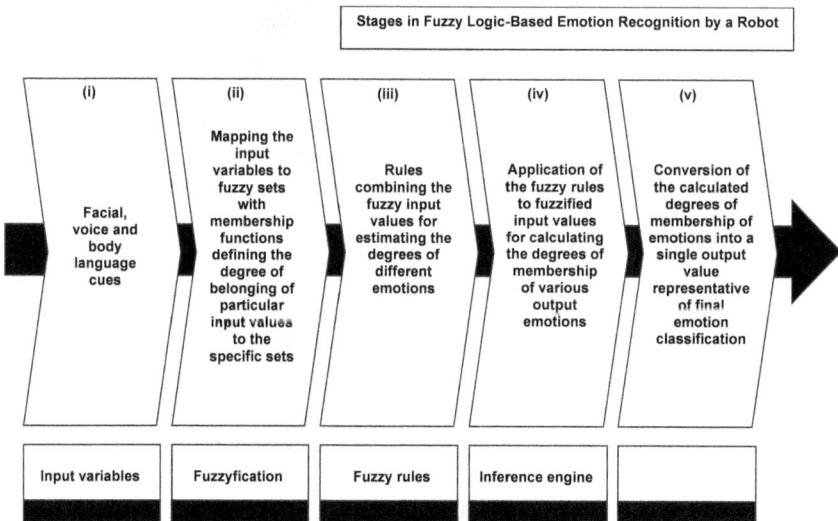

FIGURE 8.5 Stages in the application of fuzzy logic to robot emotion detection.

ambiguity and uncertainty in interpreting human emotional cues like facial expressions, tone of voice, and body language. It allows robots to identify and respond to complex emotions more accurately. Robots accomplish this by translating vague input into meaningful emotional states through the use of fuzzy rules and membership functions.

To mention a few applications of fuzzy logic in robot emotion detection, we state the following:

  i. Analysis of Facial Expression: Fuzzy logic is used to interpret subtle facial expressions. Examples are slightly raised eyebrows or slightly parted lips. They are vital to identify emotions like surprise or uncertainty.
  ii. Analysis of Voice Tone: Fuzzy logic is used to analyze variations in pitch, volume, and speech rate. Determination of emotional states like anger or sadness is done.
  iii. Social Interaction of Robots: A robot dynamically adjusts its responses by incorporating fuzzy logic. The emotional state of the user is perceived. A more natural and engaging interaction is made.
  iv. Interpretation of Vague Emotional Data: Fuzzy logic handles situations where emotional indicators are not clear-cut. A slightly furrowed brow or a slightly raised voice might represent a combination of emotions like confusion and slight annoyance.
  v. Creation of Nuanced Emotional States: Fuzzy membership functions are used by robots to represent emotions on a spectrum. This kind of function allows for a more nuanced understanding of emotions. The understanding extends beyond simple 'happy' or 'sad' categories. A robot endowed with six universal human emotions (happiness, anger, fear, sadness, disgust, and surprise) is designed and simulated (Leu et al. 2014).
  vi. Integration of Multiple Sensory Inputs: Fuzzy logic combines information from various sensors like facial expressions, speech patterns, and body movements. Amalgamating varied information creates a more comprehensive picture of a person's emotional state.
  vii. Adaptation to Individual Differences: Fuzzy rules are adjustable based on the context and individual user behavior. This adjustability allows robots to interpret emotions across different individuals in a better way.

8.4.2.3 Advantages of Fuzzy Logic

In contrast to explicit logic, which relies on clear and definite statements, fuzzy logic offers several advantages in robot emotion detection. The primary reason is its ability to handle the uncertainty and vagueness inherent in human emotions. This ability allows for more nuanced and robust emotion recognition. Traditional binary classifications fail to do so. Fuzzy logic is particularly useful for interpreting subtle facial expressions and complex emotional states in human–robot interactions.

  i. Handling of Ambiguities: Emotions are complex and often blend together. Fuzzy logic can effectively represent these blurred boundaries between different emotional states. Crisp classifications require clear-cut distinctions.

ii. Adaptability to Context: Fuzzy logic rules can incorporate contextual information about the situation or tone of voice. Better interpretation of facial expressions and body language is achieved. Therefore, more accurate emotion detection becomes possible.

iii. Intuitive Rule-Based System: Fuzzy logic rules can be formulated using natural language. This facility enables developers to incorporate expert knowledge about human emotions more easily. The inclusion of such expertise enables emotions to be incorporated into the decision-making process of a robot.

iv. Robustness to Noise and Occlusion of Image: Fuzzy logic systems can tolerate noisy or incomplete data. Such data is common in real-world scenarios. Frequently, sensor readings are imprecise. A fuzzy inference method is presented for emotion recognition from facial expression that can recognize emotions from partially occluded facial images (Ilbeygi and Shah-Hosseini 2012).

v. Gradual Response to Emotional States: Unlike binary classifications, fuzzy logic allows for a gradual transition between emotional states. The gradual transition enables robots to respond more naturally to subtle changes in human emotions.

vi. Integration of Fuzzy Logic with Other Techniques: Fuzzy logic can be conjoined with other ML algorithms to enhance emotion detection accuracy. It can be combined with deep learning for feature extraction.

An example of the application of fuzzy logic in emotion detection will make its use clear. A robot could use fuzzy logic to interpret a slightly furrowed brow and slightly downturned mouth by taking into account the surrounding context for a more accurate emotional interpretation. This might mean a state of 'mild sadness', rather than classifying it definitively as either 'happy' or sad'.

8.4.2.4   Limitations of Fuzzy Logic

The Human-to-Humanoid robot communication is particularly challenging (Mogos 2022). The main difficulty of fuzzy logic in emotion detection lies in accurately capturing complex nuances of emotions. Due to our customary overdependence on well-defined rules, there is a potential for misinterpretation when dealing with ambiguous data. Challenges are faced in handling the context. Computational complexity increases when dealing with large amounts of sensory data. These drawbacks make it less ideal for robust emotion recognition in real-world scenarios. They become dominant, especially when dealing with subtle or multifaceted emotions. Some limitations of fuzzy logic in emotion detection are outlined here.

i. Oversimplification of Emotions in Fuzzy Logic Representation: Fuzzy logic often categorizes emotions into a limited set of fuzzy states. These fuzzy states may not adequately capture the full spectrum of human emotions. Subtle variations and combinations may be difficult to express.

ii. Limitations of Fuzzy Rules in Context-Dependent Situations: Defining clear fuzzy rules for emotion detection is not a piece of cake. This happens because human emotions are often context-dependent. They are influenced by various factors beyond easily quantifiable data.

iii. Difficulty in Incorporation of Contextual Information: Fuzzy logic may not effectively incorporate contextual information. This failure is crucial for accurate emotion recognition. The same facial expression can signify different emotions depending on the context of the situation being considered.

iv. Dilemmas due to Ambiguity in Interpretation of Data: Fuzzy logic can struggle to interpret ambiguous or inconsistent sensory data. The chances of potential misinterpretation of emotions are unavoidable.

v. Requirement of a High Computational Overhead: Implementation of complex fuzzy logic systems for emotion detection requires substantial computational power. As real-time data processing is more power hungry, major issues arise with these applications.

8.4.2.5 Alternatives to Fuzzy Logic

If fuzzy logic is not useful, what are the options available? Then one can beat a retreat and solicit intervention by:

i. Resorting to ML Models: Deep learning techniques like CNNs can learn complex patterns from large datasets. They are capable of achieving higher accuracy in emotion recognition.

ii. Adopting Hybrid Approaches Based on Combination of Techniques: Combining fuzzy logic with other techniques, such as statistical analysis or sentiment analysis, is helpful. It leverages the strengths of each method to improve emotion detection accuracy.

8.5 DISCUSSION AND CONCLUSIONS

In order to achieve the symbiosis between humans and robots, the aspect of emotions must be integrated into robotic systems (Loghmani et al. 2017). The importance of emotional intelligence in robots deserves due appreciation for its multi-pronged benefits.

i. Enhancement of Human–Robot Interaction: Robots create a more positive and engaging experience for users by recognizing and responding to human emotions. The friendly behavior of robots makes humans feel more comfortable. Emotionally responsive robots that can simulate empathy increase the acceptability of users toward them (Marcos-Pablos and García-Peñalvo 2022).

ii. Improvement of Human–Robot Communication: Robots are able to interpret nonverbal cues like facial expressions and tone of voice, paving the way to more nuanced and natural communication.

iii. Delivery of Personalized Support to Users: Robots can tailor their responses based on a user's emotional state. Customized support and comfort for users are provided in situations where it is needed most.

iv. Applications in Sensitive Fields: In healthcare, robots can provide emotional support to patients. In education, robots can adapt teaching methods to meet the individual needs of students.

v. Building Trust in Users: A greater sense of trust and acceptance from users is fostered by a robot demonstrating an ability to understand and respond to emotions.

The crux of the discussion in this chapter centered on Bayesian inference and fuzzy logic methods (Table 8.2). In the context of robot emotion, we deliberated on Bayesian inference. It is a computational method that allows a robot to continuously update its understanding of a human's emotional state by incorporating new sensory data like facial expressions, tone of voice, etc., with its prior beliefs about emotions. Such robots can more accurately and dynamically interpret human emotions during interaction. Bayesian inference employs a probabilistic approach based on probability values to represent the likelihood of different emotional states. Robots can consider uncertainty and update their beliefs as new information becomes available. Starting with an initial 'prior' belief about the human's emotional state, the robot continuously refines the belief based on new sensory data. Upon receipt of new data from the human-like facial expressions, tone of voice, etc., the robot uses Bayes' rule to calculate the 'posterior' probability of each possible emotion. Thus, it effectively updates its understanding of the human's emotional state in real-time.

Bayesian inference allows a robot to analyze facial expressions, body language, and speech patterns to infer a human's emotional state. Based on the inferred human emotion, the robot adjusts and adapts its own behavior and communication style. A robot creates a more natural and engaging interaction with humans by demonstrating an ability to understand and respond to human emotions. It thus builds trust and rapport with the human user.

In the context of robotics, fuzzy logic is often employed to model and simulate a robot's emotions. It allows for a more nuanced and human-like expression of feelings.

TABLE 8.2

Takeaways from This Chapter at a Glance

Sl. No.	Takeaway	Explanation
1	Summary	Emotionally intelligent robots are able to recognize, interpret, and respond to human emotions by utilizing machine learning, natural language processing, and robot vision to analyze facial expressions, voice, and other cues. The main components of an emotional robot algorithm were outlined, highlighting the considerations and concerns that must be accounted for during algorithm formulation. Specific algorithms used in emotionally intelligent robots were discussed, namely Bayesian inference and fuzzy logic.
2	Bayesian inference	Bayesian inference provides a probabilistic perspective on emotion recognition.
3	Fuzzy logic	Fuzzy logic is a platform to handle the ambiguous nature of human emotional behavior by interpreting uncertain or vague information about emotions based on a range of input stimuli.
4	Keywords and ideas to remember	Emotional AI, emotional robot algorithm, Bayesian inference for robot emotion detection, probabilistic inference perspective, and fuzzy logic in robot emotion recognition

This becomes possible by utilizing degrees of emotions in lieu of strict binary states. Therefore, the robot can experience a range of emotions in response to various input stimuli and situations. Such expressions of feelings make a robot's responses appear more natural and adaptable.

Unlike traditional logic with clear true/false values, fuzzy logic allows for gradual transitions in terms of degrees of truth, meaning a robot can be 'moderately happy' or 'slightly frustrated'. Fuzzy logic systems often utilize linguistic rules based on human language, e.g., if the battery is low, then the robot feels anxious. These rules determine emotional states based on input data like sensor readings or user interactions. Factors like battery level, environmental conditions, user interactions, and task completion status are inputs to the fuzzy logic system that influence the robot's emotional response. The robot's emotional state generated by fuzzy logic can then be translated into observable output behaviors like facial expressions, tone of voice, or movement patterns. To give an example, when a robot assistant is tasked with performing a complex task and encounters unexpected difficulties, fuzzy logic determines that it experiences a mix of 'frustration' and 'uncertainty'. The robot will request clarification or seek assistance from the user.

Fuzzy logic allows robots to express a wider range of emotions. It makes their interactions with humans feel more natural, relatable, and realistic. These systems can be easily adjusted to accommodate different situations and user preferences by modifying the linguistic rules. They are well-suited for dealing with ambiguous or uncertain information, which is often present in real-world interactions.

The survey of algorithms for embellishing robotics with intelligence will be continued in the next chapter, starting with hidden Markov models. Fuzzy logic and hidden Markov models can be combined to build fuzzy hidden Markov models (FHMMs) for emotion recognition from recognition from various modalities, such as speech, EEG, and ECG signals enabling capturing of complex relationships between input features and emotional states with smoother transitions between states, making them more adaptable to real-world application.

REFERENCES AND FURTHER READING

Arrent K., D. J. Brown, J. Kruk-Lasocka, T. L. Niemiec, A. H. Pasieczna, P. J. Standen and R. Szczepanowski. 2022. The use of social robots in the diagnosis of autism in preschool children, *Applied Sciences*, Vol. 12, 17, 8399, pp. 1–16.

Bera A., T. Randhavane and D. Manocha. 2019. Modelling Multi-channel Emotions Using Facial Expression and Trajectory Cues for Improving Socially-Aware Robot Navigation, *2019 IEEE/CVF Conference on Computer Vision and Pattern Recognition Workshops (CVPRW)*, Long Beach, CA, USA, 16–17 June, pp. 257–266.

Bryant D. 2019. Towards Emotional Intelligence in Social Robots Designed for Children, *AIES '19: Proceedings of the 2019 AAAI/ACM Conference on AI, Ethics, and Society*, Honolulu, HI, USA, January 27–28, pp. 547–548.

Cardone B., F. Di Martino and V. Miraglia. 2023. A fuzzy-based emotion detection method to classify the relevance of pleasant/unpleasant emotions posted by users in reviews of service facilities, *Applied Sciences*, Vol. 13, 10, 5893, pp. 1–16.

Ho K. H. L. 1997. A Model of Fuzzy Emotion and Behavior Selection for an Autonomous Mobile Robot, *Proceedings 6th IEEE International Workshop on Robot and Human Communication, RO-MAN'97 SENDAI*, Sendai, Japan, 29 September to 1 October, pp. 332–337.

Hsu Y.-T., F.-Y. Leu, J.-C. Liu, Y.-L. Huang and W. C.-C. Chu. 2013. The Simulation of a Robot with Emotions Implemented with Fuzzy Logic, *2013 Eighth International Conference on Broadband and Wireless Computing, Communication and Applications*, Compiegne, France, 28–30 October, pp. 382–386.

Ilbeygi M. and H. Shah-Hosseini. 2012. A novel fuzzy facial expression recognition system based on facial feature extraction from color face images, *Engineering Applications of Artificial Intelligence*, Vol. 25, 1, pp. 130–146.

Kato S., Y. Sugino and H. Itoh. 2006. A Bayesian Approach to Emotion Detection in Dialogist's Voice for Human Robot Interaction. In: Gabrys B., R. J. Howlett and L. C. Jain (Eds.), *Knowledge-Based Intelligent Information and Engineering Systems. KES 2006.* Lecture Notes in Computer Science, Vol. 4252, Springer, Berlin, Heidelberg, pp. 961–968.

Khare S. K., V. Blanes-Vidal, E. S. Nadimi and U. R. Acharya. 2024. Emotion recognition and artificial intelligence: A systematic review (2014–2023) and research recommendations, *Information Fusion*, Vol. 102, p. 102019, https://doi.org/10.1016/j.inffus.2023.102019

Kolling T., S. Baisch, A. Schall, S.e Selic, S. Rühl, Z. Kim, H. Rossberg, B. Klein, J. Pantel, F. Oswald and M. Knopf. 2016. What Is Emotional About Emotional Robotics? In: Tettegah S. Y. and Y. E. Garcia, (Eds.), *Emotions and Technology, Emotions, Technology, and Health*, Academic Press, an imprint of Elsevier B.V., Amsterdam, Netherlands, pp. 85–103.

Leu F.-Y., J.-C. Liu, Y.-T. Hsu and Y.-L. Huang. 2014. The simulation of an emotional robot implemented with fuzzy logic, *Soft Computing*, Vol. 18, pp. 1729–1743.

Loghmani M. R., S. Rovetta and G. Venture. 2017. Emotional Intelligence in Robots: Recognizing Human Emotions from Daily-Life Gestures, *2017 IEEE International Conference on Robotics and Automation (ICRA)*, Singapore, 29 May to 3 June, pp. 1677–1684.

Marcos-Pablos S. and F. J. García-Peñalvo. 2022. Emotional Intelligence in Robotics: A Scoping Review. In: de Paz Santana J. F., D. H. de la Iglesia and A. J. López Rivero (Eds.), *New Trends in Disruptive Technologies, Tech Ethics and Artificial Intelligence. DiTTEt 2021.* Advances in Intelligent Systems and Computing, Vol. 1410, Springer, Cham, pp. 66–75.

Martin G. F.-B., F. Matia, L. G. Gómez-Escalonilla, D. Galan, M. G. Sánchez-Escribano, P. de la Puente and M. Rodríguez-Cantelar. 2023. An emotional model based on fuzzy logic and social psychology for a personal assistant robot, *Applied Sciences*, Vol. 13, 5, 3284, pp. 1–26.

Martinez-Hernandez U., A. Rubio-Solis and T. J. Prescott. 2016. Bayesian Perception of Touch for Control of Robot Emotion. *2016 International Joint Conference on Neural Networks (IJCNN)*, Vancouver, BC, Canada, 24–29 July, IEEE, pp. 4927–4933.

Mogos E. 2022. The fuzzy-based systems in the communication between a human and a humanoid robot, *Journal of Physics: Conference Series*, Vol. 2251, 012003, pp. 1–12.

Mohan S. N., F. Mukhtar and L. Jobson. 2021. An exploratory study on cross-cultural differences in facial emotion recognition between adults from Malaysia and Australia, *Frontiers in Psychiatry*, Vol. 12, pp. 1–8.

Nicolai A. and A. Choi. 2015. Facial Emotion Recognition Using Fuzzy Systems, *2015 IEEE International Conference on Systems, Man, and Cybernetics*, Hong Kong, China, 9–12 October, pp. 2216–2221.

Seyitoğlu F. and S. Ivanov. 2024. Robots and emotional intelligence: A thematic analysis, *Technology in Society*, Vol. 77, 102512, pp. 1–11.

Spezialetti M., G. Placidi and S. Rossi. 2020. Emotion recognition for human-robot interaction: Recent advances and future perspectives, *Frontiers in Robotics and AI*, Vol. 7, Article 532279, pp. 1–11.

Thilmany J. 2007. The emotional robot. Cognitive computing and the quest for artificial intelligence, *EMBO Reports*, Vol. 8, 11, pp. 992–994.

Wu J. 2024. Social and ethical impact of emotional AI advancement: The rise of pseudo-intimacy relationships and challenges in human interactions, *Frontiers in Psychology*, Vol. 15, 1410462, pp. 1–6.

Yan F., A. M. Iliyasu and K. Hirota. 2021. Emotion space modelling for social robots, *Engineering Applications of Artificial Intelligence*, Vol. 100, p. 104178, https://doi.org/10.1016/j.engappai.2021.104178

9 Emotionally Intelligent Robots
Unlocking More Opportunities

9.1 INTRODUCTION

Robot emotion models exhibit a wide heterogeneity and manifoldness in their approaches to the representation and generation of emotions. These include variations in the number of emotions considered and the underlying theoretical frameworks used in the development of the models. The triggering of emotions by external stimuli greatly differs, and so do the methods for expressing emotions through robot behavior (Zhao 2023). A spectrum of models exists that can adapt to context and user interactions. In this chapter, the discussion of algorithms related to emotional intelligence will be continued to provide readers with a deeper understanding of this extensive field. We begin with the hidden Markov models (HMMs).

9.2 HMMs FOR ROBOT EMOTION DETECTION

A Markov model for emotions is a mathematical framework that utilizes the concept of Markov chains. A Markov chain is a stochastic process unfolding a sequence of events. A stochastic process is a phenomenon in which the outcome at any given time is a random variable. It is a collection of random variables indexed by time. The future state of a stochastic process is therefore not entirely predictable, but rather depends on probabilities and randomness.

The Markov chains are used for the representation of dynamic changes in emotional states with time. The HMMs add a layer of complexity within the model (Kulic and Croft 2006; Inthiam et al. 2019a,b). This layer incorporates hidden states. The hidden states allow for a more nuanced representation of emotions.

A Markov model represents affective states of robots, e.g., relaxed, stressed, engaged, and bored. The probability of transition to a new emotional state is governed only by the current state. This characteristic property enables the modeling of shifting and evolution of emotions within a sequence of events or interactions. Figure 9.1 is a depiction of the components and working mechanism of the HMM in robotics. Figure 9.1a shows the main elements of HMM as: states (discrete emotional states represented by nodes in the diagram), transitions (arrows between states labeled with transition probabilities), and observations (inputs such as facial expressions). Figure 9.1b sketches the workflow of HMM. First, the facial muscle movements are

DOI: 10.1201/9781032695266-9

| Main Elements of the Hidden Markov Emotion Recognition Model Used by a Robot |

(i) States → Representation of discrete emotional states by nodes in the diagram

(ii)Transitions → Arrows between states labeled with transition probabilities signifying the possibility of moving from one emotion to another

(iii) Observations → Inputs such as facial expressions or speech features for updating the probability of existence in a particular emotional state at each time step

(a)

| Workflow of the Hidden Markov Emotion Recognition Model Used by a Robot |

(i) Extraction of facial muscle movements 〉 **(ii)** Calculation of probability of transitioning from the current emotional state to different possible next emotional states using the transition matrix 〉 **(iii)** Calculation of the likelihood of observation of current features considering the state to be active 〉 **(iv)** Determination of the most likely sequence of emotional states that explain the observed features over time in the best possible way by using a dynamic programming algorithm, e.g., Viterbi algorithm to find the most likely sequence of states responsible for producing a sequence of events by calculating the distance between the received signal and all the paths in a trellis diagram, a bipartite graph representing a current state and its succeeding state

(b)

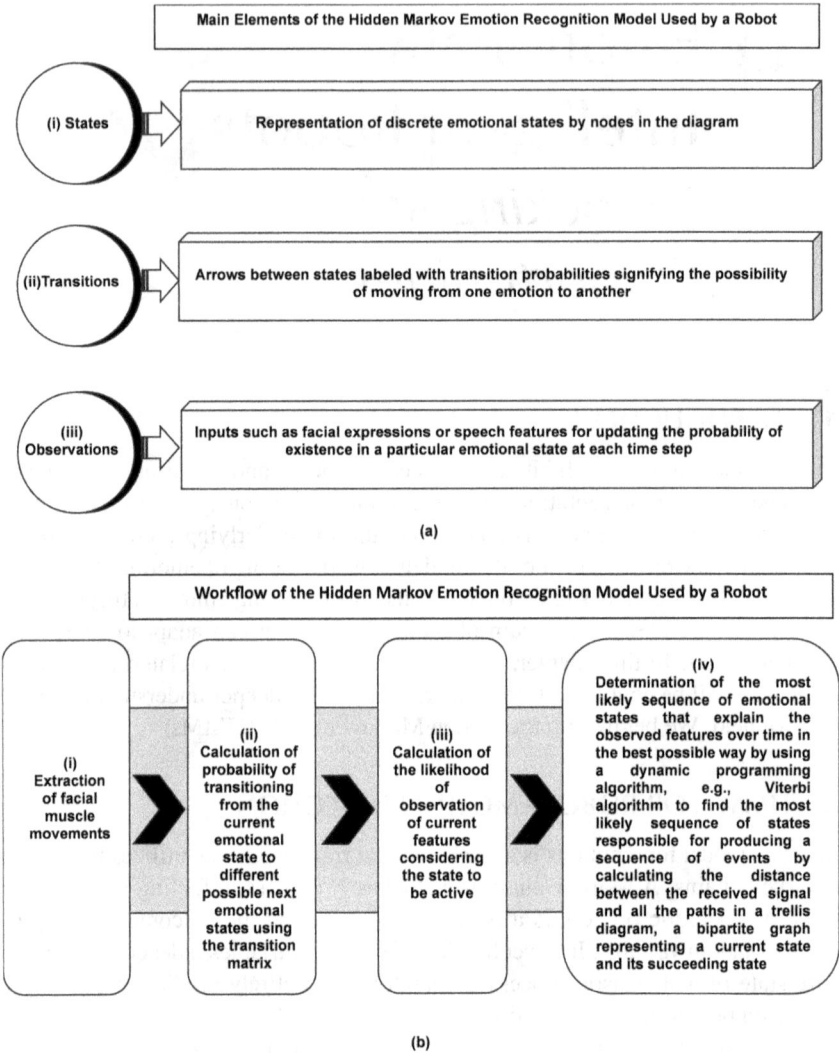

FIGURE 9.1   Hidden Markov model for emotion recognition by robots: (a) components and (b) workflow.

extracted. Second, the probability of transitioning from the current emotional state to any of the possible next states is calculated using the transition matrix. Third, the likelihood of observation of current features is ascertained. The fourth stage involves determining the most likely sequence of emotional states that explains the observed emotional features over time in the most effective way.

The staple component of a Markov model is the transition matrix (Christopher 2024). It defines the probability of transition from one emotional state to another. The transition probabilities between different emotional states are indicators of the likelihood of moving from one state to another in a Markov model. So, a transition matrix is a matrix that organizes the transition probabilities. Each entry in the transition matrix is the probability of moving from one state to another. The transition matrix is used to calculate the probability of occupying any state at any given time. So, the transition matrix is a representation of the probabilities of changing between distinct states of the robot when given its current state.

A transition diagram is a weighted directed graph. This graph represents a Markov chain. Each vertex of the graph is a state. Each directed edge signifies a transition probability.

9.2.1 Applications of Markov Models

Markov models are used in robot emotion detection to analyze sequences of sensory data like facial expressions, voice tone, or body language over time. They allow the robot to deduce the underlying emotional state of a human. Changing patterns in these observations are identified. Hence, more natural and responsive human–robot interactions are enabled through the assistance of these models. Among the applications of Markov models in robot emotion detection, mention may be made of:

i. Analysis of Facial Expressions: The transitions between different facial expressions of a person are modeled as states in a Markov chain. The modeling of transitions allows the robot to predict the current emotional state of the person. The prediction is based on a sequence of facial features observed over time, notwithstanding the fact that the expressions are subtle or partially obscured.

ii. Recognition of Speech Emotion: HMMs are used to analyze the dynamic changes in pitch, volume, and speech rate discovered in a person's voice. This analysis allows for the identification of the emotional tone in the person's speech.

iii. Multimodal Emotion Recognition: A Markov model combines information from assorted heterogeneous sources such as facial expressions, voice tone, and body language of an individual. A multi-sensor arrangement is used for collecting information from these sources. The information is coalesced by integrating data from various sensors like cameras and microphones. The combined information provides an all-inclusive understanding of a person's emotional condition.

iv. Adaptation of Robot Behavior: When a robot detects a person's emotion using a Markov model, it adjusts its responses and behaviors accordingly. The robot offers comfort if the person appears to be sad. It provides encouragement if the person seems to be frustrated. Thus, the robot's response correctly answers the person's emotion.

v. Contextual Understanding of Emotion: Markov models are able to incorporate contextual information like the current situation or past interactions. This understanding helps in the interpretation of emotional cues in a better way. More nuanced and relevant responses are then offered by the robot.

vi. Modeling of Emotional State Transitions: The modeling enables prediction of changes of a robot's emotional state based on user interactions. The prediction allows for generating more natural and adaptive responses.

vii. Estimation of Emotion Intensity: The intensity of an emotion is determined by analyzing the transition probabilities between different states within a Markov model.

viii. Personalized Emotion Recognition: A Markov model is customized for each user by learning the unique emotional patterns and behaviors of the concerned user. Model customization provides enhanced alignment with the user's needs. Consequently, its performance is improved for the intended application.

9.2.2 ADVANTAGES OF MARKOV MODELS

Markov models, particularly HMMs, offer several advantages in the detection of robot emotions. To understand this, we must be aware of the specific areas where HMMs can make a dent in processing emotions. They equip the robots with the ability to handle sequential data and model dynamic emotional states. Further, the robot can infer hidden emotional states from observed behaviors. It can adapt to changing contexts, too. Hence, real-time emotion recognition in human–robot interactions materializes. The chief advantages of using Markov models for robot emotion detection are as follows:

i. Modeling Temporal Dependencies: Unlike static classification methods, Markov models can capture the temporal relationships between different emotional states. The capturing of temporal relationships allows these models to understand the evolution of emotions over time. This understanding is based on previous observations. Natural human–robot interactions are therefore conceivable and workable.

ii. Handling Dynamic Changes in Emotions: Markov models can effectively pick up the dynamic nature of emotions. This allows processing of expressions and behaviors that change rapidly over time.

iii. Real-Time Processing of Emotions: The relatively simple structure of Markov models allows their efficient implementation for real-time emotion detection. So, robots using these models can respond promptly to changing human emotions.

iv. Probabilistic Inference of Emotions: HMMs leverage probability theory. Hence, they can provide confidence levels in emotion predictions. This capability makes them suitable for emotion recognition in ambiguous situations.

v. Inference of Hidden Emotional States: HMMs examine observable behaviors, such as facial expressions, voice tone, and body language, meticulously. By conducting a thorough examination, they recognize hidden emotional states, such as 'frustration' or 'joy'. These states might not be directly measurable.

vi. Adaptability to Contextual Information: Markov models incorporate context information into their formalism. By adopting this strategy, they can adjust their emotion recognition based on the current situation. Therefore, more accurate interpretations of human emotions in different scenarios are obtainable.

vii. Flexibility in Feature Selection: Different features like facial expressions, speech prosody, and body posture can be integrated into a Markov model. A more comprehensive understanding of human emotions is thereby achievable.

viii. Computational Efficiency: Compared to more complex models, Markov models are computationally efficient. The high efficiency allows for real-time emotion detection on robots with limited processing power.

9.2.3 LIMITATIONS OF MARKOV MODELS

What drawbacks make Markov models inadequate for emotion modeling? Markov models have limitations in modeling robot emotions due to their inherent lack of long-term memory. They can only consider the current state and not the full context or history of interactions. The history of interactions is necessary for accurately representing complex human emotions. The reason is that emotions often build over time. They depend on previous experiences. Consequently, emotion recognition and expression in robots lack accuracy.

The limitations of Markov models in robot emotion include:

i. Restrictions of Context Awareness: Although Markov models display adaptability to contextual information, they only consider the current state. They neglect the influence of past interactions or events. The past events can significantly impact emotional responses.

ii. Inability to Capture Nuanced Emotions: Human emotions are often complex and multifaceted. They show varying intensities and subtle transitions. A simple Markov model cannot accurately represent these variations faithfully.

iii. Difficulty Faced with Long-Term Emotional Dynamics: Emotions build up over time or change based on past experiences. Such building up of emotions is not properly portrayed by the short-term memory of a Markov model.

iv. Oversimplification of State Transitions: Markov models often assume discrete emotional states with fixed transition probabilities. These discrete states may not accurately reflect the continuous nature of human emotions.

v. Challenges with Complex Social Interactions: In real-world scenarios, social interactions involve multiple factors and participants. This dependence of emotions on several parameters makes it difficult to model emotions accurately using a simple Markov chain.

vi. Demand for Quantity and Quality of Data for Model Training: Training a robust Markov model for emotion detection is a conscientious job. It requires a large amount of high-quality data on human emotional expressions and behaviors, ornamented with accurate emotion labels, to ensure reliable emotion detection.

vii. Necessity of Taking Cultural Variations into Account for Model Training: Emotional expressions vary significantly across cultures. So, the data used for training Markov models should be representative of the target user population, unifying intercultural, cross-cultural, multicultural, and international aspects.

viii. Complexity of Markov Models: The Markov models are no doubt efficient for basic emotion recognition. But complex emotional states might require more sophisticated Markov models. Additional hidden states may be needed in these models, making them more difficult and muddled.

ix. Associated Privacy Concerns: Privacy concerns must be properly addressed when using facial expressions or voice analysis for emotion detection. Obtaining prior user consent is compulsory.

9.2.4 POTENTIAL SOLUTIONS TO OVERCOME THE LIMITATIONS OF MARKOV MODELS

The understanding of the limitations of HMMs prompts us to the exploration of better alternatives that can overcome their aforementioned hurdles.

i. Recurrent Neural Networks (RNNs): RNNs can learn long-term dependencies in data. This capability enables them to describe the temporal aspects of emotions and context in a better way.

ii. Emotion Appraisal Models: These models incorporate cognitive factors. Situational appraisal is included for understanding the underlying causes of emotions in detail.

iii. Multimodal Input Integration: A richer picture of emotional states is sketched by combining various sensory data. Facial expressions, speech, and body language are combined with basic parameters in building the model.

9.3 SELF-ORGANIZING MAPS IN ROBOT EMOTION DETECTION

Let us look around eagerly for other opportunities for emotion detection. From the HMMs, which prototype the probability of a sequence of events based on a set of hidden states and transition probabilities between those states, we transition to self-organizing maps (SOMs), also known as Kohonen maps. These algorithms facilitate dimensionality reduction and data clustering, offering a visual representation of the data (Simplilearn 2023).

A SOM is an artificial neural network (Jitviriya and Hayashi 2014; Jitviriya et al. 2015). The SOM neural network is an unsupervised learning model. It is able to distinguish patterns in data without any pre-labeled emotion categories. The SOM is trained to categorize and visualize emotional states by clustering data points representing different emotions. It is used for clustering of emotional data for easy understanding (Figure 9.2). Figure 9.2a shows the three layers in an emotion recognition SOM: the input layer (representing extracted facial features from an image), hidden layer (representing potential emotional states by nodes in a 2D grid of nodes) and output layer (displaying the final emotion classification based on the winning node in the SOM grid). Figure 9.2b shows the six operations in an emotion recognition SOM: feature extraction, its presentation to SOM, identifying the node in the SOM grid with smallest distance with input feature vector as the winning node, calculating the distance between the input feature vector and each node in the SOM grid, updating of neighboring nodes to the winning node in the SOM grid toward the input feature vector, and repetition of the process for many images for training.

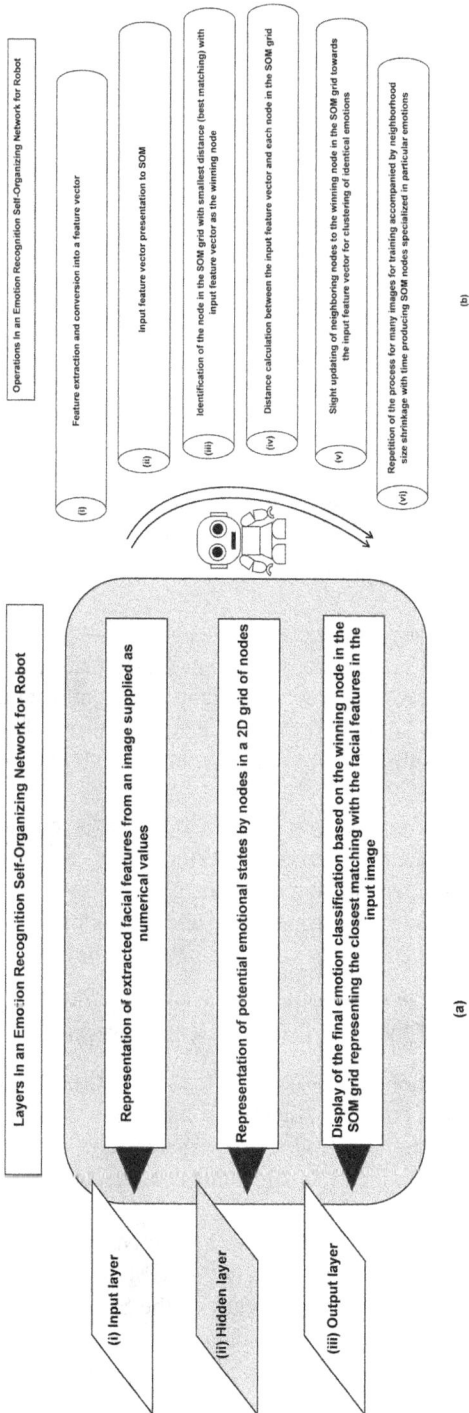

Operations in an Emotion Recognition Self-Organizing Network for Robot

(i) Feature extraction and conversion into a feature vector

(ii) Input feature vector presentation to SOM

(iii) Identification of the node in the SOM grid with smallest distance (best matching) with input feature vector as the winning node

(iv) Distance calculation between the input feature vector and each node in the SOM grid

(v) Slight updating of neighboring nodes to the winning node in the SOM grid towards the input feature vector for clustering of identical emotions

(vi) Repetition of the process for many images for training accompanied by neighborhood size shrinkage with time producing SOM nodes specialized in particular emotions

(b)

Layers in an Emotion Recognition Self-Organizing Network for Robot

(i) Input layer — Representation of extracted facial features from an image supplied as numerical values

(ii) Hidden layer — Representation of potential emotional states by nodes in a 2D grid of nodes

(iii) Output layer — Display of the final emotion classification based on the winning node in the SOM grid representing the closest matching with the facial features in the input image

(a)

FIGURE 9.2 Self-organizing network for emotional robot: (a) layers and (b) operations.

The SOM effectively reduces high-dimensional emotional data displaying facial features into a lower-dimensional space. How is this beneficial? It helps because the relationships between different emotions are easily visualized and interpreted. Complex emotional landscapes are explored with SOM. From an SOM, the researchers identify patterns and relationships between various emotional expressions. These expressions are often based on facial features or physiological data. The topological structure of the input space is maintained. Hence, similar emotions tend to be mapped to nearby nodes on the SOM grid in close proximity. Topological preservation is thereby attained.

9.3.1 APPLICATIONS OF SOM IN EMOTION RESEARCH STUDIES

SOMs are used for recognizing facial emotion as well as physiological emotion offering a two-pronged benefit. They are also applied to investigating other indicators of emotional states.

i. Recognition of Facial Emotion: Facial expressions divulged by images or videos are probed to classify emotions (happiness/sadness/anger/surprise). The facial feature points are mapped onto the SOM grid (Majumder et al. 2014).

ii. Detection of Physiological Emotion: Physiological signals, namely, heart rate, skin conductance, or electrocardiogram data, are mapped onto the SOM grid. Identification of related emotional states is sought from the maps.

iii. Analysis of Emotion in Text: Textual data are appraised. The sentiment or dominant emotion within a piece of text is identified by mapping word vectors onto a SOM.

iv. Investigation of Emotional Dynamics: Sequences of emotional data are mapped onto the SOM. An analysis is performed of the mapped data. The analysis reveals the transition and evolution of emotions over time. Thereby, fluctuations of emotions over time, their underlying processes, and downstream consequences come into view and become known.

9.3.2 LIMITATIONS OF SOMs FOR EMOTION RECOGNITION

In what ways is the utilization of SOMs restricted? The utilization of SOMs is hampered by the subjectivity of emotions, their personalization, variation from person to person, and shaping by individual, unique experiences and interpretations of situations. The quality of data supplied too determines the veracity of results.

i. Subjectivity of Emotion: Emotions are subjective and context-dependent. Hence, their accurate mapping onto a fixed SOM grid is fastidious.

ii. Data Quality Dependence: The accuracy of the SOM analysis is always at the mercy of the quality of the input data. Being grabbed in their clutches, facial expressions, physiological signals, etc., in input data play vital roles in the emotion detection of robots.

9.4   SUPPORT VECTOR MACHINES FOR ROBOT EMOTION CLASSIFICATION

While SOMs reduce dimensionality to extract meaningful features, let us investigate the utilization of a classifier machine learning algorithm in this matter. A support vector machine (SVM) is a machine learning algorithm (Tsai et al. 2009; Kang 2025). It is commonly used as a classifier to categorize different emotional states from data like speech or facial expressions. The classification is done by identifying patterns in extracted features. The identified patterns are assigned to specific emotions like happiness, sadness, anger, or neutrality.

The SVM effectively distinguishes between different emotional categories based on the input data by finding the optimal decision boundary between them. Figure 9.3 displays the main elements and workflow of a SVM for emotion recognition by a robot. Figure 9.3a shows the principal elements of the SVM: data points (extracted features representing the facial image), feature space (multidimensional space for visualizing the data distribution), hyperplane (optimal separating plane subdividing the data points), support vectors (data points most proximate to the hyperplane) and the kernel function (a mathematical function for data transformation to higher dimensional space). Figure 9.3b shows the sequential flow of the SVM algorithm: feature extraction, feature mapping, calculation for creating an optimal hyperplane, and classification of a new facial image.

9.4.1   USING SVM FOR EMOTION CLASSIFICATION

The SVM works as an emotion classifier by following a supervised machine learning approach. It learns from labeled data to create a model that predicts emotions based on input features (Alhussan et al. 2023).

i. Feature Extraction: Relevant features like pitch, energy, Mel frequency cepstral coefficients (MFCCs), or facial landmarks are extracted from the input speech or image data before classification by SVM. The MFCCs constitute a set of features representative of the spectral envelope of a sound signal. The spectral envelope is a curve of amplitude values of the signal on the Y-axis and frequency on the X-axis.

ii. Hyperplane Separation: SVM finds the hyperplane in the feature space that best separates different classes of emotions. It maximizes the margin between them. A robust classification is enabled, which is less sensitive to data uncertainty.

iii. Using Kernel Functions: SVM handles non-linear relationships between features by employing kernel functions that implicitly map the input data into a higher dimensional space. Complex emotion classification scenarios can therefore be handled.

9.4.2   APPLICATIONS OF SVM IN EMOTION CLASSIFICATION

Applications of SVM extend across a broad range, from speech emotion recognition to facial emotion recognition (FER) and text sentiment analysis. To name a few, we mention:

Principal Elements of The Support Vector Machine Algorithm for Emotion Recognition by a Robot

(i) **Data points** — Representation of a facial image with the extracted features, e.g., landmarks, pixel intensities or specific emotion-associated attributes

(ii) **Feature space** — Multidimensional space in which the extracted features are plotted for visualization of the distribution of data among various emotions

(iii) **Hyperplane** — Optimal separating plane providing the best subdivision of data points of different emotions with maximum inter-class margin

(iv) **Support vectors** — Data points in closest vicinity of the hyperplane and significantly impacting its position, thereby serving as the most critical points for classification

(v) **Kernel function** — A mathematical function that can transform the data into a higher-dimensional space, hence enabling the definition of non-linear boundaries between the classes

(a)

Sequential Flow of the Support Vector Machine Algorithm for Emotion Recognition by a Robot

(i) Feature extraction: Mouth curvature, eye shape and eyebrow position

(ii) Mapping feature to the feature space by plotting the extracted features as data points

(iii) Calculation for creation of optimal hyperplane separating the data points of different emotions with maximum margin between them

(iv) Classification of a new facial image by feature extraction, plotting it in the feature space, and determining the emotion from the side of the hyperplane into which it is found to be falling

(b)

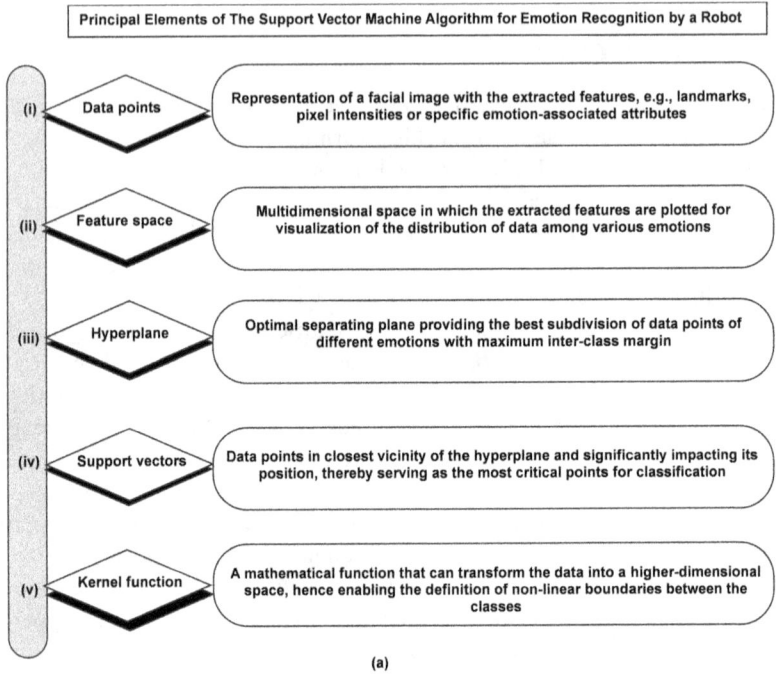

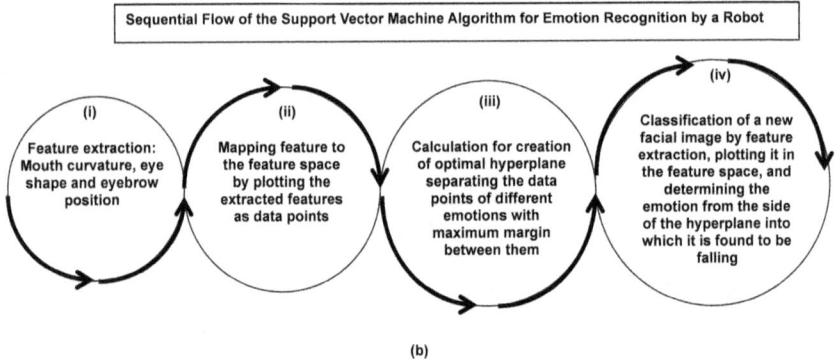

FIGURE 9.3 Using the support vector machine algorithm for emotion recognition in robotics: (a) the chief elements and (b) the process flow.

i. Recognition of Speech Emotion: Features such as the pitch and intensity of sound form the basis of classifying emotions from spoken words.
ii. Recognition of Facial Emotion: Features from facial landmarks are extracted as descriptors of emotions from facial expressions.
iii. Analysis of Text Sentiments: Emotional cues are utilized for determining the sentiment (positive, negative) underlying a written textual document.

9.4.3 Advantages of SVM for Emotion Classification

As an emotion classifier, the SVM offers many advantages for improved customer satisfaction.

i. Effective Management of Small Datasets: Training data for emotion recognition is often limited. SVM performs well even with limited training data.
ii. Potential of High Accuracy: A high classification accuracy for emotion recognition tasks is achievable by properly tuning SVM.

9.4.4 Limitations of SVM for Emotion Classification

Using SVM in emotion classification is profoundly swayed by parameter tuning and interpretability issues. So, attention must be drawn to its limitations.

i. Complexity of Parameter Tuning: Choice of the right kernel function and hyperparameters for attaining optimal performance is frequently intriguing.
ii. Issues in Interpretability: The decision-making process behind an SVM is wearisome to comprehend than other algorithms.

9.5 CONVOLUTIONAL NEURAL NETWORKS FOR ROBOT EMOTION PROCESSING

While SVMs typically require pre-extracted features to make classifications, convolutional neural networks (CNNs) are a group of widely used algorithms for emotion detection (Ghayoumi and Bansal 2006; Fuertes et al. 2023). They can learn hierarchical features from raw input data, such as images or text, through their convolutional and pooling layers in an extemporaneous manner. The CNNs are used principally in facial expression recognition. They are really top-notch at automatically extracting features from images of faces. The extracted features are used to classify emotions of happiness, sadness, anger, and surprise. Accurate emotion identification is possible from a person's face using CNNs. Figure 9.4 shows the seven layers in a CNN and describes the role of each layer. These layers are: the input layer receiving the facial image, the convolution layer performing feature extraction by applying convolution filters (small matrices sliding across an image); the activation function layer for introducing nonlinearity in the features; the pooling layer for downsampling the feature maps; the flattening layer for collapsing the spatial dimensions of data into a one-dimensional array; the fully connected layer for learning the high-level relationships between features; and the output layer delivering the predicted emotion for the supplied image.

9.5.1 Working of CNNs for Emotion Detection

What workflow do the CNNs follow? Instead of preprocessing data for deriving features, the CNN works through a layered structure as follows (Mehendale 2020; Angel et al. 2024):

Robot Emotion Recognition Using a Convolutional Neural Network

(i) Input layer: Facial image pre-processing by image resizing, normalization and alignment for ensuring the consistency of the input data fed to the network

(ii) Convolution layer: Application of filters to the image to extract features such as edges, corners and contours on the face crucial to emotion recognition

(iv) Pooling layer: Downsampling the feature maps for reduction of spatial dimensions with preservation of pertinent information

(iii) Activation function layer: Application of ReLU activation function to introduce non-linearity in the features for improving the CNN's ability of learning complex patterns

(v) Flattening layer: Flattening the pooled features

(vi) Fully-connected layer: Learning the high-level relationships between features for emotion classification

(vii) Output layer: Prediction of the emotion for the supplied input image

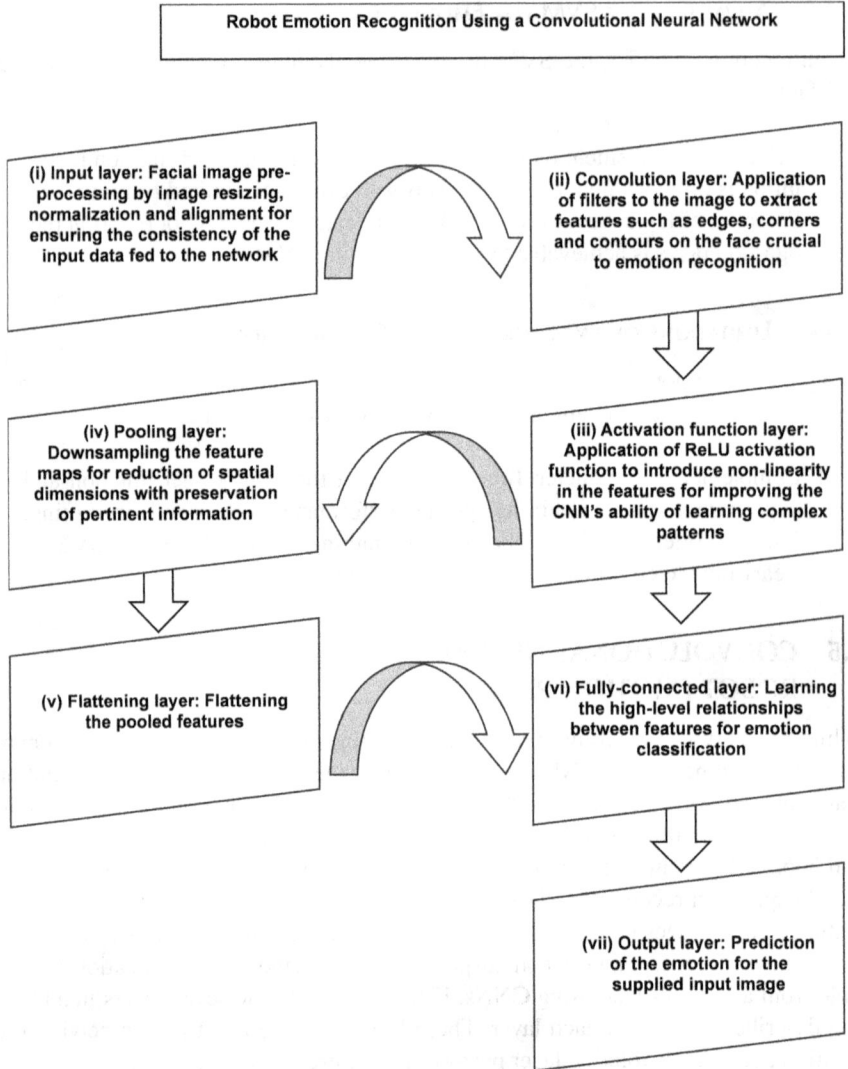

FIGURE 9.4 Layers of the convolutional neural network used for emotion recognition by a robot.

i. Convolutional Layers: These layers apply a convolution operation to the input data. This operation uses a learnable filter or kernel that slides over the input image. During sliding, it performs element-wise multiplications and sums the results. By applying these filters, local patterns like edges and textures are extracted from different parts of the face. A feature map is generated representing the presence and location of specific features in the input.

   ii. Pooling Layers: The feature maps are downsampled. The downsampling reduces the spatial dimension. Nonetheless, the most significant information is preserved.

  iii. Fully Connected Layers: These are the final layers of the network. They combine the extracted features to predict the emotion.

9.5.2 Using CNNs for Emotion Detection

Special considerations in using CNNs are as follows:

   i. Feature Extraction: Important facial features for emotion recognition are eye shape, mouth curvature, and eyebrow position. These features are spontaneously learnt by CNNs. So, manual feature engineering is hardly necessary.

  ii. Image Classification: After feature extraction, the CNN classifies the image into different emotion categories. An emotion label is assigned for the input face.

  iii. Training Data Amount and Quality: Large datasets of labeled facial images, corresponding to various emotions, are required for training CNNs. The data used for CNN training must be diverse in nature. It must represent a range of different facial expressions and demographics. Then, only high accuracy is achieved. Demographics are the statistical characteristics of a human population.

9.5.3 FER in Human–Computer Interaction

A cursory, concise description of FER will provide the background information for further discussion. FER in human–computer interaction (HCI) is a computer vision technology. It aims to analyze facial expressions to identify and classify emotional states, and adapt behavior and output accordingly for emotional interaction. It thus creates a user-friendly and engaging system that provides a more intuitive and personalized experience. Conspicuous traits of this technology are as follows:

   i. Remarkable Adaptation of System Behavior: Facial expressions are analyzed in real-time for adapting system behavior based on user emotions.

  ii. Outstanding Security and Surveillance Assistance: Facial expressions of people in public spaces are monitored to identify potential threats or suspicious behavior.

  iii. Affective Computing: Systems are developed for understanding and responding to human emotions.

9.5.4 Perplexing Situations during Use of CNNs for Emotion Detection

There are several bizarre and quaint occasions in which it is really strenuous and demanding to read people's emotions and decipher what they are actually feeling.

i. Analysis of Complex Facial Expressions: Emotion classification is difficult due to subtle variations in facial expressions.

ii. Variations of Lighting and Pose: Recognition accuracy is adversely affected by changes in lighting and head position.

9.6 DECISION TREES FOR ROBOT EMOTION DETECTION

The CNNs are particularly strong for tasks like image recognition, where they can automatically learn relevant features. Aside from CNNs, what more can be found in the portfolio of algorithms for emotion? Contrastingly, decision trees are non-parametric supervised learning methods used for classification and regression. They are capable of handling various types of data and excel in interpretability. They allow users to understand the reasoning behind the emotion predictions. A decision tree is a machine learning algorithm that uses a tree-like structure and pre-programmed rules to select appropriate responses based on detected emotions (Lee et al. 2011; Noroozi et al. 2017). The responses help to predict and classify human emotions. The basis of classification encompasses various input features, including facial expressions, voice tone, and text analysis. Each node in the tree represents a decision based on specific criteria. The tree displays a final classification of the emotion expressed. Figure 9.5 shows the four components of the decision tree: root node which is the starting point of the tree, internal nodes representing specific features of the image, branches which are the lines interconnecting the nodes, and the leaf node which are the final nodes at the ends of the branches representing the class of predicted emotion.

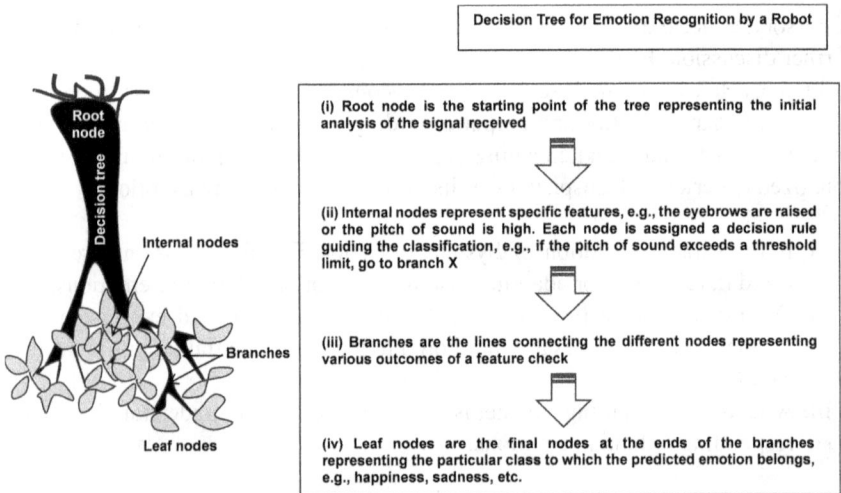

Decision Tree for Emotion Recognition by a Robot

(i) Root node is the starting point of the tree representing the initial analysis of the signal received

(ii) Internal nodes represent specific features, e.g., the eyebrows are raised or the pitch of sound is high. Each node is assigned a decision rule guiding the classification, e.g., if the pitch of sound exceeds a threshold limit, go to branch X

(iii) Branches are the lines connecting the different nodes representing various outcomes of a feature check

(iv) Leaf nodes are the final nodes at the ends of the branches representing the particular class to which the predicted emotion belongs, e.g., happiness, sadness, etc.

FIGURE 9.5   Structural organization of the decision tree structure.

9.6.1   WORKING OF DECISION TREES FOR EMOTION DETECTION

The building blocks of a decision tree are its nodes. In a decision tree, there are three different types of nodes:

i. Root Node: It is the initial node of the tree. Here, the most important feature is evaluated.
ii. Decision Nodes: Each subsequent node represents a further evaluation of a feature. It splits the data into separate branches based on the criteria.
iii. Leaf Nodes: At the end of each branch is the leaf node. It represents the final emotion classification.

9.6.2   USING DECISION TREES FOR EMOTION DETECTION

How are decision trees used in emotion analysis? Decision trees are used by learning patterns in data like facial expressions or voice tone to predict emotions. The approach followed consists of the following stages:

i. Classification: Emotion categories like happiness, sadness, anger, fear, surprise, etc. are predefined. The emotions in the input data are categorized under these headings.
ii. Feature Extraction: First, relevant features are extracted from the input data. These features are facial landmarks from an image or pitch variations from speech. Then the decision tree is applied.
iii. Hierarchical Structure: The decision tree structure allows for a step-by-step analysis. In this analysis, each node represents a question about the features. The branches of the tree lead to further decisions based on the answers.
iv. Interpretability: One can easily understand how the model reached a particular emotion classification by following the path through the tree.

9.6.3   APPLICATIONS OF DECISION TREES FOR EMOTION DETECTION

Decision trees are used in emotion detection by classifying emotions based on features extracted from various sources, like facial expressions, speech, or text, and using a tree-like structure to make predictions (Sun et al. 2019). A few application areas are as follows:

i. Emotion Recognition in AI Robotic Systems: Chatbots are built to understand the emotional tone of user interactions.
ii. Affective Computing: Emotional states are analyzed from video or audio data for applications like sentiment analysis.
iii. Healthcare: Emotional states of patients are monitored through speech analysis.

9.6.4 Challenges Confronting Decision Trees in Emotion Detection

At what point does the decision tree begin to pose problems? Difficulties arise with an increase in algorithmic complexity in dealing with large data sets and tree structure complexity. Plausibly, a tree with many nodes and branches is difficult to interpret and debug. The frustration is manifested in various ways.

  i. Detection of Complex Emotions: It is difficult to discriminate between subtle emotional nuances.
  ii. Context Dependence of Emotions: Context incorporation in a tree structure is difficult. Emotions that are influenced by context are therefore not efficiently managed.
  iii. Desired Data Quality: To achieve good performance, the data used for training must be accurate and well-labeled.

9.7 NATURAL LANGUAGE PROCESSING ALGORITHMS FOR ROBOT EMOTION DETECTION

We understand that decision trees are a flowchart-like structure that makes feature-based sequential decisions. In the decision tree algorithm, each internal node represents a test on an attribute or feature, each branch represents an outcome, and each leaf node represents a class label or decision. On the opposite side, natural language processing (NLP) algorithms generate empathetic and contextually relevant verbal responses. They analyze the text of a message or other content to determine the emotions or sentiment expressed within it. For detecting emotions in text, the most commonly used approaches include sentiment analysis algorithms, lexicon-based methods, machine learning algorithms, e.g., SVMs and Naive Bayes classifiers; and deep learning algorithms like RNNs and transformers which can analyze the context and sentiment of a piece of text to identify the expressed emotions like happiness, sadness, anger, or fear (Graterol et al. 2021; Maruf et al. 2024). Figure 9.6 shows the stages in NLP: input text; text preprocessing including tokenization, normalization and stemming/lemmatization; lexicon lookup consisting of emotion lexicon and word sentiment scoring; feature extraction comprising N-gram analysis, parts-of-speech tagging and intensity analysis; machine learning algorithm for emotion classification; and the predicted emotion as the output in two forms as emotion label and emotion intensity.

9.7.1 Emotion Detection with NLP Algorithms

Let us survey the composite structure of the NLP algorithm technology. NLP algorithmic methodology incorporates a multiplicity of techniques (Kumar and Geetha 2024), for example:

  i. Lexicon-Based Methods: Predefined dictionaries contain words associated with specific emotions. The sentiment of a text is calculated by counting the occurrences of these words.

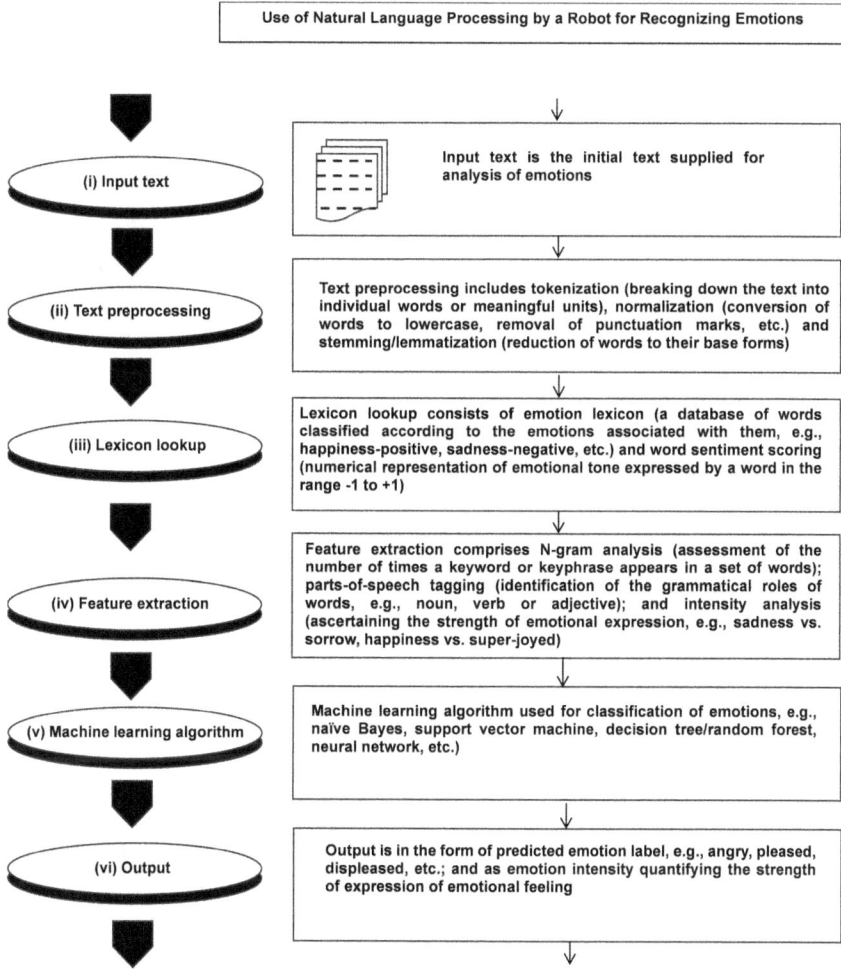

Use of Natural Language Processing by a Robot for Recognizing Emotions

(i) Input text

Input text is the initial text supplied for analysis of emotions

(ii) Text preprocessing

Text preprocessing includes tokenization (breaking down the text into individual words or meaningful units), normalization (conversion of words to lowercase, removal of punctuation marks, etc.) and stemming/lemmatization (reduction of words to their base forms)

(iii) Lexicon lookup

Lexicon lookup consists of emotion lexicon (a database of words classified according to the emotions associated with them, e.g., happiness-positive, sadness-negative, etc.) and word sentiment scoring (numerical representation of emotional tone expressed by a word in the range -1 to +1)

(iv) Feature extraction

Feature extraction comprises N-gram analysis (assessment of the number of times a keyword or keyphrase appears in a set of words); parts-of-speech tagging (identification of the grammatical roles of words, e.g., noun, verb or adjective); and intensity analysis (ascertaining the strength of emotional expression, e.g., sadness vs. sorrow, happiness vs. super-joyed)

(v) Machine learning algorithm

Machine learning algorithm used for classification of emotions, e.g., naïve Bayes, support vector machine, decision tree/random forest, neural network, etc.)

(vi) Output

Output is in the form of predicted emotion label, e.g., angry, pleased, displeased, etc.; and as emotion intensity quantifying the strength of expression of emotional feeling

FIGURE 9.6 The complete workflow from input to output stages in a natural language processing operation.

ii. Rule-Based Approaches: Linguistic rules are used as pointers toward emotional cues in text. Exclamation marks show excitement. Negative words imply sadness.

iii. Machine Learning Models:
   a. Naive Bayes Classifiers: These offer a simple and efficient method for identifying sentiment from word frequencies and their association with emotions.
   b. SVMs: These are effective for complex classification tasks, chiefly when dealing with high-dimensional data.

iv. Deep Learning Models (Guo 2022):

a. RNNs: They can handle sequential information in text. Hence, they are well-suited for understanding context and sentiment.
b. Transformers: Pre-trained models exceling in capturing complex linguistic relationships are often used for fine-tuning of specific emotion detection tasks.

9.7.2 Considerations for NLP Algorithms in Emotion Detection

Our inquisitiveness compels us to inquire about the vital factors that need attention with regard to using NLP algorithms for detection of emotions. In the development and deployment of NLP algorithms, several factors must be taken into account, e.g., data quality and quantity, understanding the context and intent behind user input, and management of ambiguities. Equally important are handling cultural and linguistic diversity such as different dialects, slang, and formal vs. informal language styles, along with privacy and security concerns.

i. Data Quality: A diverse and well-labeled training dataset is essential for NLP models for accuracy in emotion detection.
ii. Context Awareness: The context of a sentence must be understood clearly for a correct interpretation of emotions.
iii. Multi-Emotion Classification: Not only purely positive or negative sentiments, but also more nuanced emotions like joy, anger, and fear need to be identified.

9.7.3 Applications of NLP in Emotion Detection

Applications of NLP algorithms span a wide range, transcending geographical and social divisions, from analyzing social media and customer feedback for sentiment analysis to developing personalized chatbots and supporting mental health monitoring.

i. Social Media Sentiment Analysis: It helps in understanding public opinion on various topics through social media posts.
ii. Customer Service Chatbots: Customer sentiment is identified to provide better service and support.
iii. Market Research: It is a valuable tool for analyzing customer reviews and feedback.
iv. Mental Health Status Monitoring: Potential emotional distress in text-based communication is detected to ascertain the mental health status of a patient.

9.8 REINFORCEMENT LEARNING FOR ROBOT EMOTION DETECTION

How can reinforcement learning methods augment the performance of NLP algorithms? Standard NLP algorithms focus on understanding the overall emotional tone of a text. Reinforcement learning algorithms are applied to train artificial intelligence systems to recognize and respond to human emotions (Moerland et al. 2018; Akalin and Loutfi 2021). These approaches can enhance the NLP algorithms by learning from sequential interactions and feedback. Responses are adjusted based on feedback from interactions by maximizing positive outcomes and minimizing negative ones. The machines are made capable enough to learn decision-making from the emotional state of a user by providing positive or negative feedback, as determined from the outcome of their actions. Therefore, a more nuanced and context-aware understanding of emotions becomes possible, particularly in dialogues (Figure 9.7). Figure 9.7a shows the six components of reinforcement learning-based emotion recognition for a robot: agent, the core of the system; environment, the source of input data; state, the current information available to the agent about the environment; action, the decision made by the agent after examining the current state; reward, the feedback signal received from the environment indicating the correctness or inaccuracy of the action; and policy, the strategy applied by the agent to choose actions suitable for a given state. Figure 9.7b shows the five steps in the process sequence for emotion recognition: visual data input, feature extraction, emotion classification, appropriate action selection for responding to the detected emotion, and reward feedback by the environment to the agent through a reward signal.

9.8.1 Using Reinforcement Learning for Emotions Analysis

Let us ask, 'In what ways is the use of reinforcement learning beneficial for emotion analysis?' Reinforcement learning enhances emotion analysis by enabling agents to learn optimal actions through trial and error. Guidance is received through feedback from the environment. Improvements in understanding of emotions are achieved by analyzing how agents react to different situations and external stimuli. It works as follows:

    i. Emotion Recognition: A system is developed to accurately detect emotions from various input modalities, including physiological signals and body language cues, rather than relying on commonly used modalities such as facial expressions, voice tone, or text analysis.

    ii. Design of the Reward Function: A reward function is defined that impels the AI to respond appropriately to different emotions in order to guide its learning process.

    iii. Decision-Making by the AI Agent: The AI agent chooses actions that are considered suitable for the emotional context of the recognized emotion. A possible action could be offering comforting words in response to sadness. Another likely action consists of providing encouragement in situations that require motivation.

9.8.2 Applications of Reinforcement Learning in Emotion Systems

Why is it necessary to incorporate reinforcement learning into emotion systems? Reinforcement learning is integrated with emotion systems to enable agents to

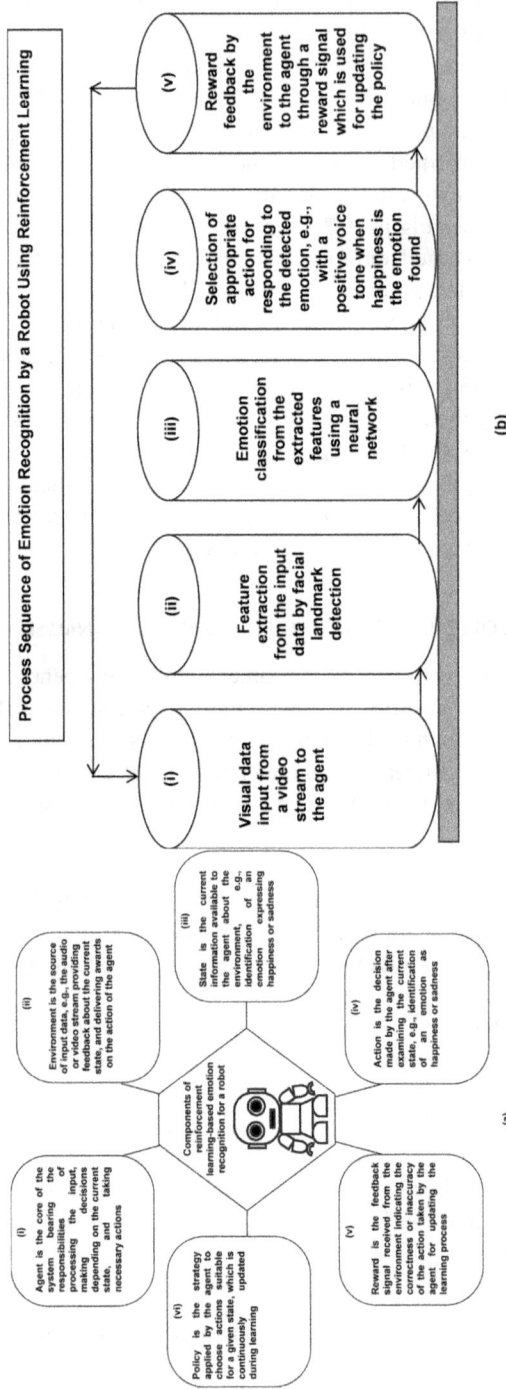

Process Sequence of Emotion Recognition by a Robot Using Reinforcement Learning

(i) Visual data input from a video stream to the agent

(ii) Feature extraction from the input data by facial landmark detection

(iii) Emotion classification from the extracted features using a neural network

(iv) Selection of appropriate action for responding to the detected emotion, e.g., with a positive voice tone when the emotion happiness is found

(v) Reward feedback by the environment to the agent through a reward signal which is used for updating the policy

(b)

Components of reinforcement learning-based emotion recognition for a robot

(i) Agent is the core of the system bearing the responsibilities of processing the input, making decisions depending on the current state, and taking necessary actions

(ii) Environment is the source of input data, e.g., the audio or video stream providing feedback about the current state, and delivering awards on the action of the agent

(iii) State is the current information available to the agent about the environment, e.g., identification of an emotion expressing happiness or sadness

(iv) Action is the decision made by the agent after examining the current state, e.g., identification of an emotion as happiness or sadness

(v) Reward is the feedback signal received from the environment indicating the correctness or inaccuracy of the action taken by the agent for updating the learning process

(vi) Policy is the strategy applied by the agent to choose actions suitable for a given state, which is continuously updated during learning

(a)

FIGURE 9.7 Reinforcement learning: (a) components and (b) the series of actions performed to achieve a desired outcome.

learn and make decisions from emotional responses (Huang et al. 2021). Its integration with emotion systems makes them more adaptable and effective in complex, real-world scenarios. A few examples are as follows:

i. Chatbots and Virtual Assistants: They enhance the conversational experience by tailoring responses to the emotional state of the user.
ii. Customer Service Systems: They identify customer frustration/dissatisfaction and provide appropriate support.
iii. Educational Platforms: Teaching methods are adapted in the light of the student's emotional engagement.
iv. Healthcare Systems: Emotional distress in patients is recognized and personalized interventions are made.

9.8.3 ONEROUS SITUATIONS FACED DURING USE OF REINFORCEMENT LEARNING IN EMOTION SYSTEMS

We would like to know: Are there any disadvantages to using reinforcement learning in emotion systems? Significant drawbacks of reinforcement learning are faced in some situations, a few of which are as follows:

i. Complex Emotional Nuances: Subtle emotions evade accurate identification and interpretation.
ii. Data Variability: Large amounts of diverse data are needed by emotion recognition models to perform well across different contexts and individuals. The data requirements are massive because reinforcement learning is susceptible to high variance and instability during the learning phase
iii. Ethical Considerations: Ethical concerns arise from the potential for misuse of emotional intelligence in AI systems.

9.9 DISCUSSION AND CONCLUSIONS

Several researchers have developed robots capable of recognizing gestures and emotions. An example is the RYAN SYSTEM (Abdollahi et al. 2023). Deep learning methods are applied for recognizing multimodal emotions. The output of this framework is integrated with RYAN's dialogue management system. The dialogue management facility is created by writing scripted conversations on many topics. Topics of science, history, nature, music, movies, and literature are covered. RYAN detects the facial expressions and analyzes the language sentiments of the user. Then it empathizes with the user via emotional conversation. It also mirrors the positive facial expressions of the user.

KISMET (a Turkish word meaning 'fate' or 'luck') is a socially assistive humanoid robot. It is located at the Massachusetts Institute of Technology (MIT) Museum, Cambridge, Massachusetts. It displays several emotions, such as calmness, anger, happiness, and sadness. It displays body postures and facial expressions, too, accompanied by voice tones.

KASPAR (Kinesics and Synchronization in Personal Assistant Robotics) is a child-sized robot. It is built by University of Hertfordshire, United Kingdom (UK). It has a range of facial gestures. The purpose is to make social interaction easier for children. This robot is especially helpful for children with autism.

PARO (Personal Robot) is a robot developed by the National Institute of Advanced Industrial Science and Technology, known as AIST. The AIST is a public research institute, Tokyo, Japan. The PARO is used for stimulating patients with dementia.

KODOMOROID {Japanese word 'kodomo' (child) + 'android'; 'android' derived from the Greek 'andro' (man) + 'eides' (shape)} is a child android robot. It is a humanoid robot with a child-like appearance interacting with people as a child will do. This robot is seen at Miraikan, The National Museum of Emerging Science and Innovation in Tokyo. It can recite news and weather reports from around the world. It is capable of speaking in various voices and languages.

The Emotional Robot

I am an emotional robot
I usually laugh and play
The whole day
Greeting everybody with hello and hi
But when I am teased by some naughty guy
I like to weep and cry.

Table 9.1 provides the sum and substance of the emotional AI algorithms discussed in this chapter. In the next chapter, we make a transition from emotion detection to task and motion planning. An emotion is a subjective feeling or mental state, e.g.,

---

TABLE 9.1
Takeaways from This Chapter at a Glance

Sl. No.	Takeaway	Explanation
1	Summary	A wide cross-section of emotionally intelligent robot algorithms was described in terms of their working mechanisms and advantages and drawbacks. Algorithms discussed included hidden Markov models, self-organizing maps, support vector machines, convolutional neural networks, decision trees, natural language processing, and reinforcement learning.
2	Hidden Markov models	These are statistical models that analyze sequences of data.
3	Self-organizing maps	These maps are artificial neural networks to project high-dimensional data onto a low-dimensional grid.
4	Support vector machines	These machines are supervised learning algorithms for classification tasks.
5	CNNs	These neural networks are deep learning architectures for image recognition and analysis.
6	Decision trees	These are tree-like structures for classification.

(Continued)

TABLE 9.1 (*Continued*)
Takeaways from This Chapter at a Glance

Sl. No.	Takeaway	Explanation
7	NLP	These algorithms analyze the sentiment of spoken or written language to interpret the emotional tone of a conversation and adjust responses accordingly.
8	Reinforcement learning	These are machine learning paradigms in which an agent learns by interacting with its environment.
9	Keywords and ideas to remember	Hidden Markov models, self-organizing maps, support vector machines, convolutional neural networks, decision trees, natural language processing algorithms, and reinforcement learning for robot emotion detection

happiness, sadness, or anger, that arises from a particular situation or experience. A task is an objectively defined operation or assignment that must be completed. It has a distinct goal. It requires some form of activity or process to be carried out. Succinctly speaking, an emotion is a feeling, whereas a task is an action. An emotion is an internal experience of an individual, while a task is an external action done by the individual. Excitement aroused by a project is an emotion. Writing its completion report is a task.

REFERENCES AND FURTHER READING

Abdollahi H., M. H. Mahoor, R. Zandie, J. Siewierski and S. H. Qualls. 2023. Artificial emotional intelligence in socially assistive robots for older adults: A pilot study, *IEEE Transactions on Affective Computing*, Vol. 14, 3, pp. 2020–2032.

Akalin N. and A. Loutfi. 2021. Reinforcement learning approaches in social robotics, *Sensors,* Vol. 21, 4, 1292, pp. 1–37.

Alhussan A. A., F. M. Talaat, E.-S. M. El-kenawy, A. A. Abdelhamid, A. Ibrahim, D. S. Khafaga and M. Alnaggar. 2023. Facial expression recognition model depending on optimized support vector machine, *Computers, Materials and Continua*, Vol. 76, 1, pp. 499–515.

Angel J. S., A. D. Andrushia, T. M. Neebha, O. Accouche, L. Saker and N. Anand. 2024. Faster region convolutional neural network (FRCNN) based facial emotion recognition, *Computers, Materials and Continua*, Vol. 79, 2, pp. 2427–2448.

Christopher V. V., Updated by B. Whitfield. 2024. Understanding the hidden Markov model, https://builtin.com/articles/hidden-markov-model#:~:text=The%20three%20basic% 20problems%20of%20a%20hidden,and%20B%20to%20maximize%20the%20 probability%20P(O'%CE%BB)

Fuertes W., K. Hunter, D. S. Benítez, N. Pérez, F. Grijalva and M. Baldeon-Calisto. 2023. Application of Convolutional Neural Networks to Emotion Recognition for Robotic Arm Manipulation, *2023 IEEE Colombian Conference on Applications of Computational Intelligence (ColCACI)*, Bogotá DC, Colombia, 26–28 July, pp. 1–6.

Ghayoumi M. and A. K. Bansal. 2016. Emotion in Robots Using Convolutional Neural Networks. In: Agah A., J. J. Cabibihan, A. Howard, M. Salichs and H. He (Eds.), *Social Robotics. ICSR 2016*. Lecture Notes in Computer Science, Vol. 9979, Springer, Cham, pp. 285–295.

Graterol W., J. Diaz-Amado, Y. Cardinale, I. Dongo, E. Lopes-Silva and C. Santos-Libarino. 2021. Emotion detection for social robots based on NLP transformers and an emotion ontology, *Sensors*, Vol. 21, 4, 1322, pp. 1–19.

Guo J. 2022. Deep learning approach to text analysis for human emotion detection from big data, *Journal of Intelligent Systems*, Vol. 31, 1, pp. 113–126.

Huang X., M. Ren, Q. Han, X. Shi, J. Nie, W. Nie and A.-A. Liu. 2021. Emotion detection for conversations based on reinforcement learning framework, *IEEE MultiMedia*, Vol. 28, 2, pp. 76–85.

Inthiam J., A. Mowshowitz and E. Hayashi. 2019b. Mood perception model for social robot based on facial and bodily expression using a hidden Markov model, *Journal of Robotics and Mechatronics*, Vol. 31, 4, pp. 629–638.

Inthiam J., E. Hayashi, W. Jitviriya and A. Mowshowitz. 2019a. Mood Estimation for Human-Robot Interaction Based on Facial and Bodily Expression Using a Hidden Markov Model, *2019 IEEE/SICE International Symposium on System Integration (SII)*, Paris, France, 14–16 January, pp. 352–356.

Jitviriya W. and E. Hayashi. 2014. Design of Emotion Generation Model and Action Selection for Robots Using a Self-Organizing Map, *2014 11th International Conference on Electrical Engineering/Electronics, Computer, Telecommunications and Information Technology (ECTI-CON)*, Nakhon Ratchasima, Thailand, 14–17 May, pp. 1–6.

Jitviriya W., M. Koike and E. Hayashi. 2015. Emotional model for robotic system using a self-organizing map combined with Markovian model, *Journal of Robotics and Mechatronics*, Vol. 27, 5, pp. 563–570.

Kang X. 2025. Speech emotion recognition algorithm of intelligent robot based on ACO-SVM, *International Journal of Cognitive Computing in Engineering*, Vol. 6, pp. 131–142.

Kulic D. and E. Croft. 2006. Estimating Robot Induced Affective State Using Hidden Markov Models, *ROMAN 2006 The 15th IEEE International Symposium on Robot and Human Interactive Communication*, Hatfield, UK, 6–8 September, pp. 257–262.

Kumar S. S. A. and A. Geetha. 2024. Emotion detection from text using natural language processing and neural networks, *International Journal of Intelligent Systems and Applications in Engineering*, Vol. 12, 14s, pp. 609–615.

Lee C.-C., E. Mower, C. Busso, S. Lee and S. Narayanan. 2011. Emotion recognition using a hierarchical binary decision tree approach, *Speech Communication*, Vol. 53, 9–10, pp. 1162–1171.

Majumder A., L. Behera and V. K. Subramanian. 2014. Emotion recognition from geometric facial features using self-organizing map, *Pattern Recognition*, Vol. 47, 3, pp. 1282–1293.

Maruf A. A., F. Khanam, M. M. Haque, Z. M. Jiyad, M. F. Mridha and Z. Aung. 2024. Challenges and opportunities of text-based emotion detection: A survey, *IEEE Access*, Vol. 12, pp. 18416–18450.

Mehendale N. 2020. Facial emotion recognition using convolutional neural networks (FERC), *SN Applied Sciences*, Vol. 2, p. 446, 8 pages.

Moerland T. M., J. Broekens and C. M. Jonker. 2018. Emotion in reinforcement learning agents and robots: A survey, *Machine Learning*, Vol. 107, pp. 443–480.

Noroozi F., T. Sapiński, D. Kamińska and G. Ambarjafari. 2017. Vocal-based emotion recognition using random forests and decision tree, *International Journal of Speech Technology*, Vol. 20, pp. 239–246.

Simplilearn. 2023. What are self-organizing maps: Beginner's guide to Kohonen map, https://www.simplilearn.com/self-organizing-kohonen-maps-article

Sun L., B. Zou, S. Fu, J. Chen and F. Wang. 2019. Speech emotion recognition based on DNN-decision tree SVM model, *Speech Communication*, Vol. 115, pp. 29–37.

Tsai C.-C., Y.-Z. Chen and C.-W. Liao. 2009. Interactive Emotion Recognition Using Support Vector Machine for Human-Robot Interaction, *2009 IEEE International Conference on Systems, Man and Cybernetics*, San Antonio, TX, USA, 11–14 October, pp. 407–412.

Zhao M. 2023. The emotion recognition in psychology of human-robot interaction, *Psychomachina*, Vol. 1, 1, pp. 1–11.

10 Robot Task and Motion Planning

10.1 INTRODUCTION

The daily routine of an individual begins with a lot of planning, either done subconsciously or consciously, starting from what to eat for breakfast, what dress to wear for going to the office, what conveyance and route to take for reaching the office, and continuing to include how to face the business meetings or participate in scholarly discussions, how to reduce stress and anxiety, and so on. Planning is an inseparable part of our everyday life, both intrinsically and inherently. It improves efficiency and productivity and increases accountability.

If robots are to be used for various jobs, they must also plan their activities in a manner similar to humans. Robot activity planning is the process of formulating a sequence of actions or a train of events necessary to be performed by the robot, i.e., creation of a list of actions that must be taken by a robot for the execution of a piece of work. The work is done by transitioning from a starting or initial state to a final goal state. The robot may be engaged in working in a static or dynamic environment. Robotic activity is a combination of two fundamental sub-activities, namely:

   i. The task to be performed by the robot, and
   ii. The necessary motion undertaken by it to complete the task.

Accordingly, there are two distinct sub-branches catering to the two sub-activities: robot task planning and robot motion planning. In this chapter, we address both of these sub-branches. Firstly, they are considered separately, showcasing their individual characteristics. Then they are dealt jointly by fusion of the two approaches to coalesce them and form a unified whole or amalgamated entity representing their inseparability. Subsequently, we systematically study the search algorithms used for robot task and motion planning (TAMP). These are used for finding optimal paths for robot movements while avoiding obstacles.

10.2 ROBOT ACTIVITY PLANNING

10.2.1 ROBOT TASK PLANNING

How are the robot tasks planned? Robot task planning works with self-reasoning by a robot using an internal AI algorithm (Hertzberg and Chatila 2008). The robot reasons with itself to finalize a sequence of actions to accomplish a given objective for reaching a desired goal in an environmental setting containing plenty of objects or items (Morecki and Knapczyk 1999). Obviously, task planning is a discrete

DOI: 10.1201/9781032695266-10

operation. The reason for its discrete nature is that it aims to select a set of actions from a range of possibilities.

10.2.2 ROBOT MOTION PLANNING

Robot motion planning is the step following its task planning. In other words, motion planning is the successor of task planning. Robot motion planning involves deciding the motions to be executed by a robot in a given environment to accomplish a certain objective for reaching a specified goal (Latombe 1991; Owen-Hill 2019). Motion planning is a continuous operation because it focuses on the constant movements of an individual robot in space. Table 10.1 explicates the aspects of task planning vis-à-vis motion planning (Marcelina 2022).

TABLE 10.1
Differences between Robot Task Planning and Robot Motion Planning

Sl. No.	Point of Comparison	Robot Task Planning	Robot Motion Planning
1	Definition	It is concerned with determining the sequence of actions needed to be performed by a robot to complete a task, irrespective of whether it is simple or complex.	It deals with the continuous movement of the robot, calculating the precise path to execute each individual action within the task plan, while avoiding obstacles and respecting the robot's physical constraints.
2	Purpose	The robot decides what work it has to do, i.e., determines its tasks.	The robot determines how to do those tasks.
2	Abstraction level	It focuses on the high-level decision-making of a robot, considering the actions involved, e.g., picking up an object or navigating to a room.	It operates at a lower level, focusing on precise joint angles and trajectories to control the robot's movement.
3	State space	It typically employs a discrete state space, where each state represents a distinct action or decision.	It usually operates in a continuous state space, considering all possible positions and orientations of the robot.
4	Input/output	Its inputs include the goal of the task and available actions, outputting a sequence of actions to achieve the goal.	Its inputs include the initial and goal positions, obstacles in the environment, and robot kinematics, outputting a collision-free trajectory.
5	Example scenario: A robot is asked to make a cup of tea for a guest	The task planner would decide the sequence of actions: go to the tea machine, press the button for a cup of tea, grab a mug, pour the tea into the mug, move the mug to the guest, and serve tea.	For each action in the task plan, the motion planner would calculate the precise movements of the robot's arm to reach the tea machine, press the button, grasp the mug, pour the tea while avoiding any obstacles in its path, and finally serve the tea to the guest.

10.2.3 ROBOT TAMP

Robot TAMP is the compounding of task and motion aspects. It is an encyclopedic term embracing the sum-total efforts necessary for the robot to create a sequence of actions to be executed by a robot in a given environment to reach a specified goal (Dantam 2021; Antonyshyn et al. 2023). The creation of sequential actions is done by taking into consideration the physical motions required for the execution of actions. So, TAMP stretches beyond task planning and has a broader scope than that. Unlike task planning, TAMP is not restricted to planning a series of actionable steps. It simultaneously considers the practicable robot movements to execute those actions. Therefore, TAMP deals with a combination of discrete and continuous operations. It is a mixed activity formed by blending a discrete operation (task planning) with a continuous operation (motion planning). Naturally, integration of discrete planning with continuous planning makes it a complicated activity comprising computationally intensive and economically expensive phases (Pan et al. 2024).

A few examples of robot TAMP are as follows:

i. Object manipulation involving moving the robot's arm to reach an object, grasping it, and then moving it to the suggested location; placing an object in the chosen location by avoiding collisions with other objects; and sorting an assortment of objects based on their shape, size, and color.
ii. Navigating a building to reach a specific room, avoiding walls and stairs; exploring unknown environments for relief and rescue operations in disaster areas; and driving cars on roads without accidents and following traffic rules.
iii. Manufacturing and assembly, e.g., welding and painting components to ensure quality and efficiency, and packaging parts on assembly lines.
iv. Robotic surgery to perform complex life-saving medical procedures.
v. Domestic chores for cleaning, laundry, and providing care for aged or crippled people.

10.3 ROBOT TASK PLANNING ALGORITHMS

10.3.1 KEY POINTS ABOUT THE ROBOT TASK PLANNING ALGORITHM

Laying down a plan leads to the formulation of an algorithm. In pursuance of the definition of robot task planning given in Section 10.2.1, a robot task planning algorithm is a computational method used to generate a sequence of actions to be undertaken by the robot. These actions, taken together, constitute a plan designed for a robot to complete a complex task. The actions are decided by considering the current state of the environment, available robotic capabilities, and the desired goals. Taking all these factors into account, a task planning algorithm begins by defining the task. The next stage is the decomposition of the task into sub-goals. This is done by breaking down the complex task into smaller, manageable actions. The decomposition process keeps in sight the capabilities of the robot. The limitations imposed by the environmental conditions must not be forgotten. A state-space representation is constructed based on the task definition and its decomposition. A search is initiated for a feasible sequence of actions. Techniques like state-space search, heuristic functions, and constraint satisfaction are utilized to optimize the plan. These are the core concepts in AI problem solving. State-space search is a technique of solving problems in which an exploration of the space of possible configurations or states is carried out to find a solution to the problem. The heuristic function gives the estimated cost or distance from a particular state to the goal state, thus helping the search algorithm in prioritization of states that lead to the goal more efficiently. Constraint satisfaction concerns finding the values to be assigned to variables for satisfying a set of constraints.

It is easy to draw an analogy between robot task planning and human task planning, namely, identifying the goal, breaking down the task into manageable steps, sequencing and scheduling the steps, allocating resources, setting timelines and deadlines, monitoring progress, and developing mitigation strategies for bottlenecks encountered.

10.3.2 Main Steps of the Robot Task Planning Algorithm

Let us draw a roadmap for planning a task for a robot mission. A roadmap is a visual way for quickly communicating a plan. Figure 10.1 shows the roadmap for planning a task by a robot. The steps in the roadmap for robot task planning are: task definition, its decomposition and state-space representation, action modeling, executing a search algorithm, state-space search, and constraint checking. These are finalized after significant brainstorming and are delineated below:

  i. Definition of Robot's Task: An explicit statement of the desired outcome of the robot's action is made. Specification of the initial and final states of the robot is unequivocally finalized. A clear mention and recognition of the relevant objects, locations, and conditions within the task environment is included.
  ii. Decomposition of Robot's Task: The complex defined task is broken down into smaller, more manageable sub-goals to reduce overwhelm. The hierarchy of actions is laid out. In this hierarchical organization, completion of the different sub-goals leads to the achievement of the overall goal.
  iii. State-Space Representation of Robot's Task: The possible configurations (states) of the robot during the task are defined. The state space is represented using appropriate data structures. Graphs or trees are used.
  iv. Modeling of Robot's Actions: A formal definition is crafted stating the available actions that the robot can perform in each state. It includes the constraints imposed on the robot. The constraints derive from joint limits, reachable workspace, and environmental limitations.
  v. Running a Search Algorithm: A search algorithm, such as A* or Dijkstra's algorithm, is utilized to find the optimal sequence of actions that constitutes the path from the initial state to the goal state of the robot. The applicable cost function, e.g., distance and time associated with each possible action, is

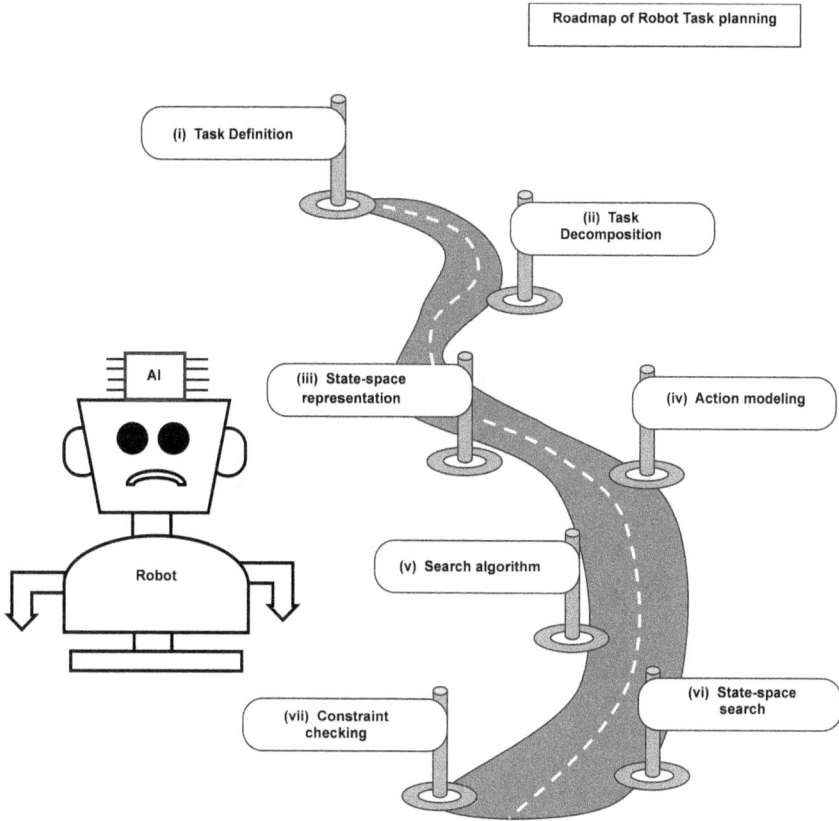

FIGURE 10.1   The robot's roadmap for task planning from definition of the task objective to plan finalization.

evaluated. The cost function quantifies the error in terms of the difference between the prediction of the algorithm and the actual value. It assesses the performance of the algorithm and guides its optimization.

vi. State-Space Search: The possible states of the robot are explored by iteratively applying actions. Algorithms used for this exploration are breadth-first search, depth-first search, or A* search to find a path to the goal state. The breadth-first search performs exploration of the graph level by level. The depth-first search explores as deeply as possible along a branch before backtracking. The A* search determines the shortest path based on the cost of traversing the edges.

vii. Checking for Constraints and Undertaking Readjustments: A verification of the adherence of the planned sequence of actions to all enforced constraints is done, e.g., collision avoidance, joint limits, etc. If any violations come to notice, replanning is done. The plan is revisited and reviewed. The path of the plan is adjusted accordingly.

10.4   ROBOT MOTION PLANNING ALGORITHMS

10.4.1   KEY POINTS ABOUT THE ROBOT MOTION PLANNING ALGORITHM

To understand the central issues of interest, we revisit Section 10.2.2. Building on the definition of robot motion planning provided in Section 10.2.2, a typical robot motion planning algorithm defines the possible configurations of the robot in the environment in terms of its positions and orientations. It checks for collisions and then iteratively builds a path from the starting point to the goal. Care is taken to avoid obstacles and blockades while building this path. Often, a search algorithm is utilized to find the best route. Figure 10.2 shows the constituent steps in the roadmap for robot motion planning: environment representation, collision detection, sampling, node connection/path building, searching for optimal path, trajectory optimization, and path execution. These steps are further expounded below.

10.4.2   MAIN STEPS OF THE ROBOT MOTION PLANNING ALGORITHM

These are conceptualized and conjured up after circumspect reflection:

i.   Representation of the Environment (Configuration Space): Possible positions and orientations of the robot in the workspace are precisely put into words to corroborate insistently that what has been said is correct. Oftentimes, these positions and orientations are represented as a configuration space. Each point in this space represents a unique pose of the robot. Configuration space (C-Space) is a mathematical space. It represents and means all possible robot configurations. For the identification of the collision areas, the obstacles within the workspace are suitably modeled.

ii. Detection of Possible Collisions: It means taking a glance to check for the intersection of a robot's configuration with any obstacles. A rigorous check is done for each potential robot configuration to find whether any part of the robot intersects with any obstacle in the environment. The cruciality of this step can be appreciated by realizing that making a guarantee that the planned path is collision-free is a compulsory requirement of motion planning that cannot be circumvented in any way because it is a safety precaution to avert accidents.

iii. Sampling of Points in the Configuration Space: Sampling-based methods are algorithms like rapidly exploring random trees (RRT). They randomly sample points in C-space to find a path. After the points in the configuration space are randomly sampled, they are vigilantly observed and checked to be collision-free. If a point is found to be collision-free, the point is accepted and added as a node in the search tree. Otherwise, it is declined inclusion.

iv. Connections of Nodes/Path Building: The newly sampled node is connected to existing nodes in the search tree. Checking is done again to make sure that there is no collision between the robot and any impediment along the connecting path. This step builds a network of potential paths within the configuration space.

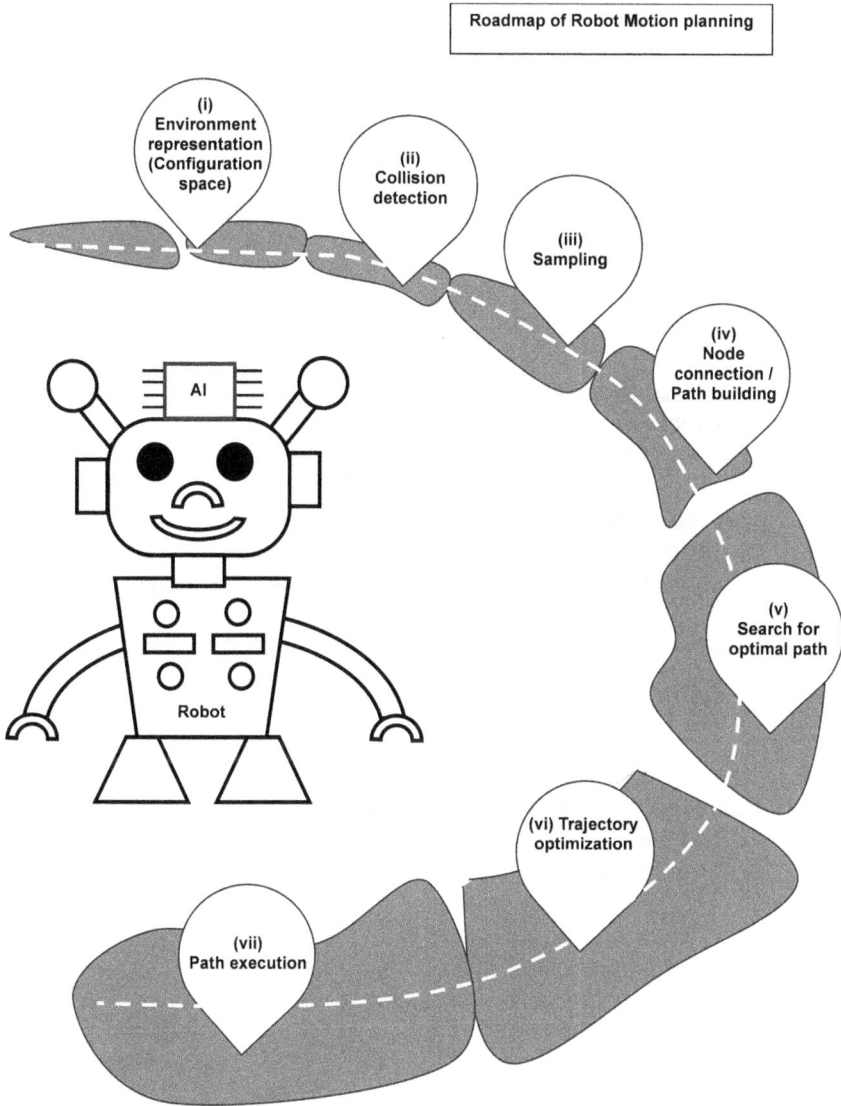

FIGURE 10.2   The roadmap for motion planning by a robot.

v. Search for Optimal Path: The graph search utilizes search algorithms like A* or Dijkstra to navigate a graph. The graph represents possible robot movements to find the shortest or most efficient path from the start configuration to the goal configuration within the constructed tree. Heuristics are used to guide the search toward the goal. They are cognitive strategies, like mental shortcuts or practical guidelines.

vi. Optimization of Trajectory (Optional): The planned path is smoothed and refined for a more even and efficient trajectory. For path smoothing and streamlining, the velocities and accelerations are modified to make sure that the robot can physically follow the trajectory.

vii. Execution of the Planned Path: The computed path is sent to the robot controller. The controller executes its responsibility to actualize the movement of the robot in reality, as decided in the plan.

10.5 ROBOT TAMP ALGORITHMS

10.5.1 KEY POINTS ABOUT TAMP ALGORITHM

We recall the discussions in Section 10.2.3. In accordance with the definition of robot TAMP given in Section 10.2.3, and unifying the procedures outlined in Sections 10.3 and 10.4 to create a comprehensive composite picture, robot task and motion algorithms refer to a set of algorithms in robotics that integrate high-level task planning (deciding what actions to take) with low-level motion planning (calculating the robot's physical movements) (Akbari et al. 2020). The intent is to achieve a complex goal through concerted effort. The goal could be assembling a product in a factory. Alternatively, it could involve navigating the robot through an environment by generating a sequence of feasible actions with corresponding robot trajectories, as noted in Section 10.2.

10.5.2 MAIN STEPS OF TAMP ALGORITHM FOR SOLVING A ROBOTIC PROBLEM

We can now combine the robot task and motion ingredients to create a cocktail. The key steps of a robot TAMP algorithm typically involve an admixture of task and motion scheduling operations: definition of the assigned task domain including states and actions, representation of the robot's configuration space, identification of constraints faced in problem solving, decomposition of the assigned task into sub-goals, generation of a sequence of actions to execute the task, planning of the corresponding motions for each action to be performed during the task, and finally, checking for collision avoidance by the robot while optimizing the overall path for task completion and motion to be undertaken. Following in the footsteps of robot TAMP, let us draw a roadmap for the same. Figure 10.3 depicts the roadmap for TAMP by a robot consisting of problem definition, TAMP, and finally optimization and refinement of the plan, as explicated below:

i. Definition of the Robotic Problem: It is a methodical and structured organization of three parts.

a. Description of the Task: The overall goal of the robot is defined. The definition includes the initial state and the desired final state. Any intermediate objectives are unambiguously spelled out.

b. Modeling of the Environment: An across-the-board representation of the robot's workspace is done. This sweeping representation consists of obstacles and relevant objects in the environment, along with their properties.

c. Robot Kinematics: The robot's configuration space is defined. The definition is inclusive of all the robot joint angles and reachable positions.

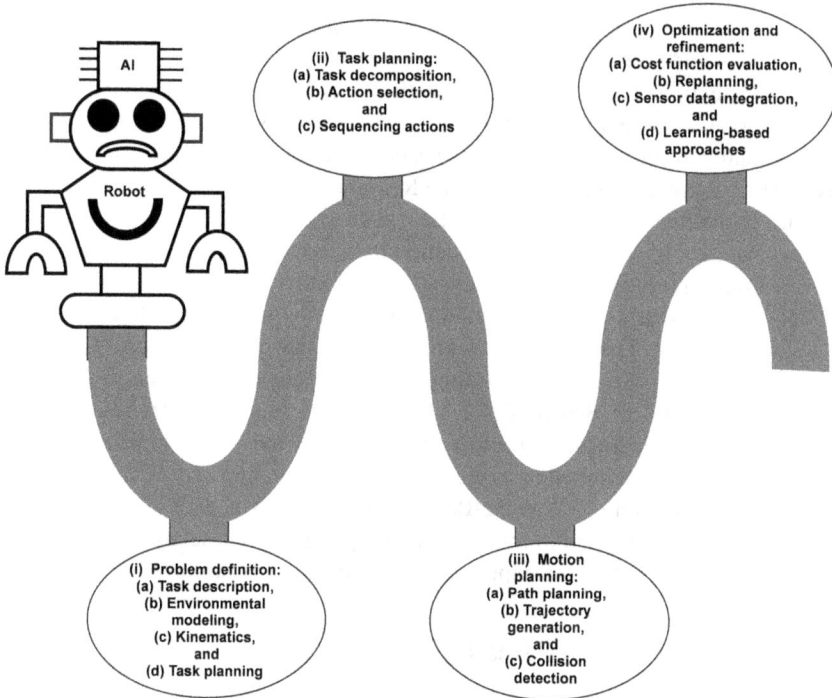

FIGURE 10.3 The roadmap for combined task and motion planning by the robot.

ii. Task Planning for Solving the Robotic Problem: Various algorithms like A* search, RRT or potential field methods are used depending on the complexity of the task and environment. The particular algorithm is decided by looking at the trials and tribulations of the case. As discussed in Section 10.3, task planning entails:

    a. Decomposition of Assigned Task: The complex task is dissevered into smaller, more manageable subtasks or actions. Each sub-task should be a single, achievable piece of work. It must be plainly and concisely stated.

    b. Selection of Actions: Appropriate actions are chosen based on the current state and desired goal. Constraints like object manipulation or environmental interactions are given due attention.

    c. Sequencing of Actions: The order of the selected actions is decided. The purpose is to produce a systematized tableau showing the logical sequential actions for performing the task.

iii. Motion Planning for Executing the Solution to the Robotic Problem: Recalling discussions in Section 10.4, it involves:

a. Path Planning: A collision-free path is generated connecting the start and goal configurations for each sub-task. Proper consideration of the robot's kinematic constraints is essential for success.

b. Generation of Robo's Trajectory: The planned path is converted into a smooth trajectory in an uninterrupted, seamless progression. The velocities and accelerations of the robot are specified.

c. Detection of Robot's Collision with Obstacles: An incessant checking is done for potential collisions of the robot with obstacles throughout the planned trajectory, considering each and every point on the trajectory.

iv. Optimization and Refinement of the Robotic Problem:

a. Evaluation of the Cost Function: The planned motion is evaluated. Metrics like path length, execution time, energy consumption, etc., are used in this evaluation process.

b. Readjustments and Replanning: The robot's plan is adjusted if found necessary. The adjustments are made to the plan based on feedback received from sensors regarding unexpected changes in the environment.

c. Integration of Sensor Data: The sensor data, e.g., vision sensor and LiDAR readings, are utilized to update the environment model. A real-time adaptation of the robot's TAMP is achieved.

d. Incorporation of Learning-Based Approaches: Machine learning techniques are utilized to improve motion planning. These algorithms benefit by learning from experience or adapting to dynamic situations.

10.6 SEARCH ALGORITHMS USED IN ROBOT TAMP

10.6.1 A* SEARCH ALGORITHM

The greatness of A* search is that it is optimal, efficient, and flexible, and so very special. It is a heuristic-based algorithm commonly used in robot motion planning and navigation (Xin et al. 2019; Ji et al. 2023). It is also used in other circumstances in which the determination of the most optimal path through a tangled and puzzling environment is imperative. A heuristic algorithm is an intuition or empirically guided rule of thumb for expeditious decision-making. It gives a feasible solution for each instance of a combinatorial optimization problem at an acceptable cost in terms of computing time and space. A feasible solution is the set of values that satisfy all the constraints of an optimization problem. The optimal solution is a feasible solution resulting in the best value of the objective function. The optimal solution is always a feasible solution. But the antithesis of this statement is not necessarily true because a feasible solution need not be an optimal solution. The objective function is the linear formula used to maximize or minimize a value.

The A* algorithm allows a robot to efficiently find the shortest path from a starting point to a goal location, which is fixed beforehand. While following this route, the robot navigates around obstacles in its environment. The navigation takes place under the supervision and watchful eye of a global cost function. The global cost function is defined as a function expressed by the equation

$$f(n) = g(n) + h(n) \tag{10.1}$$

where $g(n)$ is the cost function of the path traversed from the initial state to the node n, and $h(n)$ is the heuristic function representing the estimated cost from node n to the goal state. The cost function is a mathematical formula measuring the difference between the predicted output of an AI algorithm and the authentic expected output. Formally, we state that

$$\text{Cost Function} = \text{Predicted output} - \text{Expected output} \qquad (10.2)$$

Examination of the cost function values allows the algorithm to prioritize the nodes that are closer to the goal sate.

The key component of the A* algorithm is the heuristic function $h(n)$. The function $h(n)$ estimates the remaining cost to reach the goal from a given node. Based on the remaining cost, the algorithm lays down priority on the nodes that are likely to be nearer to the desired destination. The algorithm works by exploration of the neighboring nodes, calculating their total cost given by the equation

$$\begin{aligned} \text{Total cost function} = {} & \text{Cost from initial state to the node} \\ & + \text{Estimated cost from the node to the goal} \qquad (10.3) \end{aligned}$$

and selecting the node with the lowest cost to expand thereafter. The shortest path is calculated by taking into consideration both the costs. These costs include:

  i. the cost of reaching a node, and
 ii. an estimated cost to reach the goal from that node.

It is re-emphasized that heuristic algorithms provide computationally feasible approximate solutions. They are not accurate algorithms. Their principal merit is that they are excellent at finding 'good enough solutions' with low execution times, though not axiomatically the optimal ones. This is of course expected from their basic approach to the problem. It can be easily envisaged by noting that the heuristic-based algorithms do not conduct an exhaustive search for every possible solution. So, they are very useful when exact solutions are computationally impractical or expensive to determine. Optimality is sacrificed to gain speed for finding the solution in a reasonable time frame.

Two lists are maintained up to date in order to run the A* search algorithm. These lists are named OPEN and CLOSED (Figure 10.4).

The list OPEN is a data structure representing the set of potential paths that have not been assessed yet. It contains all the nodes awaiting exploration and evaluation by the heuristic function during the search process. These nodes have not been expanded into successors yet. Expansion of a node is the process of applying operators to the node, producing a set of nodes. The list OPEN is implemented as a priority queue. The node with the lowest $f(n)$ value is selected next to expand upon.

The list CLOSED contains the nodes that have already been visited. Its purpose is to prevent the algorithm from revisiting them unnecessarily. An already visited node is a node in the search space that has been explored and added to the list CLOSED. The algorithm has, by now, calculated the best path to reach that node from the starting point. It will not re-evaluate that node again during the search process.

This means that it is a node that has been definitively considered. Therefore, it is no longer part of the active search area.

Figure 10.4 illustrates the steps to be followed in the A* search algorithm, starting with the definition of two lists, OPEN and CLOSED. If the OPEN list is empty, failure is returned. If NO, the algorithm progresses by calculating the cost function. A node n with the smallest $f(n)$ value in the OPEN list is chosen. This node is transferred to the CLOSED list while its index is saved. It is checked whether the node n is the target end node. If NO, the node n is expanded by producing all its neighboring nodes and placing them in the OPEN list. Then the cost function is calculated for each node produced. The algorithm returns to the stage of selection of the node with the smallest value of $f(n)$. If YES, the optimal path is calculated using pointers of the

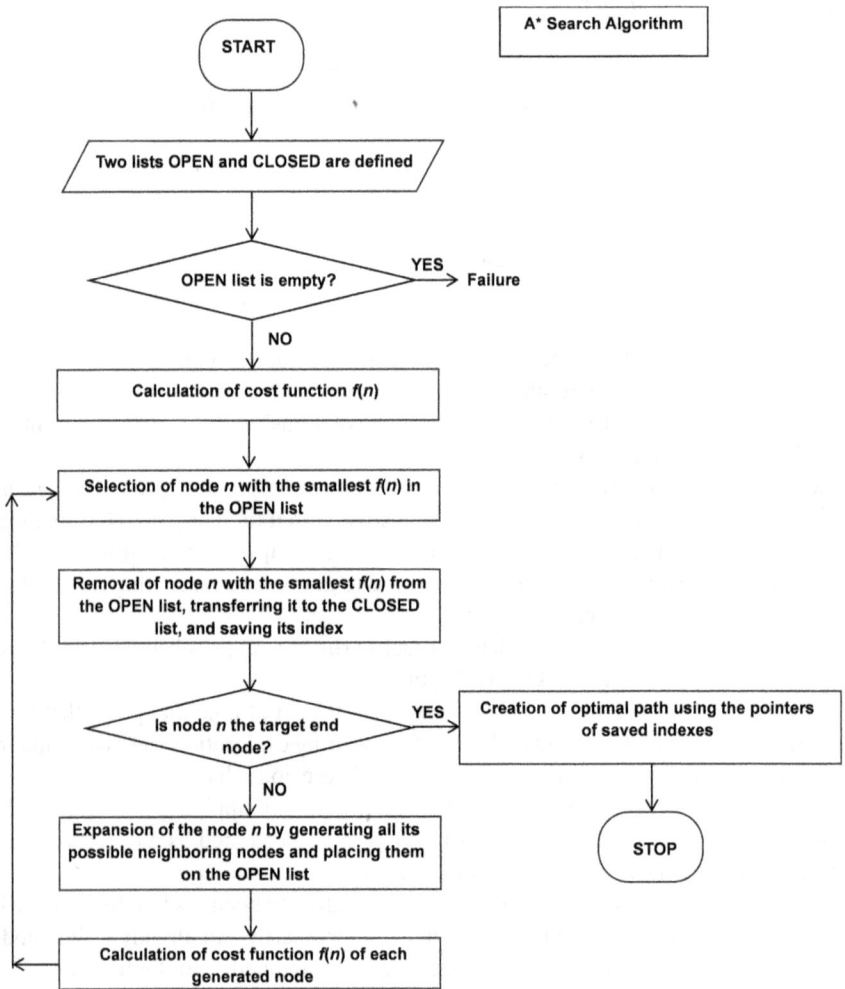

FIGURE 10.4 The A* search algorithm.

saved indices, and the algorithm stops. A thorough explanation of the procedures of the algorithm is as follows (Kumar 2024):

   i. A list OPEN is defined consisting of the nodes to be evaluated. The OPEN list is a priority queue. It stores the nodes with their estimated costs.
  ii. The list CLOSED is defined.
 iii. The node n in the list OPEN with the smallest value of $f(n)$ is selected.
  iv. The node n is removed from the list OPEN. It is transferred to the list CLOSED. Its index is saved; the index of a node in an A* search is its position in the OPEN list.
   v. If the list OPEN is empty, failure is returned, followed by exiting.
  vi. If the node n is a goal state, success is returned, followed by exiting.
 vii. The current node n in the search tree is examined and expanded by generating its possible neighboring nodes.
viii. If any successor to n is the goal node, success and solution are returned by tracing the path from the goal node to node n; otherwise, the algorithm moves to the next step.
  ix. For each succeeding node, the evaluation function is applied to the node; if the node has not been in either list, it is added to OPEN.
   x. The steps are repeated until the goal node or destination is reached.

10.6.2   Dijkstra's Algorithm

Dijkstra's algorithm is a popular algorithm used in robotics to find the shortest path between a given node and all other nodes in a graph (Sniedovich 2006; Fan and Shi 2010; He 2022). It was conceived by a Dutch computer scientist, programmer, software engineer, and mathematician, Edsger Wybe Dijkstra, in 1956, and so is named after the scientist. It is a useful tool for mobile robots to navigate warehouses and other spaces. It helps the robots to optimize their routes and avoid collisions with obstacles. While not guaranteeing efficiency, Dijkstra's algorithm provides the shortest path between nodes in a weighted graph. Its core utility is in solving the shortest-path problem.

Dijkstra's algorithm uses the weights of the edges to minimize the total distance between the source node and all other nodes. In Dijkstra's algorithm, the weights of edges are positive integers or real numbers. They represent the distance or cost between two nodes in a graph. Suppose a graph is used to represent the map of a store. The vertices of the graph are specific points in the store. Its edges are the pathways in the store. The weight of an edge could be the length of the pathway. If the graph is used to represent the cost of a robot's movement between two particular points, the weight of an edge could represent the cost of moving between those points.

The algorithm maintains two sets: one for visited vertices and another for unvisited vertices. The source vertex is the starting point. The algorithm finds the shortest path from the source vertex to all other vertices in the graph. It starts at the source vertex and iteratively selects the unvisited vertex with the smallest tentative distance from the source. It then visits the neighbors of this vertex. It persistently updates the tentative distance of the vertex if a shorter path is found. This process continues until the goal node or destination vertex is reached, or all reachable vertices have been visited and explored.

The main steps of Dijkstra's algorithm are shown in Figure 10.5: creation of a list of unvisited vertices, designation of the current vertex, and checking whether this vertex is the destination vertex. If NO, all the vertices leading to the current vertex are found. Then the distances between the source vertex and each unvisited neighbor of the current vertex are calculated. If the new distance is shorter than the previous distance, the distance is updated. If NO, the current vertex is removed from the list of unvisited vertices, and the algorithm returns to the stage of designation of the current vertex. However, if the current vertex is found to be the destination vertex, the search is deemed to be completed, and the algorithm terminates. Specific details of the steps are furnished as follows (Navone 2020):

i. Initialization and Marking of the Source Vertex: The source vertex is marked. The distance of the source vertex is set to 0. The distances of all other vertices are set to infinity.
ii. Designating the Current Vertex: The unvisited vertex with the shortest distance from the source is chosen. The unvisited vertex with the smallest distance is set as the current vertex
iii. Finding and Calculating the Distances to the Current Vertex: All the vertices leading to the current vertex are found. The distance from the source to each unvisited neighbor of the current vertex is calculated.
iv. Updating Distance: The distance is updated if the new distance is found to be shorter than the previous one.
v. Marking the Current Vertex as Visited: Once all neighbors of the current vertex have been visited, the current vertex is marked as visited. This vertex again will never be looked at again.
vi. Repetition: Steps (ii)–(v) are repeated until all vertices are visited.

Dijkstra's algorithm is based on the principle of the greedy algorithm. A greedy algorithm is a problem-solving strategy. It chooses the best option at each step without considering the future consequences or possibilities. This means that it always chooses the solution with the lowest cost. The goal is to find a globally optimal solution by making locally optimal choices.

Table 10.2 presents the similarities and dissimilarities between A* search and Dijkstra's algorithms.

10.7 RRT ALGORITHM

Dijkstra's algorithm is a deterministic algorithm operating in a static space. The RRT algorithm is a probabilistic algorithm that iteratively grows a tree to find paths in a continuous, dynamic space (Caccavale and Finzi 2022; Ding et al. 2023; Xu 2024). The tree is grown by arbitrarily sampling points in the environment. Sampling is a method of estimating the characteristics of a population by selecting individual members or a subset of the population. It helps to make statistical inferences about the whole population from these members. A sampling-based technique is a method

Dijkstra's Algorithm

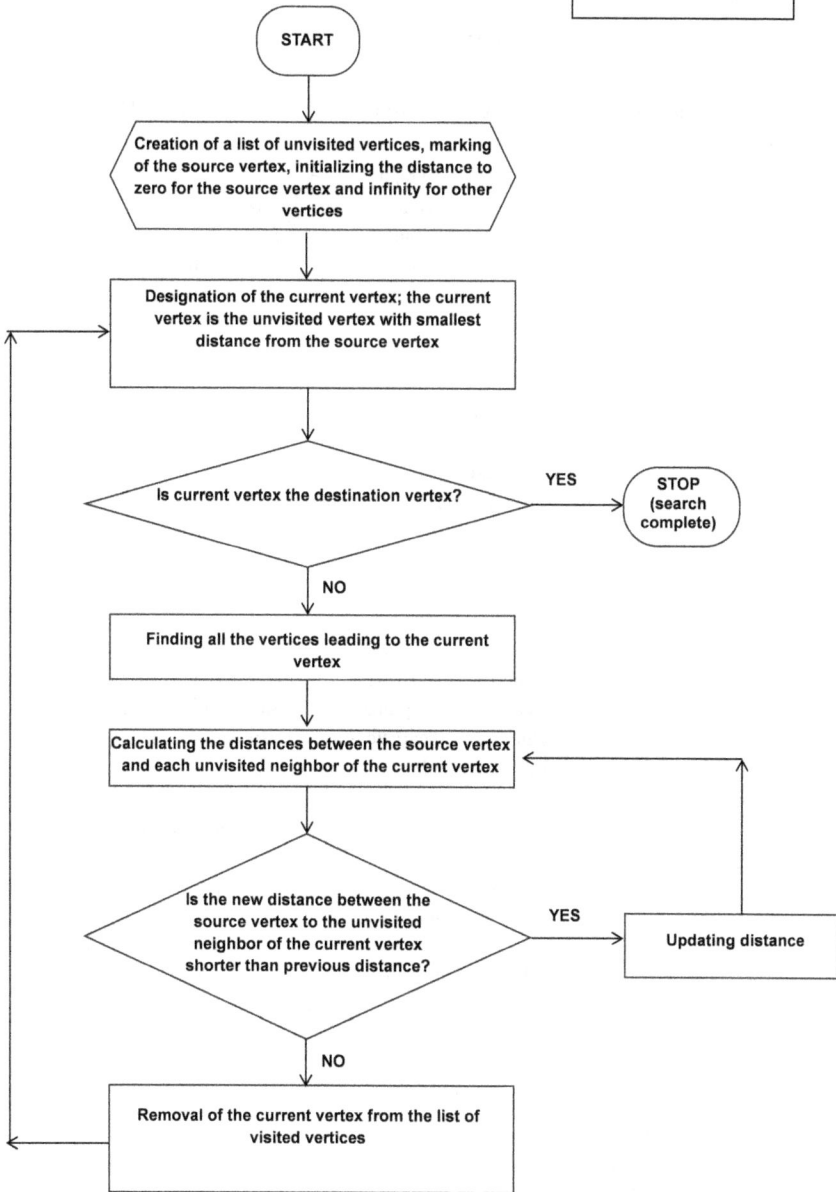

START

Creation of a list of unvisited vertices, marking of the source vertex, initializing the distance to zero for the source vertex and infinity for other vertices

Designation of the current vertex; the current vertex is the unvisited vertex with smallest distance from the source vertex

Is current vertex the destination vertex?

YES

STOP (search complete)

NO

Finding all the vertices leading to the current vertex

Calculating the distances between the source vertex and each unvisited neighbor of the current vertex

Is the new distance between the source vertex to the unvisited neighbor of the current vertex shorter than previous distance?

YES

Updating distance

NO

Removal of the current vertex from the list of visited vertices

FIGURE 10.5   Dijkstra's algorithm.

that utilizes sampling to achieve a specific outcome. It first samples the possible configurations. It then builds a graph that approximates the connectivity of the space.

In the RRT algorithm, the indiscriminately sampled points are connected to the nearest existing node in the tree. During this process, it is always ensured that the

TABLE 10.2

A* Search and Dijkstra's Algorithms

Sl. No.	Point of Comparison	A* Search Algorithm	Dijkstra's Algorithm
1	Commonality	A* search is used to find the shortest path in a graph.	Dijkstra's algorithm too is used to find the shortest path in a graph.
2	Heuristic function	It utilizes a heuristic function to estimate the remaining distance to the goal. This function allows it to prioritize paths closer to the goal. Hence, it is more likely to lead to the solution. Thus, it aids in making more informed decisions about which node to explore next.	It does not use a heuristic function. So, it explores all nodes based solely on their distance from the starting point.
3	Efficiency	It is usually faster than Dijkstra's algorithm. This is especially true for large graphs. The path prioritization feature of the algorithm is the root cause behind the fast computing.	It can be slower due to its exhaustive search strategy.
4	Optimality	It may not always find the optimal solution. Optimality is sacrificed if the heuristic is not properly designed.	It always guarantees to find the shortest path. The reason is that it explores all possible routes.
5	When preferred to be used?	It is favored when one needs to find the shortest path quickly, especially in large graphs. It is also chosen when a possibility exists for designing a reliable heuristic function to guide the search toward the goal.	It is chosen when finding the absolute shortest path is necessary, and the graph is relatively small. It is also preferred when the accuracy of the solution is more important than the speed of execution of the algorithm.

connection is collision-free. Freedom from collision is maintained until the desired goal configuration is reached.

Figure 10.6 shows the steps of the RRT algorithm, viz., initialization of the random tree, setting the starting point X_{Start} and the goal point X_{Goal}, and checking whether the tree reaches the goal point X_{Goal}. If YES, the algorithm backtracks from the node X_{Goal} to the node X_{Start} to trace the planned path. If NO, a random sampling point X_{Random} is selected. All the nodes in the tree are traversed to find the node $X_{Nearest}$ at the shortest distance from X_{Random}. Then, a new node X_{New} is generated by expansion from the node $X_{Nearest}$ along the direction of the node X_{Random}. The path from the node $X_{Nearest}$ to the new node X_{New} is checked and confirmed to be collision-free. If YES, the node $X_{Nearest}$ is added to the random tree. The algorithm returns to finding

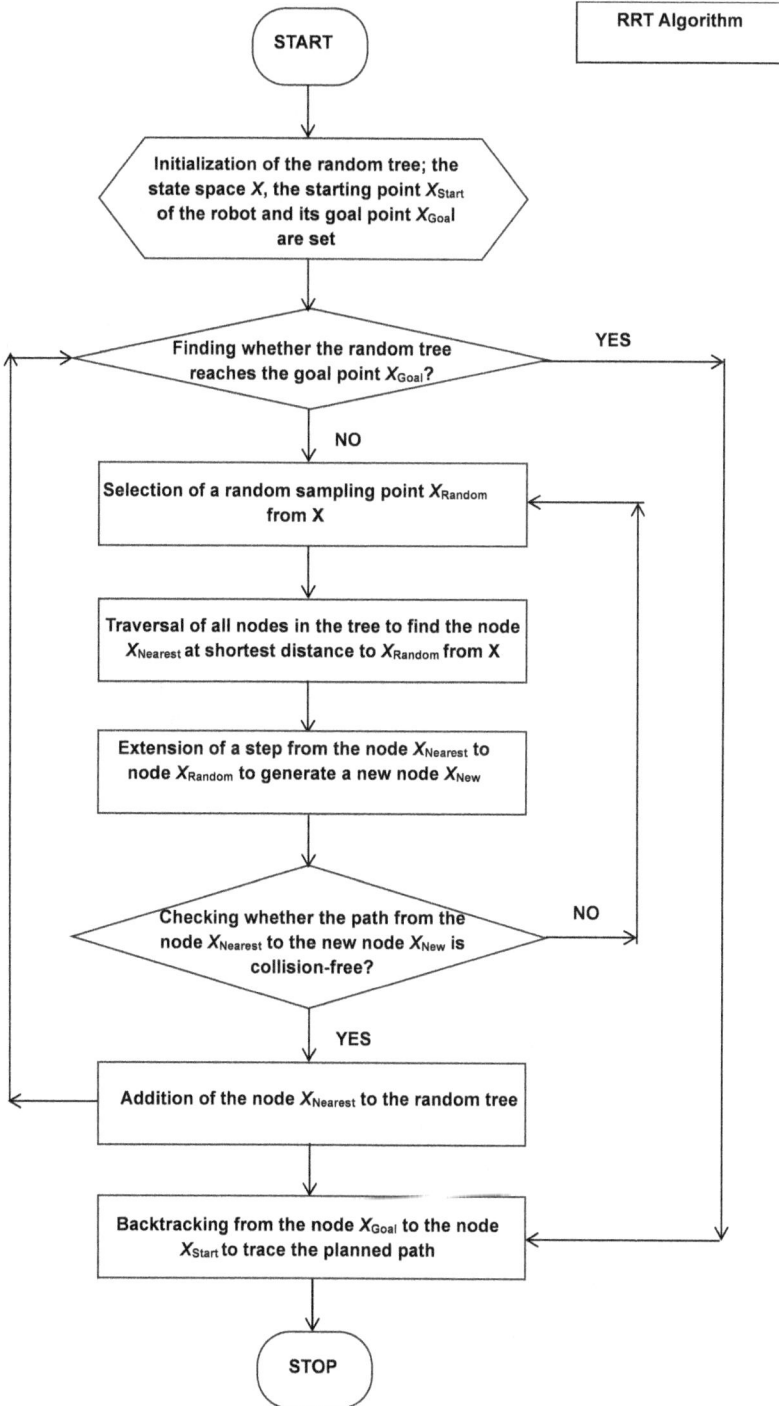

FIGURE 10.6 The RRT algorithm.

whether the random tree reaches the goal point, accompanied by backtracking from the node X_{Goal} to the node X_{Start} to trace the planned path. Then the algorithm stops. If the answer is NO for the path from the node $X_{Nearest}$ to the new node X_{New} as collision-free, then the algorithm goes back to the selection of a random sampling point X_{Random}. Further clarifications are given below (Sarkar 2024):

i. Initialization of a Random Tree at the Starting Point: The mobile robot constructs a search random tree. This random tree is based on the initial pose and target pose obtained from the scene map. It is constructed at the starting point X_{Start} of the two-dimensional state space. Then the starting point X_{Start} of the robot is set as the root node of the random tree. The goal point X_{Goal} of the robot is also stipulated.

ii. Random Generation of a Sampling Point: A random sampling point X_{Random} is haphazardly generated in the free search space. The search space is the configuration space of all possible positions and orientations of the robot. It is used to guide the expansion of the random tree.

iii. Nearest Neighbor Search: The nodes that have been generated in the whole random tree structure are traversed. After completing this traversal of nodes, the tree node that is closest to the randomly sampled point X_{Random} is found. It is selected and defined as $X_{Nearest}$.

iv. Extension of a New Node from the Nearest Neighbor Toward the Sampled Point by Taking a Predefined Step Size: The appropriate step size is expanded from the node $X_{Nearest}$ along the direction of the node X_{Random} as the extension direction. A suitable step size is set as the branch length to generate a new node X_{New}. It is the new tree node.

v. Checking Possibility of Collisions: A verification test is undertaken to validate whether or not the new node path collides with any obstacles in the environment.

vi. Expansion Cancelation on Encountering an Obstacle: If an obstacle is encountered in the expansion process, the expansion is canceled. After the cancelation, the sampling is performed again.

vii. Addition of Node in Absence of Collision: If no collision is detected, the new node is added to the tree.

viii. Iteration of Steps: Steps (ii)–(vii) are repeated until a node in the tree is close enough to the goal configuration. The algorithm repeats the above iterative process until the target node exceeds the specified number of iterations. Eventually, a fast-expanding random tree path is formed, ending the search.

10.8   PROBABILISTIC ROADMAP

Like the RRT algorithm, the probabilistic roadmap (PRM) method is a sampling-based technique (Zhang et al. 2013; Zhang 2022). While the RRT algorithm employs a local approach that starts from the initial position, the PRM algorithm adopts a global approach that encompasses the entire space. The RRT algorithm may not find optimal paths, but the PRM algorithm can do so. The RRT algorithm incurs

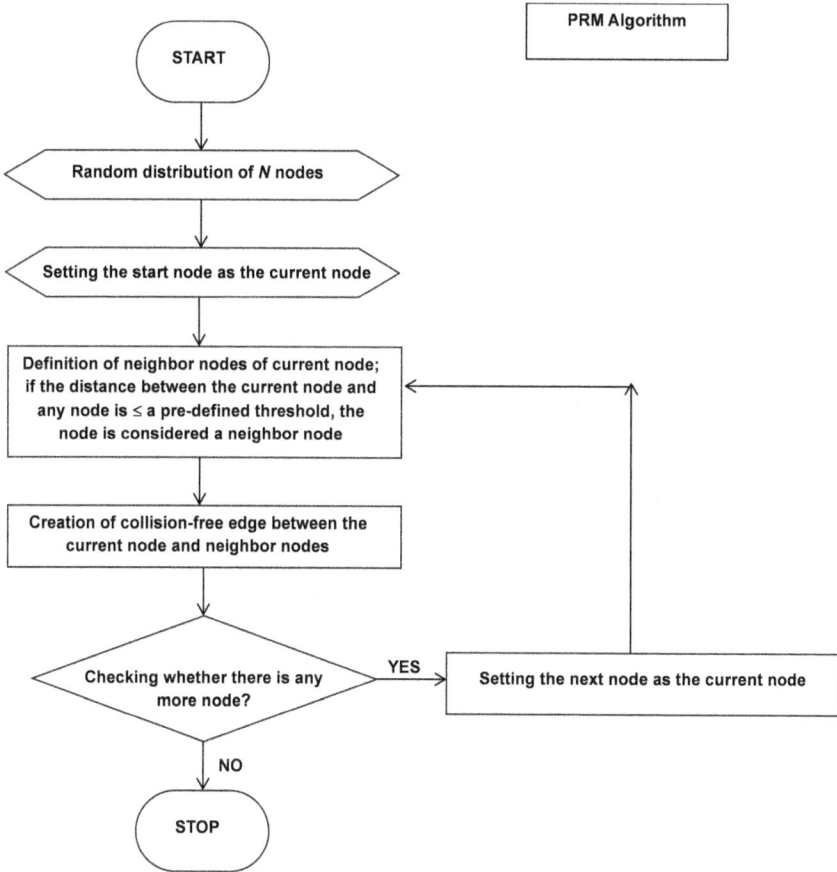

FIGURE 10.7   The PRM method.

a lower computational cost than the PRM algorithm. But the RRT algorithm can handle dynamic obstacles, whereas the PRM algorithm is not ideal for dynamic environments.

The PRM algorithm is particularly suitable for robots with a high number of degrees of freedom (DoFs). Such robots are required to perform multiple point-to-point motions in a known workspace. The DoF of a robot is the number of independent movements that the robot can make. It is a measure of a robot's motion capabilities and flexibility. The principal steps of the PRM method for robot path planning are presented in Figure 10.7. Beginning from the random distribution of N nodes, the start node is set as the current node. The neighbor nodes of the current node are defined. A collision-free edge is created between the current node and the neighboring nodes. An exhaustive check is carried out to corroborate whether there are any more nodes. If YES, the next node is set as the current node, and the algorithm returns to the definition of neighbor nodes. If NO, the algorithm stops. More details are given as follows (Khokhar 2021):

i. Preprocessing: A network of collision-free configurations (nodes) is created. The network is produced by randomly selecting configurations and testing them for collisions. This step is done only once for a given environment.

Collision detection is a process that determines if two or more objects are intersecting. The steps for collision detection generally involve checking if the objects are overlapping or lying within a certain distance of each other. The bounding volume hierarchy is a tree structure composed of a set of bounding volumes. It is wrapped around geometric objects, which form the leaf nodes of the tree. The leaf nodes are grouped as small sets. They are enclosed within larger bounding volumes. These, in turn, are grouped and enclosed within further larger bounding volumes in a recursive fashion. Ultimately, a tree structure is obtained. It has a single bounding volume at the topmost point of the tree. Collision is determined by undertaking a tree traversal starting from the root and proceeding ahead. If the bounding volume of the root does not intersect with the object of interest, the traversal is stopped. If, however, there is an intersection, the traversal proceeds further. Those branches are checked for which there is an intersection.

ii. Planning: The initial and final configurations of the robot are connected to two nodes in the network. A path is then computed through the network between these two nodes.

10.9   DISCUSSION AND CONCLUSIONS

This chapter outlined the various methods employed in robot task planning, motion planning, and TAMP all of which play a crucial role in robotics and require intensive debating and deliberation (Table 10.3). Task planning aims to build a structured plan to reach a prescribed goal (Zhang et al. 2022). It works by decomposition of the complicated long-horizon task into elementary short-duration subtasks. Hierarchical methods, heuristic searching methods, and operator planning methods have been used for task planning. Logic programming offers several advantages, such as greater expressivity and interpretability, which are helpful in making safe and reliable robots (Meli et al. 2023).

Motion planning is the extension of path planning, seeking to generate interactive trajectories in the workspace when robots interact with a dynamic environment, necessitating consideration of kinetic features and velocities of robots and moving objects nearby as they move toward the goal.

TAMP allows robots to not only plan high-level actions like picking up an object but also generate the precise movements necessary to execute those actions while avoiding collisions. It bridges the gap between abstract tasks and the physical motions required to achieve them. It enables robots to be more autonomous and adaptable in complex environments.

Task planning operates at a higher level, deciding on the overall sequence of actions, such as 'grasp the tea cup, move to the guest', while motion planning handles the lower-level details of how to physically move the robot to accomplish those actions (refer back to the example in Table 10.1). Task planning often involves discrete choices about which object to pick up and which path to take. Motion planning

TABLE 10.3

Takeaways from This Chapter at a Glance

Sl. No.	Takeaway	Explanation
1	Summary	Robot activity planning consists of task planning, motion planning, and combined robot task and motion planning. Key points and main steps of all types of algorithms for robot activity planning were described. Search algorithms used in robot task and motion planning were discussed, including the A* search algorithm, Dijkstra's algorithm, the rapidly exploring random tree algorithm, and the probabilistic roadmap.
2	A* search algorithm	It is a graph traversal and pathfinding algorithm that utilizes a heuristic function to estimate the distance to the goal from each node.
3	Dijkstra's algorithm	It is a classic, deterministic method to determine the shortest path between two points in a graph.
4	RRT algorithm	It involves randomly sampling points in the search space, connecting them to the existing tree, and gradually expanding the tree until it reaches the goal point.
5	Probabilistic roadmap algorithm	It operates by randomly sampling points in the free space to create a network of connected nodes and then searching for a path between the start and goal nodes within this roadmap.
6	Comparison of algorithms	A* and Dijkstra are 'informed search algorithms' prioritizing exploration of closer nodes to the goal, while the rapidly exploring random tree algorithm and the probabilistic roadmap are both randomized algorithms used in complex, high-dimensional spaces.
7	Keywords and ideas to remember	Robot activity planning, robot task planning, robot motion planning, robot task and motion planning, A* search algorithm, Dijkstra's algorithm, rapidly exploring random tree algorithm, probabilistic roadmap.

deals with continuous variables like joint angles and velocities to generate smooth trajectories. By integrating task and motion concepts in TAMP, robots perform intricate manipulation tasks like assembling components or navigating cluttered environments, which are difficult to program manually. Advanced TAMP systems allow real-time adaptation. The robots can react to changes in the environment or unexpected situations by adjusting the robot's actions on the fly.

Several robot path and motion planning algorithms, collision avoidance, and navigation were delved into. Path planning determines the path between the origin and destination within the workspace, which is the area where an algorithm operates or a task exists. It applies strategies based on the shortest distance or the shortest time. The traditional path planning algorithms can meet most requirements. Motion planning algorithms comprise traditional planning algorithms, and classical machine learning algorithms, including reinforcement learning (Zhou et al. 2022). Combined

TAMP is receiving extensive attention. TAMP solutions are required in situations such as self-governing robots engaged in ground drilling for the extraction of materials (Mansouri et al. 2021; Guo et al. 2023).

A self-driving car utilizes TAMP to determine when to change lanes, navigate intersections, and avoid obstacles, while generating smooth driving trajectories. Robots in industrial manufacturing can use TAMP to pick and place objects, assemble components, and perform complex manipulations while optimizing movement efficiency. A service robot assisting humans could use TAMP to plan a sequence of actions like fetching a drink, opening a door, and placing items on a table.

Integrating high-level task planning with detailed motion planning is computationally elaborate and expensive, especially in dynamic environments. Handling of uncertainties, such as dealing with sensor noise and imprecise environmental information, is extremely enervating and backbreaking for accurate motion planning. Physical constraints must be considered to ensure that the generated motions are mechanically feasible for the robot, duly considering its joint limits and dynamics.

In the next two chapters, we dedicate ourselves to the subject of autonomy in robotics. It refers to the ability of a robot to work without human control by perceiving its environment, using AI to make decisions and acting on those decisions by initiating the requisite movements voluntarily and of its own accord, in a given situation at the right moment.

REFERENCES AND FURTHER READING

Akbari A., M. Diab and J. Rosell. 2020. Contingent task and motion planning under uncertainty for human–robot interactions. *Applied Sciences*, Vol. 10, 1665, pp. 1–20.

Antonyshyn L., J. Silveira, S. Givigi and J. Marshall. 2023. Multiple mobile robot task and motion planning: A survey, *ACM Computing Surveys*, Vol. 55, 10, Article No. 213, pp. 1–35.

Caccavale R. and A. Finzi. 2022. A rapidly-exploring random trees approach to combined task and motion planning, *Robotics and Autonomous Systems*, Vol. 157, p. 104238, https://doi.org/10.1016/j.robot.2022.104238

Dantam N. T. 2021. Task and Motion Planning. In: Ang M. H., O. Khatib and Siciliano B. (Eds.), *Encyclopedia of Robotics*, Springer, Berlin, Heidelberg, pp. 1–9.

Ding J., Y. Zhou, X. Huang, K. Song, S. Lu and L. Wang. 2023. An improved RRT* algorithm for robot path planning based on path expansion heuristic sampling, *Journal of Computational Science*, Vol. 67, p. 101937.

Fan D. and P. Shi. 2010. Improvement of Dijkstra's Algorithm and Its Application in Route Planning, *2010 Seventh International Conference on Fuzzy Systems and Knowledge Discovery*, Yantai, China, 10–12 August, pp. 1901–1904.

Guo H., F. Wu, Y. Qin, R. Li, K. Li and K. Li. 2023. Recent trends in task and motion planning for robotics: A survey, *ACM Computing Surveys*, pp. 1–34, https://doi.org/10.1145/3583136

He B. 2022. Application of Dijkstra algorithm in finding the shortest path, *Journal of Physics: Conference Series*, Vol. 2181, p. 012005, *AIIM-2021 (International Symposium on Artificial Intelligence and Intelligent Manufacturing*, Huzhou, China, 26–28 November, pp. 1–4.

Hertzberg J. and R. Chatila. 2008. AI Reasoning Methods for Robotics. In: Siciliano B. and O. Khatib (Eds.), *Springer Handbook of Robotics*, Springer, Berlin, Heidelberg, pp. 207–223.

Ji X., L. Gu, Z. Fu, M. Ouyang and R. Liu. 2023. Research on Improvement and Optimization of A* Algorithm in Robot Path Planning, *RICAI '22: Proceedings of the 2022 4th International Conference on Robotics, Intelligent Control and Artificial Intelligence*, Dongguan, China, 16–18 December, Copyright © 2022 ACM, New York, USA, pp. 165–166.

Khokhar A. 2021. Probabilistic roadmap (PRM) for path planning in robotics, https://arushi-khokhar.medium.com/probabilistic-roadmap-prm-for-path-planning-in-robotics-d4f4b69475ea

Kumar R. 2024. The A* algorithm: A complete guide, https://www.datacamp.com/tutorial/a-star-algorithm

Latombe J.-C. 1991. *Robot Motion Planning*, Springer Science + Business Media, New York, 651 pages.

Marcelina. 2022. Motion planning vs path planning, Shade Robotics, https://medium.com/shade-robotics/motion-planning-vs-path-planning-e29ced760b90

Mansouri M., F. Pecora and P. Schüller. 2021. Combining task and motion planning: Challenges and guidelines. *Frontiers in Robotics and AI*, Vol. 8, Article 637888, pp. 1–12.

Meli D., H. Nakawala and P. Fiorini. 2023. Logic programming for deliberative robotic task planning, Artifcial Intelligence Review, https://doi.org/10.1007/s10462-022-10389-w, 39 pages.

Morecki A. and J. Knapczyk. 1999. Robot Task Planning. In: Morecki A. and J. Knapczyk (Eds.), *Basics of Robotics*, International Centre for Mechanical Sciences, Vol. 402. Springer, Vienna, pp. 319–377.

Navone E. C. 2020. Dijkstra's shortest path algorithm: A detailed and visual introduction, https://www.freecodecamp.org/news/dijkstras-shortest-path-algorithm-visual-introduction/#:~:text=Dijkstra's%20Algorithm%20basically%20starts%20at%20the%20node,all%20the%20other%20nodes%20in%20the%20graph.&text=Once%20the%20algorithm%20has%20found%20the%20shortest,as%20%22visited%22%20and%20added%20to%20the%20path

Owen-Hill A. 2019. Back to basics: Robot motion planning made easy, https://robodk.com/blog/robot-motion-planning-made-easy/#:~:text=Robot%20motion%20planning%20uses%20the%20same%20type,it%20is%20also%20used%20with%20industrial%20manipulators

Pan T., R. Shome and L. E. Kavraki. 2024. Task and motion planning for execution in the real, *IEEE Transactions on Robotics*, Vol. 40, pp. 3356–3371.

Sarkar A. 2024. RRT (rapidly exploring random tree) as a Python-based demonstration, https://www.linkedin.com/pulse/rrt-rapidly-exploring-random-tree-python-based-animesh-sarkar-xrqrc

Sniedovich M. 2006. Dijkstra's algorithm revisited: the dynamic programming connexion, *Journal of Control and Cybernetics*, Vol. 35, 3, pp. 599–620.

Xin W., L. Wanlin, F. Chao and H. Likai. 2019. Path Planning Research Based on an Improved A* Algorithm for Mobile Robot, *AMIMA 2019 (2nd International Conference on Advanced Materials, Intelligent Manufacturing and Automation*, Zhuhai, China, 17–19 May, *IOP Conference Series: Materials Science and Engineering*, Vol. 569, 052044, pp. 1–8.

Xu T. 2024. Recent advances in rapidly-exploring random tree: A review, *Heliyon*, Vol. 10, 11, e32451, pp. 1–29.

Zhang K., E. Lucet, J. A. D. Sandretto, S. Kchir and D. Filliat. 2022. Task and Motion Planning Methods: Applications and Limitations, *ICINCO 2022-19th International Conference on Informatics in Control, Automation and Robotics*, Lisbonne, Portugal, 2022 July, pp. 476–483.

Zhang L. 2022. Robot Path Planning based on Probabilistic Roadmaps and Velocity Potential Field in Complex Environment, *International Conference on Service Robotics (ICoSR)*, Chengdu, China, 10–12 June, pp. 96–101.

Zhang Y., N. Fattahi and W. Li. 2013. Probabilistic Roadmap with Self-Learning for Path Planning of a Mobile Robot in a Dynamic and Unstructured Environment, *IEEE International Conference on Mechatronics and Automation*, Takamatsu, Japan, 4–7 August, pp. 1074–1079.

Zhou C., B. Huang, and P. Fränt. 2022. A review of motion planning algorithms for intelligent robots, *Journal of Intelligent Manufacturing*, Vol. 33, pp. 387–424.

11 Autonomous Robots
SLAM, APF, and
PID Algorithms

11.1 INTRODUCTION

Autonomy is the quality of being self-governing. Autonomous individuals are commanded by their own personal rules. Autonomy is synonymous with self-determination, self-reliance, independence, and sovereignty.

Autonomous robots are intelligent devices that can work without recourse to human control. They are able to perceive their environment through advanced sensors. Indeed, they can also analyze the situation with AI assistance in real time, make decisions, and respond to the real world independently. These capabilities help them in performing tasks in various working and congested environments. The competencies and skills built into them render them smart enough in adapting to dynamic conditions with minimal-to-no human intervention, either through a guide or tele-operator control. Therefore, the actions of autonomous robots deeply contrast with those of traditional industrial robots. Illustrious examples of autonomous robots are self-moving vacuum cleaners, self-driving automobiles, and space probes. On the flip side, primordial and less advanced robots can only be programmed for executing repetitive tasks/movements in controlled environments (Mukhopadhyay and Sen Gupta 2007; Liu et al. 2023).

THE AUTONOMOUS ROBOT

I am an independent robot
I am my own master
A real-time decision maker
I never falter
Helping the homemaker
As a domestic aid and house caretaker
I plan my tasks meticulously
And wander in the house freely
Working silently and gracefully.
By evening, all my jobs are done
And I am praised by everyone.

This chapter describes the multi-talented autonomous robots possessing qualities for independent operation. These robots can be programmed to perform various simple and complex tasks in production environments, working tirelessly for long hours with superhuman efficiency. By surpassing human limits, they can achieve more in

DOI: 10.1201/9781032695266-11

less time. 'Superhuman' is a fictional concept that has a real-world focus on reaching high productivity and throughput in factories with effective tools and strategies.

We first compile a list of algorithms useful for autonomous robotics and then embark on a tour from one algorithm to another, gaining insights into their principles, merits, and demerits.

11.2   ALGORITHMS USED IN AUTONOMOUS ROBOTS

Algorithms play an important role in autonomous robot navigation. They hold the secret to robot action planning. They help to produce optimal paths distinguished by features such as being short in length, apart from being smooth, sturdy, and safe to tread. They are the routes that are free from obstacles and hurdles.

Autonomous robots function through the agency of various algorithms. All these algorithms work together in unison to enable a robot to perceive its environment, navigate, and make decisions on its own. Autonomous robot algorithms are listed in Figure 11.1. These algorithms are: pathfinding algorithms, short path finder algorithm, simultaneous localization and mapping (SLAM) algorithm, artificial potential field (APF) algorithm, PID (proportional-integral-derivative) control algorithm, decision matrix algorithm, bug algorithm, vector field histogram (VFH) algorithm, generalized Voronoi diagram (GVD), perception algorithms, and reinforcement learning algorithms. A cursory description of these algorithms is given below; it will be followed by an in-depth treatment in impending sections of this chapter:

   i. Pathfinding Algorithms: They use data to predict and pre-scan the paths of robots from their current positions to destinations. They aid in discovering the best driving route on a map, taking into account the state of the traffic. Traffic condition refers to the status at a given location at a particular instant of time. The A* and Dijkstra's algorithms are two indelible instances that have stood the test of time.
  ii. Short Path Finder Algorithm: It is used to find the shortest and easiest way of traversing a maze, a confusing network of passages with twists and turns. Dijkstra's algorithm works by iteratively selecting the node with the smallest distance from the starting point. Then it updates the distances to its neighbors. Thus, it effectively creates the shortest route from the source to all reachable nodes.
 iii. SLAM Algorithm: It is used by a robot for sketching a map of an unknown environment and tracking down its own location inside the map (Jain et al. 2021).
  iv. APF Algorithm: It is a magnetic field-inspired method for path planning in robotics that employs a notion akin to magnetic forces to guide a robot toward a target point (Al Jabari et al. 2022).
   v. PID Control Algorithm: It maintains the desired behavior of the robot by adjusting the control signals fed to it. The adjustments are based on the error measured by the difference between the current state and the wanted or wished-for state of the robot (Wang 2025).

Path finding algorithms

Short path finder algorithms

Simultaneous localization and mapping algorithm

Artificial potential field algorithm

Proportional-Integral-Derivative algorithms

Decision matrix algorithm

Bug algorithm

Vector field histogram algorithm

Generalized Voronoi diagram algorithm

Perception algorithms

Reinforcement learning algorithms

Autonomous robot algorithms

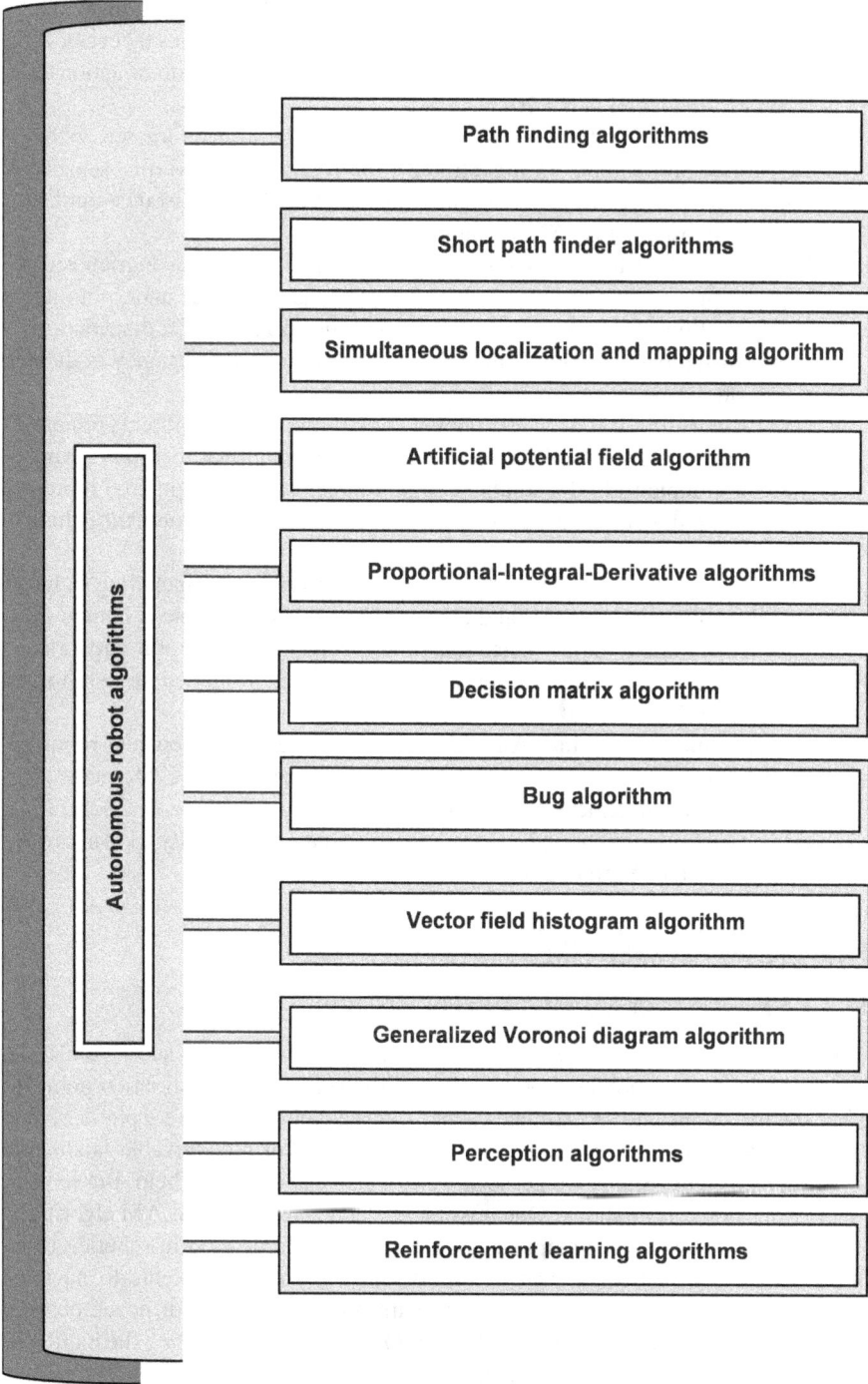

FIGURE 11.1   Algorithms used in autonomous robotics.

vi. Decision Matrix Algorithm: Here, a robot analyzes the sensor readings and environmental parameters in light of possible necessary actions by organizing them in a matrix format to make a selection of the best path of action to reach its final location (Li 2023).

vii. Bug Algorithm: This is a simple obstacle avoidance strategy known as the wall-following method. In this strategy, the robot uses proximity sensors (ultrasonic or infrared) to consistently follow a wall from a distance until it finds a path to the goal (Sivaranjani et al. 2021).

viii. VFH Algorithm: It is an obstacle avoidance algorithm. A histogram representation of the surrounding environment is the main idea underpinning this algorithm; the histogram is a graphical visualization of a distribution of data using bars with data values on the *X*-axis and the frequency of data points on the *Y*-axis (Yim and Park 2014).

ix. GVD Algorithm: It is used for robot path planning in environments replete with complicacies of multiple obstacles, where traditional methods struggle and grapple to find a solution. It guarantees safe collision-free routes by defining the closest areas to different points in the environment (Chen et al. 2022).

x. Perception Algorithms: These algorithms use sensory data, including sight, sound, and touch, to understand the environment. An example is computer vision, specifically object detection, where convolutional neural networks are utilized for image processing. Robot vision was discussed at length in Chapters 5–7.

xi. Reinforcement Learning Algorithms: These algorithms enable robots to learn optimal actions through a trial-and-error practice of repeated attempts and refinements in complex environments (Wen et al. 2025). In this hit-and-miss or cut-and-try method, solutions are cyclically tried at random and refined until one works.

11.3   SLAM ALGORITHM

11.3.1   Principles of Mapping and Localization

What do we do when we navigate an unknown environment? We access tools like Google Maps on our mobile phones and enter the starting and destination points to get travel directions and an estimated time. Alternatively, if we have a physical map and a compass, we can find directions and keep track of recognizable landmarks to stay on the right path. If we get lost, we can also ask locals for help. But how do robots work in such situations? One possible method is to use the SLAM algorithm.

The SLAM algorithm is employed as a computational method in robotic vision. It allows moving platforms such as robots and autonomous systems to navigate unknown spaces and environments, such as uncharted territories or novel, obscure places, safely and effectively (Liu et al. 2021; Qiao et al. 2024). During the motion of the robot through an unfamiliar and outlandish environment, it takes sensor readings from its camera or laser scanner to identify landmarks. Landmarks are the features that are easily noticeable in the landscape from a distance. The robot then uses this information to create a map of its surrounding environment. The map is continuously updated and regularly refreshed by integrating new landmark information with the existing map in a consistent and accurate manner. Robotic activity is not restricted to map building only. We know that a job half done is as good as none. So, the robot concurrently determines its own position within that map. Further, it regularly updates both the map of the environment and its own estimated position in real time. In this way, the usefulness of the algorithm is demonstrated in allowing robots to navigate without reliance on pre-existing maps.

11.3.2 MAIN STEPS OF THE SLAM ALGORITHM

The typical SLAM robot algorithm follows a pipeline consisting of many steps of which the principal ones are: data acquisition by the robot's sensors, extraction of landmarks in the acquired data, data association by linking sensor measurements to landmarks, state estimation by calculating robot's current position and orientation (pose) from sensor data and updating it through new measurements, map building, and loop closure detection when the robot revisits a previously mapped area (loop closure is the process of determining that a robot has visited a previously investigated location). These are divided into front-end and back-end operations.

Figure 11.2 shows the steps involved in SLAM. The algorithm begins when the robot's camera and LiDAR sensors start functioning. The front-end operations are composed of: data acquisition, feature identification, landmark extraction, data estimation, and position determination, while the back-end operations entail the calculation of the robot's current position and pose. Any of the SLAM variants are used, viz., EKF-SLAM, FastSLAM, GraphSLAM. Accordingly, the robot's position and pose are updated, followed by exploration, map building, and updating. After this process, the algorithm stops. Loop closure detection and map refinement are completed, as indicated. Further minutiae of steps in the algorithm are presented below:

i. Sensor Data Acquisition: The robot continuously senses its environment. It uses sensors like LiDAR, cameras, or a combination of sensors to collect environmental data. The sensors capture information such as distances to obstacles and visual features of the robot's surroundings. The data is in the form of images or laser scans about the robot's circumjacent areas.

Different types of sensors used for acquiring data have their assets and flaws. The accuracy and complexity of the SLAM algorithm are impacted by the individual, specialized, and dedicated sensors used. This means that the algorithm is sensor-dependent.

ii. Feature Identification and Landmark Extraction from the Sensor Data: Depending on the sensor type, distinctive features in the sensor data, notably corners or edges, or specific patterns, monuments, or prominent and distinctive constructions, are identified to serve as landmarks. The algorithm uses these landmarks to refine the robot's position and orientation.

iii. Data Association and Position Determination by Matching with Landmarks: The robot matches landmark observations across different viewpoints. It

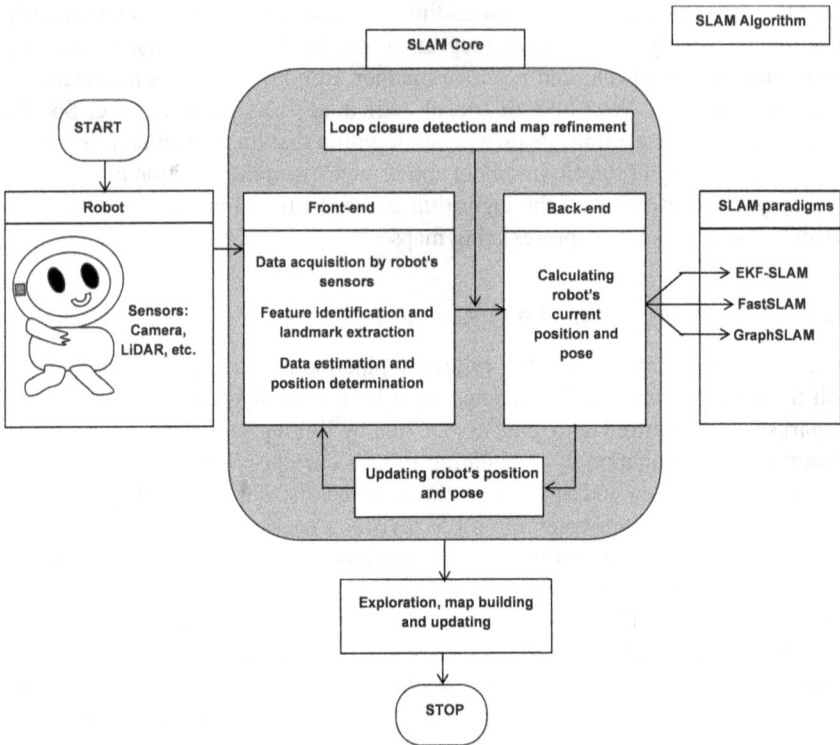

FIGURE 11.2   The simultaneous localization and mapping algorithm.

attempts to match newly detected landmarks to previously observed ones (existing features in the map) to determine correspondences between them. The correspondences are applied to find the robot's position relative to the landmarks and update its position and map. Incorrect matches are filtered out based on distance or other criteria; this process is termed outlier rejection. Noise and potential ambiguities render landmark matching really laborious and exhausting.

iv. State Estimation or Pose Update: The algorithm calculates the robot's current poses within the map. These calculations are done using wheel encoders, motion sensors (odometry) to estimate relative movement between robot positions and orientations (poses), and a filtering technique like a Kalman filter or particle filter. Based on its current estimated pose and the sensor data, the robot updates its estimated position.

Note 1: Odometry is a method that applies motion sensor data to estimate a robot's position from a starting point.

Note 2: The Kalman filter is a popular mathematical algorithm. It utilizes noisy data and a predictive mathematical model of the system to estimate the system's state, enabling real-time process monitoring. A series of measurements is performed over time to calculate the state of a system, and a recursive filtering method is applied to minimize the mean squared error (MSE). The recursive filtering method is a computational approach that continuously updates its approximations of the robot's state and the map of the environment using new sensor data by reusing the results from the prior calculations. A combination of linearity (where the change in output is proportional to the change in input) and Gaussian noise (normal or random noise) is considered for predicting and correcting real values. Thus, mathematically tractable calculations are done to form an impression of real-world values with noisy data. The noise is assumed to follow a Gaussian or normal distribution characterized by a bell-shaped curve. It is a common archetype for casual and unplanned fluctuations.

v. Exploration, Map Building, and Updating: The location of identified features on the map is stored as 3D points or other suitable data structures. The robot persists in its exploration of the area until it has adequate landmarks to create a map of the environment. It then builds a map by integrating new landmark information with the existing one, ensuring consistency and accuracy.

The robot continuously updates the map based on new information by adding new landmarks and their relative positions with sustained exploration, incorporating information from the state estimation. To maintain computational efficiency, optimization algorithms such as bundle adjustment – a nonlinear least-squares method for refining visual reconstructions – are often used to refine the map and robot pose estimates.

vi. Loop Closure Detection: To correct for accumulated errors, the algorithm detects loops in the path, i.e., situations where the robot reenters a formerly mapped area. This is done by making a similarity identification in which current sensor data is compared to previously stored map information. Whenever a loop closure is detected, the robot's estimated pose and map are adjusted to minimize inconsistencies and variabilities. Thus, loop closure detection corrects for potential errors by allowing for map refinement.

11.3.3 Different Versions of the SLAM Algorithm

There are several approaches to the SLAM algorithm, as well as various types of SLAM algorithms.

11.3.3.1 Visual SLAM (vSLAM) Algorithm

In this SLAM, cameras are primarily used to capture visual data of an environment. The visual data is unscrambled and interpreted to build a map of the environment. Computer vision is used for identification, confirmation, and cataloging of features and patterns in the images. It is an economical method. Detailed color and texture information is obtained. It is suited to well-illuminated scenes and augmented reality applications, e.g., medical training, education, entertainment, manufacturing, and retail. It is used in indoor robotics. But changes in lighting conditions, natural light or artificial illumination, or presence of textureless or featureless surfaces lacking any notable characteristics create difficulties to sabotage the algorithm functionality.

11.3.3.2 LiDAR SLAM Algorithm

It utilizes laser scanners to acquire depth information for more precise mapping of the environment (Malik 2023). A 3D map is created by measuring distances to various objects and generating a 3D point cloud, a dataset containing numerous points in a 3D X, Y, Z coordinate system with each point representing a specific spatial measurement on the surface of an object. The method is highly accurate. Soft, low lighting-based dimly illuminated scenes evoking serene and tranquil to mysterious and somber moods, as well as environments containing several obstacles, are mapped. The hardware used is more expensive than that of vSLAM. Difficulties are experienced with reflective surfaces, e.g., curved mirrors, which cause waves to bounce off in different directions. It is useful for autonomous vehicles, industrial robotics, dark spaces, and outdoor terrains.

11.3.4 COMMON SLAM ALGORITHMS

  i. Extended Kalman Filter-SLAM (EKF-SLAM) Algorithm: The EKF-SLAM algorithm is an extension of the standard Kalman filter, which is enhanced for a definite purpose. It is designed to handle nonlinearity by linearizing the equations of the system around its estimated current state. Its need arises because the robotic motion is not always necessarily linear. The robot's position, velocity, and acceleration are modeled as a nonlinear system.
  ii. Rao-Blackwellized Particle Filter-SLAM (FastSLAM) Algorithm: It provides an improvement over the EKF-SLAM algorithm. Here, particle filters are used for handling uncertainty and nonlinearity. Hence, mapping is efficiently done in complex environments
  iii. GraphSLAM Algorithm: The map and robot path in the environment are graphically represented. The nodes of the graph are the landmarks. Its edges denote the movement of the robot between the nodes. Therefore, map updating, its optimization, and loop closure detection can be efficiently done. Particle filters are recursive Bayesian filters. They constitute a sequential Monte Carlo method for estimating the state of a dynamic system when confronted with nonlinearities and non-Gaussian noise. Here, a set of weighted samples or particles is used for approximating the posterior probability distribution.

11.3.5 APPLICATIONS OF THE SLAM ALGORITHM

The SLAM algorithm is the foundation of robot navigation. It allows computers to perform computationally intensive tasks much faster than humans. A few examples of situations of its utilization are:

  i. Navigation of Autonomous Robots: SLAM is used to plot the robot's trajectory and steer its mobility along the plotted course. It can be applied to unknown indoor or outdoor environments.

ii. Navigation of Self-Driving Cars and Other Autonomous Vehicles: Maps of the road and surroundings are sketched using SLAM. These maps aid in planning routes for vehicle transportation.

iii. Navigation of Drones: Maps of difficult terrains are built with SLAM with regard to their physical features. Aerial exploration of the mapped region is conducted using drones, which seek guidance from these maps.

iv. Augmented Reality: Virtual objects are computer-generated digital images. These are accurately positioned in the real world with the help of SLAM by allowing a device to build a map of the surroundings and at the same time understand its location. Precise appearance of virtual overlays on real-world surfaces is thereby ensured with the user's movement.

11.3.6 Advantages of the SLAM Algorithm

The advantages of an algorithm refer to the situations, qualities, or opportunities that result in a positive outcome. Let us see the ways in which the SLAM algorithm proves beneficial in robotics.

i. Provision of Self-Governing Navigation to Robots: Robots can navigate without pre-existing maps, permitting robot activity for exploration of dynamic environments. Non-requirement of a map is a boon in many situations.

ii. Offering Real-time Mapping Facility: Maps of the environment are continuously generated as the robot moves around. The map generation enables the planning of a secure path for robot's motion avoiding obstacles.

iii. Affording Capability of Sensor Fusion: Data from multiple sensors such as cameras, LiDAR, and inertial measurement unit (IMU) are integrated for increasing accuracy and robustness of the algorithm. The IMU is a device that measures and reports the acceleration, angular rate of motion and acceleration of an object. It contains accelerometer, gyroscopes and magnetometers.

iv. Furnishing Adaptability: Changes in environment of robot can be handled by SLAM. Their handling allows the robots to adapt to new situations.

11.3.7 Limitations of the SLAM Algorithm

Reliance on an algorithm without understanding its limitations leads to unforeseen and potentially negative outcomes. We would like to mention the following drawbacks of the SLAM algorithm.

i. High Computational Cost Demands: Real-time data processing using SLAM algorithm needs significant power consumption. The power requirement increases especially when dealing with large spaces and complicated lighting conditions.

ii. Loop Closure-Related Issues: Identification of errors and making the necessary corrections is difficult when revisiting previously mapped areas. As a consequence, drifts are noticed in the map. The issue is intensified with increasing complication of environment because cumulative errors creep in. They accumulate over time leading to significant deviation from true value, which is observed as an underestimation or overestimation of the value.

iii. Environmental Dependences and Effects of Circumstantial Variability on SLAM: Poor lighting, textureless monotonous surfaces, or cluttered environments are detrimental to SLAM performance impacting its utilization adversely.

iv. Sensor Noise-Induced Errors: SLAM is extremely sensitive to noise from sensors, e.g., undesired fluctuations or variations in the output that do not reflect the true state of the measurand. The noise introduces inaccuracies in the map and robot localization.

v. Maintenance of Correct Calibration of Sensors: A precise calibration of sensors is warranted to guarantee accuracy in SLAM. As the sensors are prone to drift with time, maintenance of correct calibration is a painstaking necessity that cannot be oversighted. Corrections for temporal drifts in sensor characteristics must be invariably applied.

11.4 APF ALGORITHM

The APF algorithm is a robot path planning algorithm. The APF algorithm can be used in conjunction with SLAM to provide local path planning capabilities for robots that already utilize SLAM for localization and mapping. APF focuses on the specific task of path planning and obstacle avoidance, whereas SLAM concentrates on the overall problem of mapping and localization.

In the APF algorithm, the location of the robot's goal is represented by an attractive potential and the obstacles in its path are represented by repulsive potentials. The attractive and repulsive forces are used to create a virtual potential field. The motion of the robot is controlled within this virtual potential field. The algorithm guides the robot toward a goal location while simultaneously avoiding obstacles (Rimon and Koditschek 1992; Tao 2024; Bharali et al. 2025).

11.4.1 APF ALGORITHM TERMS

The algorithm primarily involves creating a potential field around a robot. It works by impersonating a magnetic field concept through simulation of the attractive and repulsive forces. As already indicated, the attractive forces pull the robot toward the goal whereas the repulsive forces push it away from obstacles. Before going into details of the algorithm, the main terms of the algorithm are defined in the context of robotics and navigation. Figure 11.3a shows a robot and a target on the opposite sides of an obstacle which is located in the middle. The directions of the attractive and repulsive forces between the robot and the target are shown by arrows. With reference to the diagram, we introduce the algorithm terms and then explain its working.

APF: This field is a mathematical representation of the attractive and repulsive forces acting at each point in the environment surrounding the robot.

Attractive force: This force is a representation of pull exerted from the robot toward the goal. It pulls the objects together. It is generated by the distance and direction

Potential Field Around a Robot

Attractive force between the robot and the target (Full lines)

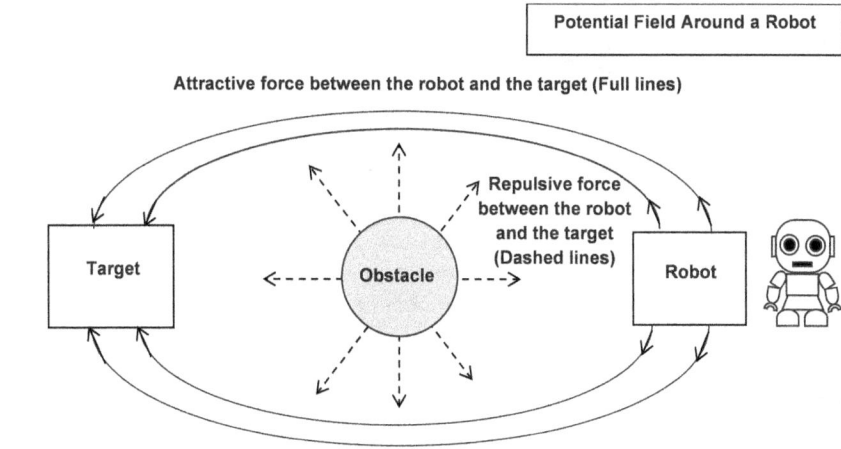

(a)

APF Algorithm

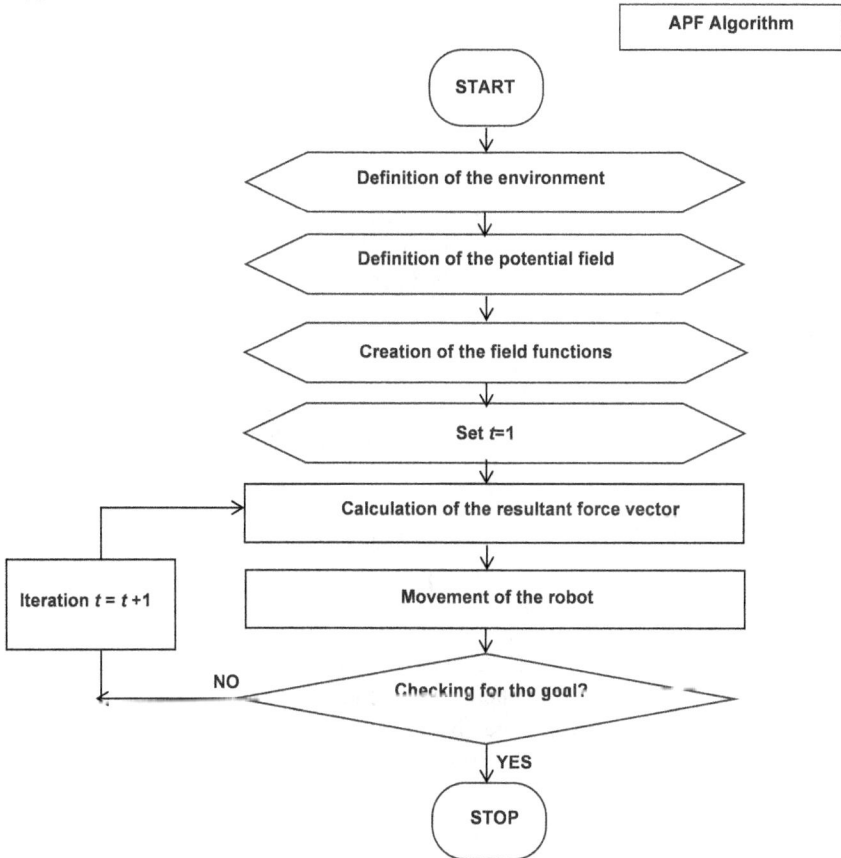

(b)

FIGURE 11.3 The artificial potential field algorithm: (a) key terms and (b) execution steps.

to a designated goal point. Its direction points directly toward the goal. Therefore, the closer the robot is to the goal and the more directly it is aligned with the goal, the stronger is the attractive force. The goal acts as a source of attraction (positive potential).

Repulsive force: This force is a representation of the push of the robot away from obstacles. It pushes the objects away. It is generated by the distance and direction to obstacles. Hence, the closer the robot is to an obstacle and the more directly the robot is aligned with the obstacle, the stronger is the repulsive force pushing it away from the obstacle. The obstacle is a source of repulsion (negative potential).

Working of the Algorithm: An environment is modeled as a landscape in which obstacles are represented as high-potential areas represented as hills, and the goal is a low-potential area at the bottom of a valley. The robot is guided toward the goal by following the gradient of the potential field.

The algorithm proceeds as the robot navigates by following the gradient of this potential field. Effectively, the robot moves toward the goal while avoiding any impediments on the way. The gradient attracts the robot toward the target but repels it away from obstacles. During the course of its motion, the velocity of the robot is determined by summing up all the forces acting on it (Rostami et al. 2019).

11.4.2   Main Steps of the APF Algorithm

The algorithm formulation consists in formalizing the status of the various terms mentioned in the preceding subsection for the robot whose motion is being studied (Xia et al. 2023). Figure 11.3b illustrates the steps in the algorithm. It starts by defining the robot's environment and the potential field. Then the field functions are created. Setting $t = 1$, the resultant force vector acting on the robot is calculated and the robot movement takes place. It is checked whether the robot has reached the goal. If NO, iteration $t = t + 1$ is set and the algorithm returns to the force vector calculation stage. If YES, the algorithm stops. The steps of the algorithm are elucidated below to dispel any doubts (Figure 11.3b).

  i. Definition of the Environment: The definition involves formal identification of the important positions regarding the robot's movement and in reference to the robot. These are the starting and goal positions of the robot as well as locations of all obstacles in the workspace.
 ii. Definition of the Potential Field: The definition entails the creation of a mathematical function that generates an attractive potential around the goal and a repulsive potential around obstacles.
iii. Creation of Potential Field Functions: These are the two parts comprising the potential field.
     a. Attractive Potential Field: This field is created by introducing a function that generates an attractive force toward the goal. It decreases with distance to the goal, and is designed to have a minimum at the goal.
     b. Repulsive Potential Field: This field is created with a function that generates a repulsive force away from obstacles. It increases as the robot

gets closer to an obstacle, and is designed to be inversely proportional to the distance from an obstacle.

c. Calculation of the Potential at Each Point: The calculation is performed to determine the combined potential by summation of the attractive potential from the goal and the repulsive potential from all obstacles. Regarding the calculation points, the calculation is done for each point in the environment or workspace of the robot.

iv. Computation of the Resultant Force Vector Acting on the Robot: Computation is performed to find the gradient of the potential field at the current position of the robot. The gradient of the potential represents the resultant force acting on the robot. This computation is done at each iteration.

v. Movement of the Robot: For motion of the robot, the position of the robot is updated based on the calculated resultant force vector. It is always ensured that the robot moves toward the goal while sidestepping any obstacles stopping it from moving.

vi. Repetition: Iteration is continued through the previous steps until the robot reaches the goal location or encounters a situation where it cannot move further. Such a situation arises when the robot is trapped in local minima where the gradient is zero. The trapping of robot causes it to oscillate around that point.

11.4.3 IMPORTANT CONSIDERATIONS ABOUT THE APF ALGORITHM

Correctness, clarity, and efficiency of the algorithm and the optimal use of resources are ensured by adopting various measures:

i. Selection of Potential Function: For ensuring a smooth navigation of the robot, one must stay away from becoming cemented in local minima. Hence, the choice of appropriate functions for the attractive and repulsive potentials is an important consideration. They must be chosen after careful, appropriate judgments made by thoughtful evaluation and discernment.

ii. Detection of Obstacle: Accurate detection of obstacle is necessary for creation of a trustworthy repulsive potential field. This is made achievable by using a reliable sensor and associated unswerving instrumentation system. An ultrasonic sensor, a camera providing depth information, or a LiDAR are employed for this intent. The instrument is equipped with sophisticated data processing techniques. The instrument's methods must be such that they provide precise data about the distance and location of obstacles through accurate identification and modeling of the boundaries of the obstacle. The data processing techniques of interest must include:

a. Segmentation of Obstacle: The pixels or points in sensor data attributed to obstacles must be correctly identified. They must be unmistakably disassociated from background noise.

b. Filtration of Obstacle: False positives that are likely to interfere with the potential field calculation must be eliminated. So must be the noisy data points.

   c. Modeling of the Obstacle Shape: The shapes of obstacles should be approximated with geometric primitives. Circles, rectangles, or more complex mathematical models are used depending on the situation at hand.
 iii. Issues about Local Minima: APF algorithm sometimes gets fastened to local minima. In these situations, the resultant force becomes zero. Zeroing of the force happens even when the goal is not reached. Techniques are available to alleviate this issue. Addition of random noise or modification of the potential field are beneficial. The current position of the robot and the robot's environment must be taken into account. Then the parameters of the potential field must be dynamically adjusted. Espousing this procedure helps in avoiding the robot's getting trapped in local minima. Application of a smoothing algorithm to the calculated potential field further blunts or softens the abrupt and sharp gradients. Consequent to this rounding and evening out, the path becomes smoother, making it easier for the robot to follow.

11.4.4   APPLICATIONS OF THE APF ALGORITHM

The APF algorithm helps to automate several operations in robotics, such as:

   i. Navigation of Mobile Robots: It is used for guiding robots through cluttered or disorganized and congested or jammed areas while avoiding hindrances. This is done by identifying suitable pathways and adjusting movements of the robot based on real-time obstacle detection, and adapting to dynamic environmental changes.
  ii. Planning of Autonomous Vehicle Path: It is used for creating collision-free trajectories for self-driving cars. On these trajectories, cars can navigate from place to place without bumping into each other and dashing into the pedestrian crowd.
 iii. Navigation of Unmanned Aerial Vehicle (UAV): The flying of drones in gorged environments bursting at the seams is successfully regulated with APF.

11.4.5   ADVANTAGES OF THE APF ALGORITHM

The APF algorithm offers the advantages of efficiency and clarity in robotic problem-solving. Among its potential benefits, we would like to give prominence to the following:

   i. Intuitive and Easy Implementation of the Algorithm: The concept of attractive and repulsive forces is relatively simple to understand in principle. It is also easy to implement in coding.
  ii. Real-Time Capability: The calculation of force is computationally efficient with minimal computational overhead requirement. The computational efficiency allows for near-instantaneous decision-making and adaptation to changing conditions. Therefore, its recommendation for real-time path planning is obvious and incontrovertible.

11.4.6   Disadvantages of the APF Algorithm

Among the main downsides of the APF algorithm, the following stand out clearly:

    i. Problem of Local Minima: In complex environments, the robot might get jammed in local minima (Zhu et al. 2006). Here, the forces from obstacles annul each other, thus preventing the robot from reaching the goal. As the robotic vehicle does not reach the destination, there is no assurance of success of the algorithm in such configurations. A hang-up ensues.

    ii. Oscillation of Robot Path: The robot might oscillate around obstacles due to rapidly changing forces. It appears to be locked in a repetitive back-and-forth bouncing in the form of an oscillatory or vibratory motion. The design efforts made for potential field play a vital role in producing such oscillations. An ill-chosen potential function leads to sharp gradients. The sharp gradients coerce the robot to make overreactions to small changes in distance to obstacles or the goal. Non-balancing of the attractive and repulsive forces is another cause of occurrence of oscillations. Then one force dominates over the other, and oscillations are instigated. Local minima too are the likely responsible factors for swaying and swinging behavior.

11.5   PID ALGORITHM

While the APF algorithm is devoted to robot path planning, the PID algorithm is a feedback control system that adjusts the state of a system to match a desired setpoint or target value (Figure 11.4). It is used to control the movement of a robot precisely

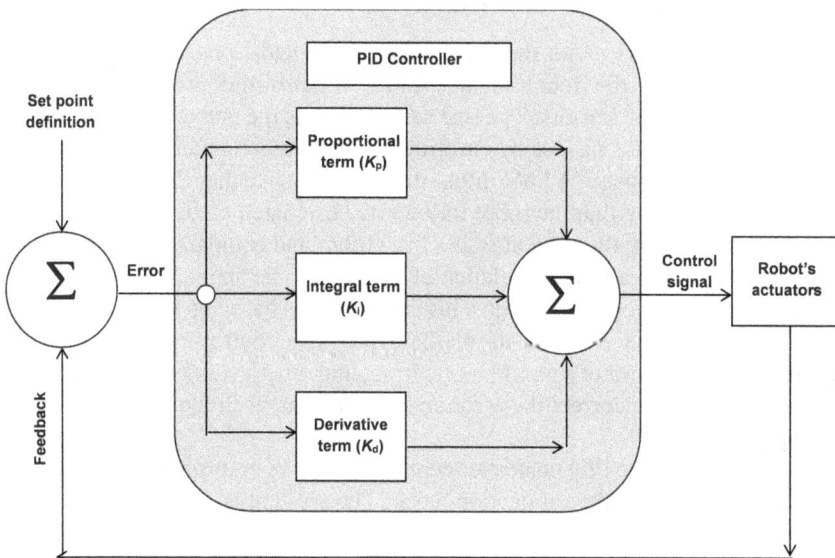

FIGURE 11.4   The proportional-integral-derivative controller.

through the regulatory action. The regulatory action is calculated by combining three parameters, namely, the current error, the accumulated error over a duration of time, and the rate of change of the error. In this manner, the algorithm allows the robot to maintain a desired position or trajectory. The algorithm is frequently used for robotic tasks such as tracking a defined path by adjusting the speeds of robot's motors based on the feedback signals received from on-board sensors (Carmona et al. 2018; Minh Nguyet and Ba 2023).

11.5.1 Components of the PID Algorithm

As its name suggests, the PID algorithm has three principal components (Waseem 2023; Smith 2024):

i. Proportional (*P*) Component: This component reacts and responds directly to the current error. It provides an immediate correction proportional to the magnitude of error.
ii. Integral (*I*) Component: This component accumulates the error over time. Building up the mistake with time helps in elimination of steady-state errors. A gradual adjustment of the control signal based on the past errors is utilized.
iii. Derivative (*D*) Component: This component measures the rate of change of the error to make a prediction of the future behavior of error. The pace at which the error changes with time helps in reducing oscillations. It improves the response time by allowing to counterpoise any sudden changes rapidly and damping overshoots.

11.5.2 Steps of the PID Algorithm

The PID algorithm works with the help of sensors, calculations, and adjustments. The sensors used by the robot include ultrasonic/infrared proximity sensors, encoders, GPS, and line sensors. The line sensor detects the presence of a contrasting black line on a white surface by emitting infrared radiation and measuring the reflected radiation intensity. A phototransistor indicates whether the line is present or absent, thus ensuring that the robot follows its designated path.

The sensors measure the current state of the robot and compare it to the desired state. This comparison allows calculation of the error to be applied. The PID algorithm then calculates a control output. This calculation is based on the proportional, integral, and derivative components of the error. The control output is used to adjust the speed of the robot's motor, its steering, and other actuators. These adjustments help the robot to correct the error and enable it to reach the desired position satisfyingly.

Figure 11.4 shows the PID controller containing blocks on proportional, integral and derivative terms and the summation block. The setpoint is defined. The control signal is fed to the robot's actuators. The feedback signal is compared with the setpoint to calculate the error. The steps of the algorithm are detailed below:

i. Definition of the Robot Setpoint: This is done by specification of the desired position, velocity, or other parameters of the robot that are to be maintained by the algorithm.

ii. Measurement of the Feedback Signal: Sensors acquire information about the current state of the robot, e.g., its actual position and velocity.

iii. Calculation of Error Signal: This signal is found by computing the difference between the setpoint and the measured feedback value.

iv. Proportional Term (K_p) Computation: A control output signal is calculated. It is directly proportional to the current error.

v. Integral Term (K_i) Computation: The error signal is accumulated over time. It provides a correction for persistent errors.

vi. Derivative Term (K_d) Computation: The rate of change of the error is calculated. It helps to anticipate future locomotion behavior of the robot. It intends to avert potential overrun. So, the robot is prevented from any likely overshoot.

vii. Combination of Three Terms: The proportional, integral, and derivative terms are added together to determine the final control output.

viii. Application of Control Output Signal: The calculated control signal is transmitted to the actuators, usually motors of the robot to adjust its movement.

11.5.3 IMPORTANT ASPECTS OF PID CONTROL IN ROBOTICS

To make the PID algorithm more accessible and useful, it is necessary to lay emphasis on the aspects that cannot be ignored. Some of these are:

i. Tuning of PID Controller Gains (K_p, K_i, K_d): Choosing suitable values for the proportional, integral, and derivative gains is an essential prerequisite for attaining responsive and steady robot control serving as a necessary precondition for robot's stability.

ii. Selection of Suitable Sensors Fulfilling Specifications: The performance of the PID controller is greatly impacted by the accuracy and precision of the sensors used to measure the error. Therefore, they should be selected after careful thought.

iii. Comprehension of Dynamics of Robot Motion: Effectively designing and tuning of a PID controller is highly reliant on understanding the mechanical characteristics of the robot with which the designer must be fully conversant.

11.5.4 APPLICATIONS OF THE PID ALGORITHM

The PID algorithm finds widespread usage in robotics. The following applications merit special attention:

i. Tracking of Robot's Path and Accurate Adherence to its Moving Line: It is used for following and maintaining a predefined path for the robot. The direction and speed of the robot are continuously adjusted to force it to stay exactly on the intended trajectory line by varying motor speeds based on the sensor readings from a line-following sensor. It makes the robot capable of detecting lines by measuring reflected light emitted by its own infrared LED.

ii. Avoidance of Obstacles: The movement of the robot is controlled on the basis of proximity sensor data for maintaining a safe distance of the robot from obstacles. Maintaining safe distances from nearby objects is a reliable collision avoidance precaution.

11.5.5   ADVANTAGES OF THE PID ALGORITHM

The PID algorithm offers several advantages for autonomous robots, including its implementation simplicity and ability to provide precise control in various situations. It is reiterated that the algorithm is often a favorite choice in robotics in view of the following plus points and privileges (Yuldashev and Solovev 2024):

i. Simple and Easy Implementation: From the software viewpoint, PID controllers are relatively straightforward to understand and implement. They allow rapid prototyping and deployment on robots.

ii. Broad Range of Applicability: PID controllers are effectively used for a wide range of robotic motion control tasks. The variety of tasks include the controlling of position, velocity, and acceleration of robots.

iii. Precision of Control: PID algorithms achieve accurate and stable robotic control by combining proportional, integral, and derivative actions. Steady-state errors are minimized in this multipart process.

iv. Tunability of Gains: Based on the particular robot system, the gains of the P, I, and D components are altered to fine-tune and tailor the response of the controller to achieve desired performance characteristics of the robot.

v. Robustness to Environmental Disturbances: PID controllers are able to handle external disturbances to the system. They can maintain control even in dynamic environments.

11.5.6   LIMITATIONS OF THE PID ALGORITHM

The limitations of this algorithm include the possibility of instability in case of improper tuning of gains, trouble in managing highly nonlinear systems, and sensitivity to noise, particularly in the derivative component. These restrictions degrade its utility for monitoring discombobulated, dynamic environments. Exact robotic modeling is difficult in these cases.

i. Complexity of Tuning: Although appearing to be conceptually simple at the first sight, determination of optimal PID gains is often perplexing. This is done by trial and error manifestly for complicated robotic systems. Sometimes advanced tuning techniques are resorted to.

ii. Sensitivity to Noise: The derivative component of PID amplifies noise in the system. Instability ensues if noise is not properly filtered to improve the quality of the signal.

iii. Limitations Concerning Nonlinear Systems: PID controllers are designed for linear systems. Naturally, these controllers do not perform optimally in scenarios with highly nonlinear dynamics. Additional control strategies are necessary for tackling these situations.

iv. Possibility of Overshooting Response: Improper tuning causes significant overshooting of the response of a robotic system. Consequently, the accuracy and stability of the system are negatively affected.

v. Limited Adaptability to Changing Environments: In general, the PID controllers are not designed to automatically adapt to significant changes in the environment or robot dynamics. Re-tuning of the controller is required to cater to such variations.

Overall, PID algorithms serve as invaluable tools for controlling autonomous robots. Their simplicity and effectiveness make them favorites of design engineers in many scenarios. Notwithstanding these benefits, careful consideration of their drawbacks and making appropriate tuning are crucial for optimal performance. Expressly, dynamic environments must be handled with caution.

11.6 DISCUSSION AND CONCLUSIONS

Autonomous robots can work in hazardous conditions inside nuclear reactors or aero-engines where humans obviously will not even dare to imagine getting entry. They can improve efficiency, safety, and productivity in many industries by consistently handling repetitive tasks with precision avoiding errors associated with manual operations and reducing production downtime. Humans can focus on more complex work. Autonomous robots can assemble parts. They can weld parts and paint finished products. They can help with inspection, monitoring, and quality assurance in manufacturing. They can be engaged in logistics and warehousing to optimize picking, sorting, and storing items. They can prepare orders and transport heavy payloads in the supply chain. They can be employed in healthcare to work for disinfection, and delivering medical supplies. They can assist in agriculture with harvesting, weeding, crop monitoring, and optimizing irrigation systems. They can reduce labor costs wherever applicable. Owing to their adaptability to changes in their environments, they can perform myriad other operations in dynamic and unpredictable environments of several sectors to usher in a new revolution in robotized, computerized, and mechanized manufacturing.

In this chapter, the SLAM, APF, and PID algorithms in autonomous robotics were reviewed (Table 11.1). This discussion of autonomous robotic algorithms will be continued in the next chapter to unearth some of its boundless potentialities.

TABLE 11.1

Takeaways from This Chapter at a Glance

Sl. No.	Takeaway	Explanation
1	Summary	Algorithms used in autonomous robots were listed. The advantages and limitations of these algorithms were outlined.
2	SLAM algorithm	It is an algorithm that enables the robot to build a map of its environment and locate its position within that map. Different versions of the SLAM algorithm are discussed, viz., visual SLAM and LiDAR SLAM. Common SLAM algorithms are mentioned, e.g., extended Kalman filter (EKF-SLAM), FastSLAM, and GraphSLAM.
3	APF algorithm	This algorithm simulates a potential field in which attractive forces pull the robot toward the goal while repulsive forces push it away from obstacles. Main steps, important considerations, and applications of artificial potential field algorithm are elaborated.
4	PID algorithm	It is a feedback control algorithm which works by adjusting a controller output. Components of the proportional-integral-derivative algorithm, its steps, important aspects and applications are reviewed.
5	Keywords and ideas to remember	Autonomous robots, simultaneous localization and mapping algorithm, visual SLAM, LiDAR SLAM, artificial potential field algorithm, proportional-integral-derivative algorithm

REFERENCES AND FURTHER READING

Al Jabari H., A. Alobahji and E. A. Baran. 2022. A New Artificial Potential Field Based Global Path Planning Algorithm for Mobile Robot Navigation, *2022 IEEE 17th International Conference on Advanced Motion Control (AMC)*, Padova, Italy, 18–20 February, pp. 444–449.

Bharali M., S. Das, K. Nath and M. K. Bera. 2025. Modified Artificial Potential Field Algorithms for Mobile Robot Path Planning. In: Stroe D. I., D. Nasimuddin, S. H. Laskar and S. K. Pandey (Eds.), *Emerging Electronics and Automation. E2A 2023.* Lecture Notes in Electrical Engineering, Vol. 1237, Springer, Singapore, pp. 137–148.

Carmona R. R., H. G. Sung, Y. S. Kim and H. A. Vazquez. 2018. Stable PID Control for Mobile Robots, *15th International Conference on Control, Automation, Robotics and Vision (ICARCV)*, Singapore, 18–21 November, pp. 1891–1896.

Chen D., Q. Xu, J. Liu, M. Zou, W. Chi and L. Sun. 2022. A Generalized Voronoi Diagram based Robot Exploration Method for Mobile Robots, *2022 IEEE International Conference on Robotics and Biomimetics (ROBIO)*, Jinghong, China, 5–9 December, pp. 1029–1035.

Jain S., U. Agrawal, A. Kumar, A. Agrawal and G. S. Yadav. 2021. Simultaneous Localization and Mapping for Autonomous Robot Navigation, *2021 International Conference on Communication, Control and Information Sciences (ICCISc)*, Idukki, India, 16–18 June, pp. 1–5.

Li X. 2023. Autonomous Robot Motion Decision Algorithm Based on Decision Tree Algorithm and Deep Learning Technology, *2023 2nd International Conference on 3D Immersion, Interaction and Multi-sensory Experiences (ICDIIME)*, Madrid, Spain, 27–29 June, pp. 259–262.

Liu B., Z. Guan, B. Li, G. Wen and Y. Zhao. 2021. Research on SLAM Algorithm and Navigation of Mobile Robot Based on ROS, *2021 IEEE International Conference on Mechatronics and Automation (ICMA)*, Takamatsu, Japan, 8–11 August, pp. 119–124.

Liu L., X. Wang, X. Yang, H. Liu, J. Li and P. Wang. 2023. Path planning techniques for mobile robots: Review and prospect, *Expert Systems with Applications*, Vol. 227 (C), 120254, pp. 1–30.

Malik S. 2023. Lidar SLAM: The ultimate guide to simultaneous localization and mapping, https://www.wevolver.com/article/lidar-slam

Minh Nguyet N. T. and D. X. Ba. 2023. A neural flexible PID controller for task-space control of robotic manipulators, *Frontiers in Robotics and AI*, Vol. 9, 975850, pp. 1–10.

Mukhopadhyay S. C. and G. Sen Gupta (Eds.). 2007. *Autonomous Robots and Agents*, Springer-Verlag, Berlin, Heidelberg, 267 pages.

Qiao J., J. Guo and Y. Li. 2024. Simultaneous localization and mapping (SLAM)-based robot localization and navigation algorithm, *Applied Water Science*, Vol. 14, 151, pp. 1–8.

Rimon E. and D. E. Koditschek. 1992. Exact robot navigation using artificial potential functions, *IEEE Transactions on Robotics and Automation*, Vol. 8, pp. 501–518.

Rostami S. M. H., A. K. Sangaiah, J. Wang and X. Liu. 2019. Obstacle avoidance of mobile robots using modified artificial potential field algorithm, *EURASIP Journal on Wireless Communications and Networking*, Vol. 2019, Article Number 70, pp. 1–19.

Sivaranjani S, D. A. Nandesh, R. K. Raman, K. Gayathri and R. Ramanathan. 2021. An investigation of Bug Algorithms for Mobile Robot Navigation and Obstacle Avoidance in Two-Dimensional Unknown Static Environments, *2021 International Conference on Communication information and Computing Technology (ICCICT)*, Mumbai, India, 25–27 June, pp. 1–6.

Smith G. M. 2024. What is a PID controller? https://dewesoft.com/blog/what-is-pid-controller

Tao S. 2024. Improved artificial potential field method for mobile robot path planning. *Applied and Computational Engineering*, Vol. 33, pp. 157–166.

Wang Z. 2025. Application of the PID Algorithm in Robot, *ITM Web of Conferences, 73, 01025 International Workshop on Advanced Applications of Deep Learning in Image Processing (IWADI 2024)*, Kuala Lumpur, Malaysia, 27–29 December, 7 pages.

Waseem U. 2023. PID controller & loops: A comprehensive guide to understanding and implementation, https://www.wevolver.com/article/pid-loops-a-comprehensive-guide-to-understanding-and-implementation

Wen S., Y. Shu, A. Rad, Z. Wen, Z. Guo and S. Gong. 2025. A deep residual reinforcement learning algorithm based on soft actor-critic for autonomous navigation, *Expert Systems with Applications*, Vol. 259, p. 125238, https://doi.org/10.1016/j.eswa.2024.125238

Xia X., T. Li, S. Sang, Y. Cheng, H. Ma, Q. Zhang and K. Yang. 2023. Path planning for obstacle avoidance of robot arm based on improved potential field method, *Sensors (Basel)*, Vol. 23, 7, 3754, pp. 1–15.

Yim W. J. and J. B. Park. 2014. Analysis of Mobile Robot Navigation Using Vector Field Histogram According to the Number of Sectors, the Robot Speed and the Width of the Path, *2014 14th International Conference on Control, Automation and Systems (ICCAS 2014)*, Gyeonggi-do, Korea (South), 22–25 October, pp. 1037–1040.

Yuldashev T. and A. Solovev. 2024. Basics of PID controllers: Design, applications, advantages and disadvantages, https://www.integrasources.com/blog/basics-of-pid-controllers-design-applications/

Zhu Q., Y. Yan and Z. Xing. 2006. Robot Path Planning Based on Artificial Potential Field Approach with Simulated Annealing, *Sixth International Conference on Intelligent Systems Design and Applications*, Jian, China, 16–18 October, pp. 622–627.

12 Autonomous Robots
Broadening the Perspective

12.1 INTRODUCTION

As technologically advanced robots, including ground robots, underwater robots, and unmanned aerial vehicles, are being increasingly utilized in industry, security, and military applications, a multitude of autonomous robot algorithms have been developed to achieve robots' autonomy. The technical literature has a virtual deluge of research papers on this topic. Each algorithm demonstrates its problem-solving ability, mettle, and fortitude in a distinct area, and finds befitting applications depending on the robot's environment and the desired task. The autonomous robot algorithms primarily focus on the themes of path planning, obstacle avoidance, and decision-making for a robot within a dynamic environment. Some algorithms rely heavily on accurate sensor data for decision-making. Others are based on pre-programmed paths and behaviors, and therefore, can operate independently of sensors. Plain and modest environments benefit from easier algorithms, while intricate and dynamic settings require advanced methods. Algorithms with faster computation times are essential for applications that require quick, reactionary responses to changing conditions. Therefore, it will be expedient to expand our coverage of autonomous robot algorithms in Chapter 11. Building on this, we present numerous ingenious algorithms in this chapter to enable the reader to gain a holistic understanding of the *status quo* in autonomous robotics.

12.2 GENERAL ASPECTS OF THE DECISION MATRIX ALGORITHM FOR ROBOTS

Effective decision-making underpins all management processes. In fact, it is the cornerstone of successful management, leading to the sustainability of individuals and organizations. A decision matrix within the framework of autonomous robotic algorithms refers to a structured approach for analyzing problems that a robot will encounter (Venkata Rao and Padmanabhan 2006; Ralfs et al. 2022). In this approach, a robot analyzes various germane factors related to it. These factors include the data recorded by its sensors, conditions of the environment, and potential actions taken by the robot. The prime aspect of this analysis is the use of a matrix format as a rectangular arrangement of numbers or symbols laid out in rows and columns to systematically evaluate and choose the best course of action based on predefined criteria. In essence, it facilitates the process of making well-informed decisions by a robot in intricate circumstances. A well-informed decision-making involves arriving at decisions after gathering all the relevant circumstantial information about a problem and considering numerous possibilities and options for redressal.

DOI: 10.1201/9781032695266-12

12.2.1 PURPOSE OF A DECISION MATRIX

A weighted decision matrix is polyonymous, being known by various names. Some of these names are grid analysis, Pugh matrix, decision grid, or problem selection matrix. The decision matrix is a powerful method for assessing and selecting the optimal choice from a range of opportunities. It is particularly useful if one has to hand-pick many options with several different factors involved in influencing the outcome. It is relatively easy to use and is most effective when deciding between a few comparable choices. Using a decision matrix is strongly endorsed when one is presented with several comparable options. It is also recommended when an individual must select only one choice from many given alternatives. It is persuasively suggested in cases where a rational decision should be made rather than one based on an emotional standpoint. It is instrumental for decision-making in robotics.

12.2.2 CREATION OF A DECISION MATRIX

For generating a decision matrix, one must thoroughly comprehend the issues that arise when handling a given situation, as well as their ramifications and relative significance in determining the solution to address the situation. After all these issues are properly understood, one can frame an analytical table or matrix containing rows and columns. In this table, decision alternatives are listed as rows of the matrix. The columns of the table list the relevant factors such as effectiveness, ease, and costs related to these alternatives. An evaluation scale is set up. This scale assesses the value of individual alternatives and combinations. Normally, the scale has the following form: the highest importance is assigned a value of 10, and the lowest importance is equated to 0. This scale must be consistent and unwavering throughout the matrix. To appraise the score of an entry in the matrix, the original ranking of that entry is multiplied by the corresponding weight, which is a numerical value expressing its importance in relation to other entries in determining the consequence. Then all the factors under each option are added together to get a weighted sum for that option.

12.2.3 STEPS IN MAKING A DECISION MATRIX

Creating a decision matrix algorithm is a multi-phase process comprising seven primary steps (Figure 12.1): conceptualization, parameterization, organizing the decision matrix, filling in the entries, assigning weights, calculating weighted scores, and aggregating them to obtain the total score. If the desired criteria are satisfied, the algorithm is stopped. Otherwise, we return to the step of constructing the decision matrix, unless we arrive at the best concept. These steps are elaborated below (UMass 2025):

    i. Identification of Alternatives from Which a Selection Is to Be Made: The available options catering to the problem in hand are found and listed in order to choose between similar choices.

    ii. Formulation of Criteria for Making Decisions: The vital factors that influence decisions are clearly defined and laid down. Outlining the crucial

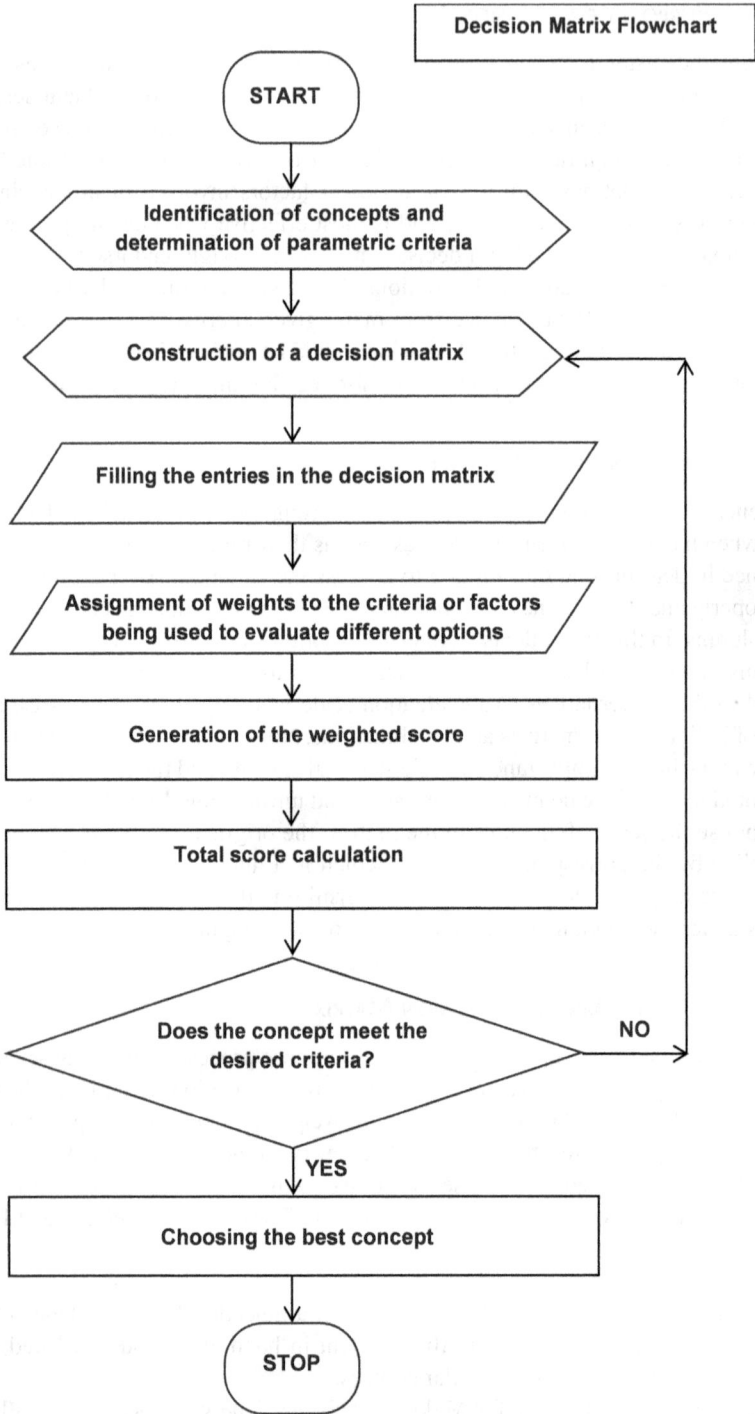

FIGURE 12.1 Flowchart of the decision matrix algorithm.

factors aids in focusing on the best decision while steering clear of subjectivity. Subjectivity refers to the unique and personal perspective, feelings, opinions, and experiences of an individual. These qualities shape the individual's understanding and interpretation of the world. They contrast with objective facts or universally agreed-upon truths.

iii. Creation of a Decision Matrix in a Grid Format: A grid is constructed to evaluate and compare the multiple considerations and options that are visible.

iv. Filling the Entries in the Decision Matrix. A predetermined scale is agreed upon for rating the considerations on a single benchmark. A 1–3 scale suffices if variations between options are limited. But a 1–5 or, 1–10 scale becomes necessary if there are several options.

v. Assignment of Weights to the Criteria: There is a hierarchy of importance of the criteria in the decision-making process. Some criteria or variables need to be prioritized over others. Therefore, numerical weights are assigned to each criterion. The allotted weights reflect their relative impacts on the decision in order to indicate the best option.

vi. Generation of the Weighted Scores of the Options: The more significant an option is, the higher its weight. As more important criteria are assigned higher values of weights, the weighted score assists in ranking the options for selecting the optimal choice.

vii. Calculation of the Total Score for Each Option: As a last action in the decision matrix, the total score is calculated. The total score provides a clear picture of the problem, allowing for the best decision to be made. It is easy to select the choice that best fits the desired criteria by merely looking at this picture.

12.2.4 ADVANTAGES OF DECISION MATRIX

As already mentioned in the beginning of Section 12.2, perhaps the most formidable task in everyday life is making correct decisions, especially those that affect an entire team and their performance. Various aspects must be observed. The technique aids in making difficult and complex decisions, particularly in cases of a team of people working together to achieve a target. When stakeholders participate in this process, several skewed viewpoints are involved. So, one cannot rely on everybody. The decision matrix promotes introspection among team members. It makes them analyze their decisions impartially. In such cases, the decision matrix technique is regarded as the most effective tool for making decisions for intricate situations plagued by perplexities.

Following a decision matrix approach, one is able to give precedence to tasks in order of their significance, one can construct arguments, and solve problems. Therefore, a decision matrix is a perfect instrument when one encounters several quantitative criteria. It helps in selecting among seemingly comparable solutions to dispel the confusion and bewilderment originating from the blurred similarities between them.

12.2.5   Disadvantages of Decision Matrix

The disadvantages arise from the likely errors introduced during the evolution of a decision matrix. The criteria alternatives for framing the decision matrix are chosen randomly. This arbitrariness means that there is no way to know whether the list is complete. It is likely that some important criteria have been overlooked. It is equally probable that some less important criteria are included or given more weight. The less important ones distract the decision-maker from making the right choice. Ultimately, the values that are attributed to guesses are derived from quantitative measurements. So, the decision matrix sometimes gives a deceptive and incredibly illusory appearance of being scientific without providing any quantitative measures.

12.2.6   Specialized Aspects of a Decision Matrix in Autonomous Robot Artificial Intelligence (AI)

The main aspects of a decision matrix for an autonomous robot are (Medrano-Berumen and İlhan Akbaş 2020):

   i. Collection of Data: The robot collects data from the sensors installed on it, e.g., the cameras, LiDAR, and ultrasonic sensors. Raw data is not fed into the decision matrix. Translation of the gathered data into relevant parameters contributing to decision-making for the problem is done for the preparation of decision matrix.
   ii. Matrix Structuring and Organization: The matrix is organized into a row-and-column format. As already mentioned, the rows of the matrix represent potential actions. The columns in the matrix signify different factors or criteria, e.g., the distance of the obstacle, the type of terrain, and the level of safety.
   iii. Assignment of Weighting Factors for Criteria in the Matrix: Each criterion in the matrix is assigned a weight. The weight assignment is based on the importance of the criterion for the decision in order that the considerations are correctly prioritized. After weights have been allocated, the vital considerations stand out clearly among the less influential ones.
   iv. Evaluation Process of Input Data for Decision-Making: The robot compares the sensor data to the matrix. It calculates a score for each possible action. The assistance of weighted criteria is sought for this calculation. Ultimately, the robot chooses the action with the highest score and implements the same.

12.2.7   Common AI Algorithms Used with Decision Matrices in Autonomous Robots

An instance of the use of a decision matrix is understood from the example of robot navigation in a chaotic setting, where a robot uses a decision matrix to choose the best path. Factors like distance to obstacles, type of terrain, and potential risks are preconceived and premeditated by the robot. The robot assigns higher weights to factors that significantly contribute to safe navigation. It utilizes several AI algorithms for acting independently, notably (IIoT World 2018):

i. Bayesian Inference Algorithm: This algorithm enables the robot to incorporate uncertainty into its analysis. The robot updates its beliefs about the environment based on new sensor data. This makes it possible for the robot to judge the situations more deeply. Hence, it can decide correctly and congruously, rendering the right verdicts.

ii. Reinforcement Learning (RL) Algorithms: These algorithms allow the robot to learn through a trial-and-error approach. In these algorithms, the robot is rewarded for its positive actions and penalized for any negative actions performed by it. Through this reward-and-penalty procedure, the robot's decision-making process undergoes continuous refinement over time.

iii. Deep Learning (DL) Algorithms: Decision trees and neural networks are used to train the robot in two different ways. First, the robot is trained on the critical factors that are most important for decision-making. Second, it is trained to assign weights to these factors, keeping their criticality levels in mind.

12.2.8 Chief Considerations When Using Decision Matrices

The following considerations ought to be given careful thought:

i. Management of Complexity of Decision Matrix: The larger the number of factors involved in making decisions, the higher is the complexity of the decision matrix. The more complications introduced, the greater is the need for careful design and optimization of the decision matrix.

ii. Examination of Unforeseen Cases: Unexpected events and potential setbacks are likely to be encountered during the use of a decision matrix. The matrix must, therefore, be designed to manage such issues by furnishing clear-cut answers.

iii. Explainability of the Decision-Making Process: The decision matrix should provide a perspicuous exposition of the process by which the robot reached a particular decision. The decision-making process must be crystal clear from the structure of the decision matrix and weight assignment considerations. It must be understandable with as little effort as feasible.

12.2.9 Decision Matrix for a Self-Driving Robotic Vehicle

A decision matrix for a self-driving vehicle is structured in a row-and-column format, similar to a normal matrix, with horizontal and vertical lines (Umbrello and Yampolskiy 2022). This structure presents an assessment of several possible courses of action in light of the environmental characteristics detected by the sensors of the vehicle. On the basis of this assessment, the vehicle can determine the safest and best-suited line of action in real time. Factors such as road conditions, status of traffic congestion, presence of pedestrians and potential roadblocks are envisaged and properly accommodated in the computation. Each cell in the matrix represents a possible decision. This decision is worked out from the combination of input parameters.

12.2.9.1   Key Elements of a Self-Driving Vehicle Decision Matrix

The main elements of a matrix are its constituent rows and columns. These are ascribed separate roles in the following ways:

i. Rows of the Matrix: The entries in the rows of the matrix illustrate many possible environmental conditions that the vehicle will face during traveling, e.g.,
   a. The road is clear or jammed,
   b. The vehicle is getting closer to an intersection,
   c. The vehicle is approaching a pedestrian crossing,
   d. The vehicle is nearing a spot where lane changing is needed, or
   e. There is a sharp turn ahead.
ii. Columns of the Matrix: The entries in the columns of the matrix show the possible courses of action that the vehicle will take, e.g., whether it will maintain its speed, apply acceleration, or press its brakes to slow down; other possibilities are that the vehicle will change lanes, turn left/right, or go straight.

Vital factors included in the decision matrix are as follows:

i. Real-time sensor information on a range of topics:
   a. Object Detection: Type of object, whether pedestrian, vehicle, or bicycle; distance of the object from the vehicle; speed of the object if it is moving,
   b. Lane Markings: Transverse/longitudinal, lane width, text/symbols, present/absent, clear/vague,
   c. Traffic Signals: Red (stop), yellow (caution), green (go), whether present/absent,
   d. Weather conditions, including whether it is sunny, cloudy, raining, foggy, windy, or snowing, along with temperature and humidity levels.
   e. Visibility or other atmospheric conditions limiting the sightline, whether clear or poor,
   f. Road Geometry: These could be specified in different forms, e.g., a bend or curve in the road or a sharp turn where one cannot see around the corner; a sloppy/bumpy road or a smooth, flat road with no incline.
ii. Vehicle State: Current speed of the vehicle, its acceleration and the angular direction of its steering determine the state of the vehicle
iii. Ethical Considerations: They must be categorically complied with. Notable among them are:
   a. Reduction of the chances of injury to the passengers and driver of the vehicle, as well as other road users, must be rigorously followed.

b. Road users vulnerable to unintentional injuries by accidents, e.g., children, pedestrians, cyclists, etc., must be prioritized.

12.2.9.2 Decision-Making Process

The process from sensory input to actuation of the vehicle mechanism consists of:

i. Sensor Inputs: The self-driving vehicle uninterruptedly seeks data from its sensors about the present status of the surrounding environment.
ii. Analysis of Sensor Data: The system processes the sensor data. Relevant information for driving a vehicle is extracted. This includes the location and movement of other vehicles, the presence of any pedestrians or perambulators on/near the road, and the indications of markings on the road.
iii. Evaluation of Decision Matrix: The data is analyzed. Based on the analysis, the system looks up at the applicable cell in the decision matrix. From the cell, it determines the best action to be initiated.
iv. Execution of Action by the Vehicle: The vehicle executes the selected action. The action could involve braking or accelerating the vehicle, or changing lanes, as required during the finalization of the decision.

12.2.9.3 Challenges in Vehicle Decision Matrix Design

Many difficult situations arise during the designing of a decision matrix, viz.

i. Handling Complex Scenarios on Roads: There are occasions that require negotiating rare or unexpected situations regarding which no clear rules, transparent policies, and defined guidelines exist. What happens when a person talking on a mobile phone and carelessly crossing the road without looking at the traffic suddenly comes in front of the vehicle? Then emergency braking is the only solution.
ii. Dealing with Borderline Cases: Suitable answers and responses to abstruse or edge situations are not described. These responses require detailed discussion and clarification.
iii. Getting Caught in Ethical Dilemmas: These quandaries arise when decisions are to be made in grave, life-threatening situations where there is no definite outcome. A decision is to be made from multiple options. None of these options might be completely morally right, thus forcing the robot to choose between conflicting ethical principles. They might potentially cause harm to different parties. Negative consequences are likely to occur regardless of the decision made. Careful attention and amendments are therefore necessary.

12.3 BUG ALGORITHM

While a decision matrix helps robots make choices based on various factors, the bug algorithm is a path-planning strategy that enables them to reach their destination. The bug algorithm is an effective and efficient method for autonomous robots to avoid obstacles on their paths (Yufka and Parlaktuna 2009; McGuire et al. 2019).

Its utility varies with context. It is mostly utilized by autonomous robots that have local sensor information to guide them toward a target goal. It is particularly useful when the robot does not have a complete map of its environment beforehand. It is highly useful in scenarios such as indoor robotics, where it controls robots for cleaning jobs and those that move around in crowded areas. Therefore, circumstance-based benefits for the algorithm can be availed by the user.

The bug algorithm is a path-planning algorithm based on the principle of a robot following a wall. It enables a robot to navigate effectively around obstacles in an environment by essentially following the wall of an obstacle. The robot follows the wall until it reaches a point on the boundary of the obstacle that is closest to its goal (in the simplest Bug0 variant of the algorithm). Then it continues its motion toward the goal until it encounters another obstacle. The wall-following robot repeats the process until it reaches its prescribed destination (Buniyamin et al. 2011; Liu 2024).

We shall look into further details about the departure point of the robot because it depends on the particular variant of the bug algorithm. So, we shall talk about this point further when we come to the discussion of variants.

12.3.1 Main Features of the Bug Algorithm

What are the characteristics of the bug algorithm? Let us give a rejoinder to this query.

   i. Basic principle of the Algorithm: Suppose an obstacle is detected by a robot while it is moving toward the goal. Immediately upon detecting the obstacle, the robot begins to follow the edge of the obstacle. How long does the robot do so? The robot follows the edge of the obstacle until it reaches a point on the boundary that is closest to the goal (in the Bug0 variant). Thereafter, a resumption of the robot's movement toward the goal takes place.
  ii. Local Sensing of Obstacle and Non-requirement of Complete Environmental Map: A striking feature of the bug algorithm is that, unlike some other path-planning algorithms, the bug algorithm does not require a pre-existing map of the environment. What is the significant advantage of this feature? The feature makes it suitable for situations where a robot needs to navigate in unknown or dynamically changing environments. Then the robot relies solely on local sensor data, like proximity sensors, to detect obstacles. Such reliance on sensor readings makes it possible to dispense with the need for a pre-drawn map of the environment. The sensor data is its sole guide.
 iii. Variants of the Algorithm: Several variations of the bug algorithm have been developed. Examples of variants are Bug0, Bug1 and Bug2. These variants have varying levels of elaborateness, involvement and memory requirements. They are designed to deal with specific challenges. They improve algorithm efficiency thereby allowing its adaptation to specialized needs.

Bug0 variant is the most basic type in the series of bug algorithm versions. In this variant, the robot simply follows the boundary of the obstacle in one direction.

It moves in the clockwise direction until it reaches the closest point to the goal. In the Bug1 variant, the robot first completes a full circle around the obstacle. Then it exits from the location on the obstacle's boundary that is closest to the goal. In the Bug2 variant, the robot follows the contour of the obstacle but says goodbye to it no sooner than it can move directly toward the goal along the line connecting the current position of the robot to the goal. The Bug0 algorithm does not guarantee that the robot will reach the goal in all scenarios, whereas the Bug1 algorithm does, albeit at the cost of inefficiency. Bug2 results in shorter distances of traveling than Bug1. However, it can be inefficient in some situations. Let us take up the Bug1 algorithm.

12.3.2 STEPS OF THE BUG1 ALGORITHM

Taking the Bug1 version as a case study, let us survey its main steps. The steps of the Bug1 algorithm (Figure 12.2) are as follows (Kurtipek 2020): robot movement, obstacle detection, activation of obstacle avoidance, and following the boundary of the obstacle (Kurtipek 2020). If the robot does not reach the point on the obstacle boundary nearest to the goal, the obstacle avoidance behavior is repeated. If it reaches that point, it continues moving toward the goal. At this stage, it is examined whether the robot has reached the goal. If YES, the algorithm is stopped. If NO, the algorithm returns to the step from which the robot's movement started.

i. Movement of the Robot toward the Goal: The robot starts by moving in a straight line. This line points directly in the direction the intended goal position.
ii. Detection of Obstacle by the Robot: As soon as the robot detects an obstacle on its path, it initiates the obstacle avoidance behavior built into its machinery.
iii. Movement of the Robot while Following the Obstacle Boundary: The robot follows the edge of the obstacle in a chosen direction, which is usually clockwise. The robot continues moving until it reaches the point on the boundary that is closest to the goal. This persistence of motion ensures that the robot can move directly toward the goal again.
iv. Storage of the Closest Point to the Goal by the Robot: While following the obstacle, the robot stores in its memory the coordinates of the point on the boundary that is closest to the goal. It thus remembers the specific location on the edge of the obstacle that is located at the shortest distance away from the goal. The remembrance of this point is necessary for the robot, as it marks the correct position for the robot to later navigate around the obstacle. The robot will access it when deciding to leave the obstacle and resume its straight-line path toward the goal. Hence, this point serves as the departure point or leave point for the robot.
v. Return of the Robot to the Closest Point to the Goal: Once the robot has navigated around the obstacle, it recalls the leave point. Thus, it moves back to the previously stored closest point on the obstacle boundary.
vi. Resumption of Robot Movement toward the Goal: From the closest point, the robot continues moving toward the goal. The motion persists, and the robot keeps moving until it encounters another obstacle on the way. Then the obstacle avoidance steps stated above are repeated.

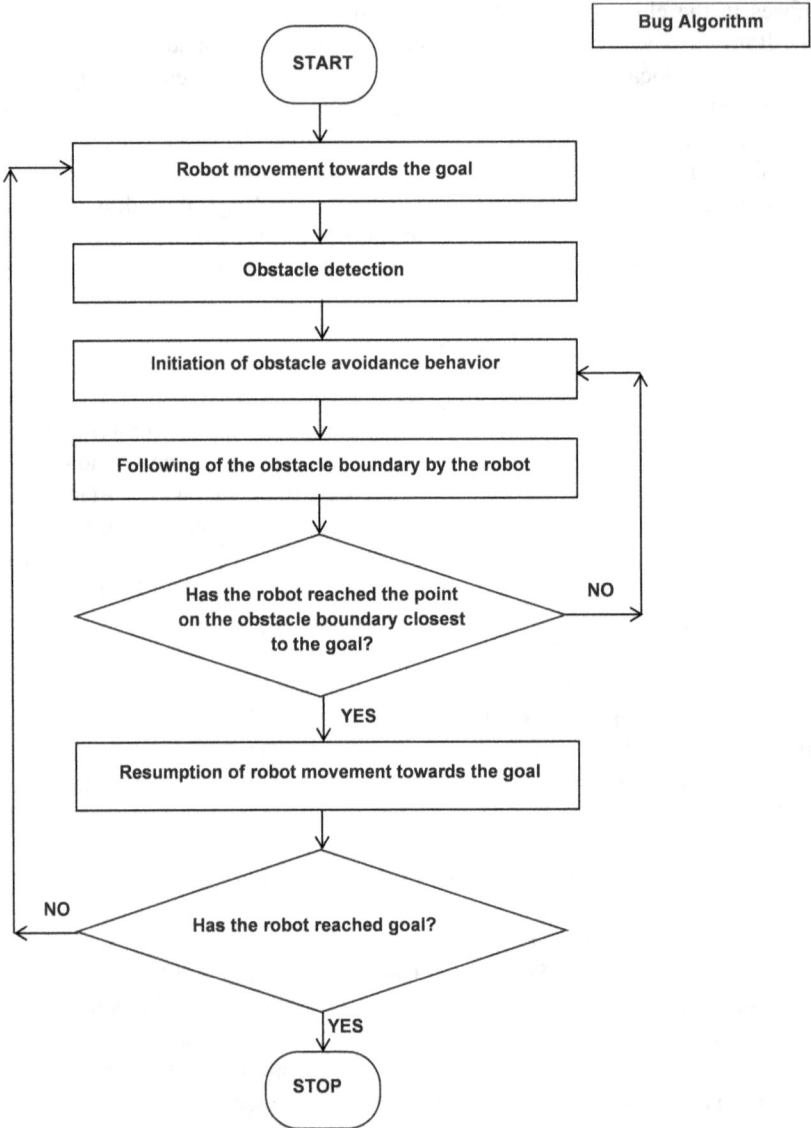

FIGURE 12.2 The bug algorithm.

12.3.3 APPLICATIONS OF THE BUG ALGORITHM

The bug algorithm streamlines the daily activities of many types of robots, a few of which captivate our attention:

  i. Robot Vacuum Cleaners: These robots navigate around furniture, other domestic items, or moving persons in a room to reach the desired area to be cleaned. The furniture, domestic items, and persons act as obstacles to robot movements.

ii. Warehouse Robots: These robots navigate a passageway in a warehouse while dodging obstructions like bundles, bags, baskets, containers, and other packages in the warehouse.

iii. Service Robots: These robots circumnavigate a home environment with fixed furniture and people. They move around in the home providing the services to people, e.g., healthcare, hospitality, and logistics.

iv. Autonomous Vehicles Negotiating Complex Environments: These robots are useful in situations where a full map of the robot's surroundings is not available to seek guidance for its locomotion. An example is a robot navigating through a crowded outdoor space.

12.3.4 ADVANTAGES OF THE BUG ALGORITHM

Recognizing and utilizing the advantages of the bug algorithm drive innovation and progress.

i. Comfort of Comprehension and Application: It is an easy-to-understand and easy-to-implement algorithm on robots having limited computational power. Its straightforward nature makes it a good starting point for robot navigation.

ii. Robustness in Dealing with Unforeseen Conditions: It is capable of handling unexpected obstacles in real time.

iii. Low computational Cost: It does not require computationally intensive and expensive operations.

12.3.5 DISADVANTAGES OF THE BUG ALGORITHM

Understanding potential disadvantages of the algorithm helps us anticipate problems well in advance and take proactive steps to avoid them.

i. Non-optimality of Robot Operation: The robot may wander about irregularly. During its dawdling and rambling, the robot may take longer paths than those obtained by optimal solutions.

ii. Arrogance of Following Boundaries on Getting Stuck in Certain Situations: The robot gets stuck in some environments. Depending on the obstacle layout, the robot may end up traveling a considerable distance by following the boundaries of obstacles. Such insistence on adhering to the boundaries renders the exercise inefficient and futile, making it an unproductive and wasteful activity.

12.4 VECTOR FIELD HISTOGRAM ALGORITHM

Another algorithm used for path planning in autonomous robotics is the vector field histogram (VFH) algorithm, a real-time obstacle avoidance algorithm in robotics (Borenstein and Koren 1991). It is used by autonomous robots predominantly for local path planning. It differs in approach from the bug algorithms, which follow obstacle boundaries until reaching the goal or a point close to it. Unlike the bug algorithms, the VFH algorithm identifies obstacle-free paths from a polar histogram of sensor data.

For robot path planning, the VFH algorithm calculates the steering directions based on sensor data supplied in the form of range readings to navigate around obstacles. The size of the robot and its turning radius are taken into account during navigation. In effect, the robot is securely guided toward a desired target direction while eschewing collisions with objects on the track.

12.4.1 CHIEF POINTS ABOUT THE VFH ALGORITHM

The main ideas of the VFH algorithm are (Chen et al. 2019):

  i. Sensor Input: The algorithm primarily relies on range sensors, such as ultrasonic or LiDAR. Using these sensors, the obstacles around the robot are detected. From the information gathered by its sensors, a realistic representation of the robot's environment is evolved.
 ii. Data Representation by Histogram: The sensor data is converted into a 2D polar histogram or density heatmap in which data points specified by their (x, y) rectangular Cartesian coordinates are grouped into bins in polar coordinates (r, θ) where the radius r represents the distance of the data point from the origin and the angle θ represents its angle. Such grouping of data points in terms of radius r and angle θ produces a circular or radial grid. Application of an aggregation function to each bin, like counting the number of points in each bin or summing a value associated with each point, produces a density plot of the data distribution where the color or intensity of each bin represents the aggregated value. In the 2D polar histogram thus generated, each cell of the histogram represents a direction and distance from the robot. Hence, it enables easy visualization of potential obstacles that the robot may encounter in different directions as it moves toward the goal.
iii. Logic for Obstacle Avoidance: By analyzing the histogram, the algorithm identifies the directions with minimal obstacle density. These directions are called openings. Then it performs calculations to find a steering direction that directs the robot toward the most suitable open space to avoid collisions.
 iv. Considerations for Robot Geometry: VFH takes into account the physical dimensions of the robot and its turning radius. Therefore, the calculated steering commands are easily abided by the robot to execute its operation.

12.4.2 MAIN STEPS OF THE VFH ALGORITHM

The principal steps of the VFH algorithm for autonomous robots are (Figure 12.3): sensor data acquisition by sensing the environment surrounding the robot using a range of sensors; creating a polar histogram graphically representing obstacle density around the robot in polar coordinates that is subdivided into slices (single wedge-shaped sections of the circular graph in the form of angular sectors or bins representing specific directions or angles around the robot with the heights of the

VFH Algorithm

START

Sensor data acquisition

Polar histogram creation

Obstacle detection by identification of high-density points in the histogram

Steering angle calculation for obstacle-free direction

Robot movement in obstacle-free direction

NO

Has the robot reached goal?

YES

STOP

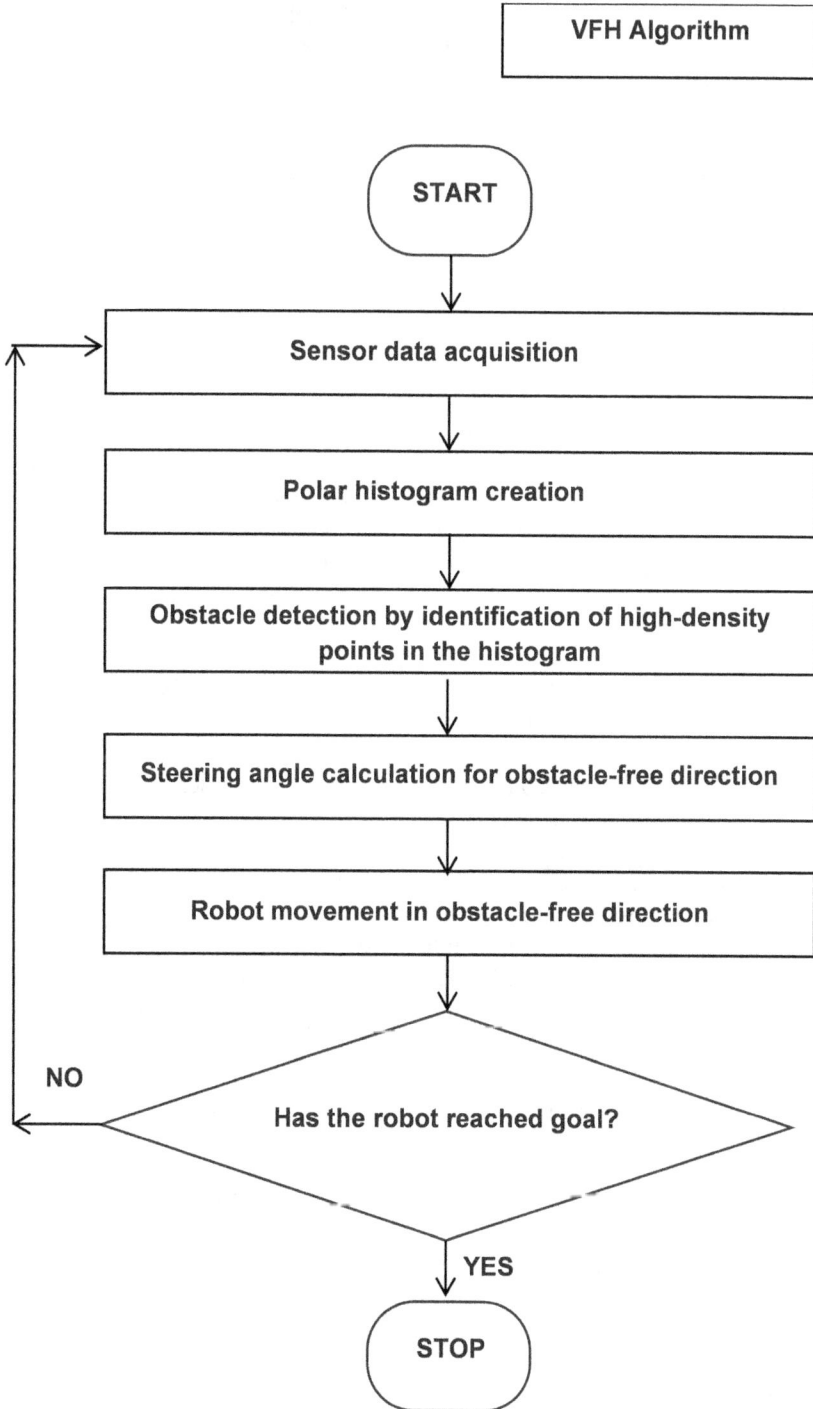

FIGURE 12.3 The VFH algorithm.

slices showing the likelihood of encountering obstacles in those directions); identifying high-density points in the histogram representing obstacles; recognizing valleys, defined as low-obstacle-density areas in the histogram; selecting the steering direction corresponding to the most desirable valley to avoid obstacles; and moving toward the obstacle-free target direction (Babinec et al. 2012; Kumar and Kaleeswari 2016; Alagić et al. 2019). If the robot reaches the goal, the algorithm is stopped. If it does not reach the goal, the algorithm goes back to the stage of acquiring data from sensors. All these steps are carried out within a two-stage data reduction process to calculate the desired control commands for the robot, as detailed in point (iii) in the description of operating procedure given below:

i. Acquisition of Data about the Robot and Its Environment by Sensors: The robot is equipped with range sensors like sonar or LiDAR. The robot's sensors perform an all-round scanning operation to collect distance information from the surrounding environment. The information is about any obstacles to the robot movement present in its neighborhood.

ii. Creation of a Polar Histogram: The range data from the sensors is transformed into a 2D polar coordinate system centered on the robot forming a polar histogram. Each cell in the histogram represents a direction and distance from the robot. It corresponds to a specific angle around the robot, creating a circular view of the surrounding environment. The value in each cell typically represents the number of obstacles detected in that direction. It is an indicator of obstacle density in that direction, with higher values showing the presence of more obstacles.

iii. Detection of Obstacles: The algorithm identifies high-density areas in the histogram. The high-density areas are the regions in which potential obstacles are found.

    a.   Data Reduction (Stage A): Obstacle density values in the histogram are smoothed to reduce noise. Hence, a more continuous representation is created. This step involves filtering or averaging neighboring cells.

    b.   Data Reduction (Stage B): The algorithm searches for openings in the histogram. These are directions with low obstacle density. Valleys in the histogram are identified, which correspond to directions with minimal obstacle density. The valley that is closest to the desired target direction is selected while considering the robot's current heading.

iv. Calculation of the Steering Command: Based on the identified openings and the chosen valley, the algorithm calculates the necessary steering angle giving the direction to navigate the robot toward the obstacle-free path.

12.4.3 VARIANTS OF VFH ALGORITHM

The basic VFH algorithm has been improved with several variants:

i. VFH+: It is an improved version of VFH algorithm that takes into account the robot's physical size and incorporates a cost function for more refined direction selection. It has additional features such as better handling of narrow openings and improved robustness to sensor noise (Ulrich and Borenstein 1998).

ii. VFH*: It is a further enhanced version of VFH, embellished with additional considerations beyond VFH⁺, like a look-ahead verification mechanism to anticipate potential future collisions of the robot. Such refinement of the algorithm provides more reliable robot path planning (Ulrich and Borenstein 2000).

12.4.4 APPLICATIONS OF THE VFH ALGORITHM

The VFH algorithm is primarily used in autonomous robotics for real-time obstacle avoidance in dynamic environments. It allows the robots to navigate around obstacles by calculating a preferred steering direction based on sensor readings, particularly from range sensors like LiDAR or SONAR. The need for detailed environmental maps is avoided. These qualities make it an ideal algorithm for applications like mobile robot navigation, cleaning robots, and autonomous vehicles moving in muddled and disarranged spaces.

  i. Navigation of Mobile Robots: The VFH algorithm is commonly employed in mobile robots to navigate environments with numerous obstacles. It helps by providing steering directions to avoid collisions while reaching a desired destination.
 ii. Controlling the Robots Used for Indoor Cleaning: Cleaning robots utilize the VFH algorithm to detect furniture, walls, and other obstacles in a room. The detection of obstacles enables them to maneuver around objects while cleaning the room efficiently.
iii. Guidance of Autonomous Vehicles in Cluttered Environments: The VFH algorithm helps autonomous vehicles to navigate around obstacles like pedestrians, pallets, or parked cars in scenarios such as warehouses or crowded streets.
 iv. Performing Robotic Manipulation Tasks: The VFH algorithm is used for obstacle avoidance during robotic arm movement for robots engaged in manipulation tasks. Thus, it prevents their collisions with surrounding objects.

12.4.5 ADVANTAGES OF THE VFH ALGORITHM

Knowing the advantages leads to greater success and positive outcomes when using the algorithm.

  i. Real-Time Processing Capability: VFH is a computationally efficient algorithm. It is designed for fast obstacle avoidance in real-time applications. The underlying reasons are its quick data reduction process and steering direction calculation, critical for dynamic environments.
 ii. Flexibility of Operation: VFH is adaptable to various robot geometries and sensor configurations.

iii. Simplified Practical Realization: The concept of using a histogram for obstacle representation is relatively straightforward to understand and implement without difficulty. It can be integrated with various robot control systems.

iv. Use of Sensor-Based Navigation: The algorithm relies on sensor data. This advantage makes it adaptable to changing environments without requiring a pre-built map.

v. Following a Robot-Centric Approach: The algorithm takes into account the robot's physical properties in order to proclaim realistic steering commands.

12.4.6 LIMITATIONS OF THE VFH ALGORITHM

Understanding limitations of the algorithm helps us develop resilience and the ability to overcome difficulties.

i. Restriction to Local Planning Only: VFH is primarily a local planner, focusing on immediate obstacle avoidance. It may not lead to the most optimal path in complex environments. It is not designed for long-term global path planning. Furthermore, it may not always account for the dynamic constraints of the robot, such as its maximum turning radius.

ii. Possibility of Becoming Stuck: VFH might not always find a clear path in complex environments. It gets trapped in tight spaces. It may wiggle through narrow passages or environments characterized by closely located obstacles.

iii. Potential for Oscillatory Behavior: VFH might lead to oscillations in tight spaces as the robot continuously struggles to avoid obstacles.

12.5 GENERALIZED VORONOI DIAGRAM ALGORITHM

Like the VFH algorithm, the generalized Voronoi diagram (GVD) is a crucial member of the family of algorithms employed in autonomous robotics for path planning and obstacle avoidance. While the VFH is a local, sensor-based algorithm, the GVD algorithm follows a global, graph-based approach. While VFH is a real-time algorithm, GVD is a pre-planned algorithm. The VFH algorithm is based on an implicit sensor-based environmental model. The GVD algorithm uses an explicit, map-based environmental model. The VFH algorithm is less complex. The GVD algorithm is relatively more complex. The VFH algorithm provides simple navigation in a dynamic environment. The GVD algorithm is suited to complex navigation in a static environment.

12.5.1 FUNCTIONAL MECHANISM OF THE GVD ALGORITHM

Let us now explain the functional details of the GVD algorithm. A GVD is a computational geometry structure used in autonomous robotics (Garrido and Moreno 2015; Li et al. 2020; Chi et al. 2022). It is a roadmap that provides all possible path homotopy classes in an environment containing obstacle regions, offering maximum clearance from these regions. 'Homotopy' is a concept from topology (the investigation of the fundamental properties of a robot's configuration space and their impact on robot motion planning and control) for the classification of trajectories that a robot can follow. This classification is done by taking into consideration which paths can undergo continuous deformation into one another without encountering obstacles. The homotopy classes constitute a way to categorize robot trajectories based on their ability to be continuously deformed into one another without intersecting obstacles. Thereby they significantly reduce the search space for finding a valid path.

The GVD algorithm is a path-planning algorithm that leverages a GVD to navigate a robot through an environment. For smooth robot navigation, it divides an environmental space into regions based on the distance to multiple obstacles. Each region represents the area closest to a specific point, such as a robot's potential position. The robot's shape and movement constraints are duly considered. Not only is the center point of the robot kept in sight, but also its full geometry and possible orientations. By keeping an eye on the robot's center, navigation of non-point-like robots is rendered possible. Therefore, large vehicles or robots with articulated limbs are automatically taken care of. In this manner, the algorithm defines the safe zones for a robot to navigate through successfully without meeting any obstruction. In brief, the algorithm accomplishes efficient path planning and obstacle avoidance by creating a roadmap of safe paths within a complex environment based on the identification of the safest or most accessible areas and routes with maximum clearance from obstacles, and delineating safe corridors between obstacles, allowing for an efficient and collision-free movement of the robot.

12.5.2 MAIN STEPS OF THE GVD ALGORITHM

A review of the main steps clarifies how the algorithm performs its operation. Figure 12.4 presents the steps of this algorithm. These steps are (Özcan and Yaman 2019; Lee et al. 2023): representing the obstacle and building its environmental map, GVD construction, distance metric selection, Voronoi cell generation, graph production, path planning, and deciding about the optimality of the determined path. If the path found is optimal, the algorithm stops. Otherwise, it reverts back to path planning and continues unless satisfactory results are obtained. Details of the steps are given below.

   i. Representation of the Obstacle: Obstacles are defined as polygons or other geometric shapes. Their full extent in the environment is considered.
  ii. Mapping of the Environment: The robot builds a map of its environment. Sensors like LiDAR are used for creating a representation of obstacles.
 iii. Construction of GVD: Based on the map, the algorithm calculates the GVD. The GVD essentially creates a network of interconnected points. This network represents the most accessible paths within the environment.
  iv. Selection of the Distance Metric: GVDs utilize a distance metric that takes into account the robot's geometry and orientation. Instead of the simple

GVD Algorithm

START

Obstacle representation

Building the environmental map

Construction of GVD

Selection of distance metric

Generation of the Voronoi cell

Production of the graph structure

Robot path planning by A* search algorithm

Is the path to reach the goal optimal?

NO

YES

Choosing the optimal path

STOP

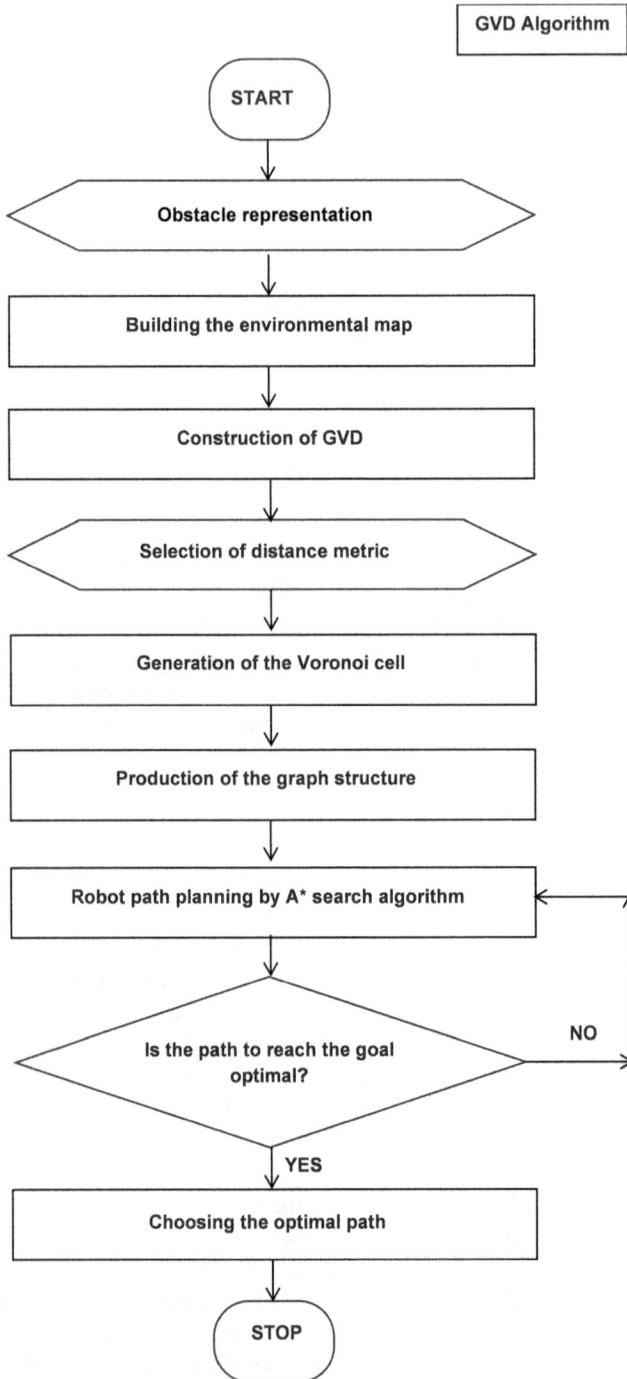

FIGURE 12.4 The GVD algorithm.

Euclidean distance, it often uses the Minkowski distance. The Minkowski distance, a generalization of several well-known distance measures, is calculated by adding the absolute differences between two points raised to a power or parameter p. The value of the power, or parameter p determines the type of distance metric used. Various values of p represent different distance measures. The $p = 1$ value represents the Manhattan distance, $p = 2$ gives the Euclidean distance and $p = \infty$, the infinite norm.

v. Generation of Voronoi Cell: The algorithm calculates the set of points closer to a specific robot configuration than any other. Hence, it creates a Voronoi cell for each potential robot pose.

vi. Construction of Graph: The boundaries of the Voronoi cells are connected to form a graph structure. The graph represents the accessible paths within the environment.

vii. Planning of the Robot Path: After the GVD is generated, the robot uses a simple graph search algorithm like A* to find the optimal path from its current location to the desired goal. The robot navigates along the GVD points to avoid obstacles.

12.5.3 Applications of the GVD Algorithm

Applications of the GVD algorithm make routine robot tasks easier. Let us elaborate on these tasks.

i. Robot Path Planning: Collision-free paths are generated for robots navigating through environments filled with numerous obstacles.

ii. Robot Motion Planning: Feasible motions for robots with non-holonomic constraints are calculated. The non-holonomic constraints refer to the path-dependent constraints on the velocity of a robot's mechanical system that are not derivable from position constraints, e.g., those faced with wheeled robots with limited turning radius.

iii. Exploration of Unknown Environments: The GVD algorithm guides autonomous robots to explore unknown environments by identifying areas with high information gain indicating a more effective splitting of data.

iv. Autonomous Navigation in Indoor Environments: Robots, such as cleaning robots or delivery bots, can utilize GVDs to navigate around furniture and other obstacles.

v. Coordination of Multi-robot Teams: The GVD algorithm is utilized to coordinate the movement of multiple robots by providing a shared understanding of the environment, considering their mutual interference and ensuring safe distances.

vi. Industrial Robotics: The GVD algorithm is used for robot path planning in manufacturing settings with layouts of various types.

12.5.4 Advantages of the GVD Algorithm

The benefits of the GVD algorithm enable better outcomes in robot navigation, giving it a competitive edge over other algorithms:

i. Efficient Planning of Robot Path and Selection of Route: The GVD algorithm provides a high-level representation of the environment. Such a representation allows for faster path calculation compared to raw sensor data.
ii. Efficient Route Selection: The GVD structure allows for quick identification of the most accessible paths between start and goal points. The robot's size and limitations are taken into consideration, which is an obvious advantage.
iii. Robustness to Complex Environments: The GVD algorithm can handle a variety of obstacle layouts and is particularly useful for navigating complex environments with multiple obstacles or tight spaces where traditional Voronoi diagrams may not be sufficient.
iv. Avoidance of Collisions: The GVD algorithm enables efficient path planning to avoid collisions by identifying the closest point to an obstacle for a given robot configuration.
v. Flexibility to Different Situations: The GVD algorithm can be easily adapted to different robot sizes and motion constraints.

12.5.5 LIMITATIONS OF THE GVD ALGORITHM

Limitations of the GVD algorithm are important for ensuring risk avoidance during its application.

i. Difficulties faced in Dynamic Environments: The GVD algorithm needs to be recalculated frequently if the environment changes significantly, which can be a troublesome and tedious activity.
ii. High Computational Cost: Generating a GVD can be computationally expensive, especially in large or highly dynamic environments.

12.6 DISCUSSION AND CONCLUSIONS

This chapter dealt with the algorithms used for designing autonomous robots (Table 12.1). The most well-known example of the autonomous robot is the self-driving robot, a self-sufficient decision-making system which processes data inputs from various sensors, and models it using DL algorithms (Mogaveera et al. 2018; Reda et al. 2024). A perception, localization, prediction and decision-making approach is adopted for path planning and motion control.

A dataset of path following behavior is constructed by manually driving a robot along steep mountain trails and recording video frames from the camera mounted on the robot along with the corresponding motor commands (Hwu et al. 2017). This dataset is used to train a deep convolutional neural network. The neural network module, which was mounted on the robot and powered by the robot's battery, leads to a self-driving robot that could successfully traverse a steep mountain path in real time.

After consideration of robotic speech, vision, emotional intelligence, robot task and motion planning, and autonomous robots in the foregoing chapters, all of which involve a single robot, it is high time now to divert our attention to a collection of robots working together as a swarm. Union is strength. When robots are organized as a disciplined team of workers, they can perform tasks involving heavy loads and toxic substances, thereby preventing many accidents and saving human lives, time,

TABLE 12.1
Takeaways from This Chapter at a Glance

Sl. No.	Takeaway	Explanation
1	Summary	This chapter described four algorithms used for making autonomous robots, namely the decision tree, the bug, VFH, and GVD algorithms. These four algorithms operate with different approaches to obstacle avoidance and path calculation.
2	Decision tree algorithm	The general purpose of a decision matrix and the procedure for its creation were explained, together with steps in making it, indicating its advantages/disadvantages. Specialized aspects of a decision matrix in autonomous robot AI were dealt with. Common AI algorithms used with decision matrices in autonomous robots are Bayesian inference, reinforcement learning, and neural networks. Chief considerations when using decision matrices are complexity management, examination of unforeseen cases, and the explainability of the decision-making process. Key elements of a self-driving vehicle decision matrix, including its decision-making process and challenges involved, were outlined.
3	Bug algorithm	The bug algorithm is a simple, reactive obstacle avoidance strategy where a robot follows the edge of an obstacle until it can resume its path toward the goal.
4	VFH algorithm	The vector field histogram is a real-time motion planning algorithm that utilizes a polar histogram to represent the density of obstacles in different directions, thereby identifying obstacle-free directions for steering a robot based on sensor data.
5	GVD algorithm	The generalized Voronoi diagram algorithm works by partitioning a space into regions based on proximity to multiple seed points, such as landmarks and waypoints, for planning a path that finds the most efficient route around obstacles.
6	Keywords and ideas to remember	Decision Matrix in autonomous robot AI and for a self-driving robotic vehicle, bug algorithm, vector field histogram algorithm, generalized Voronoi diagram algorithm

and money. The concluding three chapters of the book will explore the opportunities, prospects, and technical snags of swarm robotics.

REFERENCES AND FURTHER READING

Alagić, E., J. Velagić and A. Osmanović. 2019. Design of Mobile Robot Motion Framework Based on Modified Vector Field Histogram, *International Symposium ELMAR*, Zadar, Croatia, 23–25 September, pp. 135–138.

Babinec A., M. Dekan, F. Duchoň and A. Vitko. 2012. MMaMS, Modifications of VFH navigation methods for mobile robots, *Procedia Engineering*, Vol. 48, pp. 10–14.

Borenstein J. and Y. Koren. 1991. The vector field histogram-fast obstacle avoidance for mobile robots, *IEEE Transactions on Robotics and Automation*, Vol. 7, 3, pp. 278–288.

Buniyamin N., W. A. J. Wan Ngah, N. Sariff and Z. A. Mohamad. 2011. Simple local path planning algorithm for autonomous mobile robots, *International Journal of Systems Applications, Engineering & Development*, Vol. 5, 2, pp. 151–159.

Chen W., N. Wang, X. Liu and C. Yang. 2019. VFH Based Local Path Planning for Mobile Robot, 2019 *2nd China Symposium on Cognitive Computing and Hybrid Intelligence (CCHI)*, Xi'an, China, 21–22 September, pp. 18–23.

Chi W., J. Wang, Z. Ding, G. Chen and L. Sun. 2022. A reusable generalized Voronoi diagram-based feature tree for fast robot motion planning in trapped environments, *IEEE Sensors Journal*, Vol. 22, 18, pp. 17615–17624.

Garrido S. and L. E. Moreno. 2015. Mobile Robot Path Planning Using Voronoi Diagram and Fast Marching. In: Luo, Z. (Ed.), *Robotics, Automation, and Control in Industrial and Service Settings*, IGI Global Scientific Publishing, Hershey, PA, pp. 92–108.

Hwu T., J. Isbell, N. Oros and J. Krichmar. 2017. A Self-Driving Robot Using Deep Convolutional Neural Networks on Neuromorphic Hardware, *International Joint Conference on Neural Networks (IJCNN)*, Anchorage, AK, USA, 14–19 May, pp. 635–641.

IIoT World. 2018. Machine learning algorithms in autonomous driving, https://www.iiot-world. com/artificial-intelligence-ml/machine-learning/machine-learning-algorithms-in-autonomous-driving/#:~:text=to%20avoid%20this.-,Decision%20Matrix%20 Algorithms,boosting%20(GDM)%20and%20AdaBoosting

Kumar J. S. and R. Kaleeswari. 2016. Implementation of Vector Field Histogram Based Obstacle Avoidance Wheeled Robot, *International Conference on Green Engineering and Technologies (IC-GET)*, Coimbatore, India, 19 November, pp. 1–6.

Kurtipek S. 2020. Robot motion planning: Bug algorithms, https://medium.com/@sefakur-tipek/robot-motion-planning-bug-algorithms-34cf5175ab39#:~:text=Bug%2D1%20 Algorithm:,It%20is%20a%20complete%20algorithm

Lee J. T., T.-W. Kang, Y.-S. Choi and J.-W. Jung. 2023. Clearance-based performance-efficient path planning using generalized Voronoi diagram, *International Journal of Fuzzy Logic and Intelligent Systems*, Vol. 23, 3, pp. 259–269.

Li L., X. Zuo, H. Peng, F. Yang, H. Zhu, D. Li, J. Liu, F. Su, Y. Liang and G. Zhou. 2020. Improving autonomous exploration using reduced approximated generalized Voronoi graphs, *Journal of Intelligent and Robotic Systems*, Vol. 99, pp. 91–113.

Liu K. 2024. A comprehensive review of bug algorithms in path planning, *Applied and Computational Engineering*, Vol. 33, 1, pp. 259–265.

McGuire K. N., G. C. H. E. de Croon and K. Tuyls. 2019. A comparative study of bug algorithms for robot navigation, *Robotics and Autonomous Systems*, Vol. 121, p. 10326, https:// doi.org/10.1016/j.robot.2019.103261

Medrano-Berumen C. and M. İlhan Akbaş. 2020. Validation of decision-making in artificial intelligence-based autonomous vehicles. *Journal of Information and Telecommunication*, Vol. 5, 1, pp. 83–103.

Mogaveera A., R. Giri, M. Mahadik and A. Patil. 2018. Self Driving Robot Using Neural Network, *2018 International Conference on Information, Communication, Engineering and Technology (ICICET)*, Pune, India, 29–31 August, pp. 1–6.

Özcan M. and U. Yaman. 2019. A continuous path planning approach on Voronoi diagrams for robotics and manufacturing applications, *Procedia Manufacturing*, Vol. 38, pp. 1–8.

Ralfs L., N. Hoffmann and R. Weidner. 2022. Approach of a Decision Support Matrix for the Implementation of Exoskeletons in Industrial Workplaces. In: Schüppstuhl T., K. Tracht and A. Raatz (Eds.), *Annals of Scientific Society for Assembly, Handling and Industrial Robotics 2021*, Springer, Cham, pp. 165–176.

Reda M., A. Onsy, A. Y. Haikal and A. Ghanbari. 2024. Path planning algorithms in the autonomous driving system: A comprehensive review, *Robotics and Autonomous Systems*, Vol. 174, 104630, pp. 1–45.

Ulrich I. and J. Borenstein. 1998. VFH+: Reliable Obstacle Avoidance for Fast Mobile Robots, *Proceedings. 1998 IEEE International Conference on Robotics and Automation (Cat. No.98CH36146)*, Leuven, Belgium, 20 May, Vol. 2, pp. 1572–1577.

Ulrich I. and J. Borenstein. 2000. VFH*: Local Obstacle Avoidance with Look-Ahead Verification, *Proceedings 2000 ICRA. Millennium Conference. IEEE International Conference on Robotics and Automation. Symposia Proceedings (Cat. No.00CH37065)*, San Francisco, CA, USA, 24–28 April, Vol. 3, pp. 2505–2511.

UMass. 2025. Decision-making process: 7 steps to effective decision making, https://www.umassd.edu/fycm/decision-making/process/

Umbrello S. and R. V. Yampolskiy. 2022. Designing AI for explainability and verifiability: A value sensitive design approach to avoid artificial stupidity in autonomous vehicles, *International Journal of Social Robotics*, Vol. 14, pp. 313–322.

Venkata Rao R. and K. K. Padmanabhan. 2006. Selection, identification and comparison of industrial robots using digraph and matrix methods, *Robotics and Computer-Integrated Manufacturing*, Vol. 22, 4, pp. 373–383.

Yufka A. and O. Parlaktuna. 2009. Performance Comparison of the Bug's Algorithms for Mobile Robot, *INISTA 2009 International Symposium on INnovations in Intelligent SysTems and Applications*, 29 June to 1 July, Trabzon, Turkey, 5 pages.

13 Robotic Swarms
Preliminaries

13.1 INTRODUCTION

Hitherto, our attention has been concentrated on the functioning of the single robot, a solitary machine designed to work independently, diligently, and indefatigably, applying forces and controlling movements of various forms to perform actions. It is abundantly clear that a single robot will fail to solve large-scale problems, despite carrying an extensive gadgetry of sophisticated sensors, actuators, and processing electronics. While a single robot excels in precise, repetitive tasks, tasks distributed over large areas are not within its domain of implementation. We know that a large group of insects, e.g., bees, wasps, ants, termites, or locusts, moving together constitutes a swarm. Motivated by the swarms of gregarious insects observed in natural settings, a robotic swarm is contemplated as a large population of simple, small, and inexpensive robots. The robots in a swarm are its members. The member robots work collectively in a decentralized manner through local interactions and sensing among themselves and with the environment to accomplish confounding tasks without access to global information. In a decentralized system, the control, power, or activities are not concentrated in a single, central authority. Rather, they are distributed among many separate entities or locations.

A few instances of commendable tasks performed by robotic swarms are rescue missions in times of catastrophe and mayhem, surveillance and defense activities, warehouse automation, logistics, oil spill response, precision agriculture, and environmental monitoring. Swarm robotics is a cohesive strategy to coordinate several relatively simple robots working collegiately sharing liabilities and accountability. It can scale up to the inclusion of hundreds or thousands of robots (Şahin 2005).

SINGLE ROBOT VS. ROBOTIC SWARM

Robotic teamwork
Can make many dreams work
Two robot minds are better than one
To finish the job and get the work done
If one robot fails, its partner takes over immediately
And the work progresses uninterruptedly
Some jobs are big, others are small,
Together, robots can do them all.

Due to their simpler design, robotic swarms can be significantly cheaper to manufacture than a single, highly complex robot. Moreover, swarms can adapt to changing environments and unexpected, often surprising situations more effectively than a single robot. Additionally, the damage and crippling of one individual robot in the swarm

DOI: 10.1201/9781032695266-13

has minimal impact on the overall system. Thus, multiple, often simpler robots collectively complete a compound task through coordinated interactions and distributed intelligence, offering increased adaptability and redundancy. The collective behavior of the robot swarm duly compensates for the failure of individual robots. Essentially, a swarm leverages the power of a large number of robots and local communication. It achieves seemingly impossible and ambitious goals that a single robot might struggle with. Therefore, from this chapter onward, we shift our focus to swarms of robots in lieu of the single robot that has been in the spotlight in the foregoing chapters. Swarm robotics can do wonders and become a game-changer. It is a paradigm-shifter that will bring a transformative change, leading to improved performance and innovation.

13.2 BIO-INSPIRED ALGORITHMS USED IN SWARM ROBOTICS

Nature is collaborative, with members of a species working together for survival, fully knowing that union is strength. Collective action and unity build a powerful community. We can learn valuable lessons from the nature. Therefore, in swarm robotics, various algorithms inspired by biological phenomena are employed (Hereford and Siebold 2010). Apart from genetic algorithm (GA), which originates from the process of natural selection in genetics, these algorithms are mainly derived from a sense of enthusiasm and excitement that nature provides through the collective behaviors of animals like birds, bees, ants, etc. The motive of all these algorithms is to solve tortuous problems by coordinating a large group of robots with simple rules. The coordination compels them to work in an interwoven and integrated fashion for achieving their pursued outcome (Bhowmick et al. 2024).

Each algorithm mimics the collective behavior of a natural swarm. The connotation of word 'mimicking' can be negative depending on the context and intent. So, let us clarify. 'Mimicking' here is not done for mocking but as a part of the learning process. This caricature is not any playful or derisive attempt but a respectful, solemn, and praiseworthy activity for a beneficial purpose. It is called biomimicry. It is the science of learning from nature, imitating natural processes, and emulating natural ecosystems to create more sustainable and efficient solutions to real-world challenges faced by humans. It is a powerful approach to designing swarm robotics methods, building upon the principles of collective intelligence and the behaviors of natural swarms. Biomimicry draws its lessons from nature's solutions to problems, encouraging a hands-on, experiential approach that fosters creativity and deeper understanding. Through billions of years of evolution, nature has developed incredibly effective ways to address various types of problems, and we, too, can gain from this expert knowledge base gifted by nature. Sustainability is a cardinal ingredient and the lifeblood of this knowledge base because nature works in ways that meet the demands, necessities, and exigencies of the present generation without compromising the needs of future generations.

In a natural-like swarm, the individual robots make local decisions like ants or bees based on limited information. Such local decisions translate for culmination into an emergent global behavior. These algorithms are often used to optimize numerous tasks, e.g., activities of path planning, target searching, foraging, and obstacle avoidance within a swarm of robots. A few commonplace examples of these algorithms are given in Figure 13.1. The diagram shows a swarm of robots and conscripts the seven bio-inspired swarm robotic algorithms as given below:

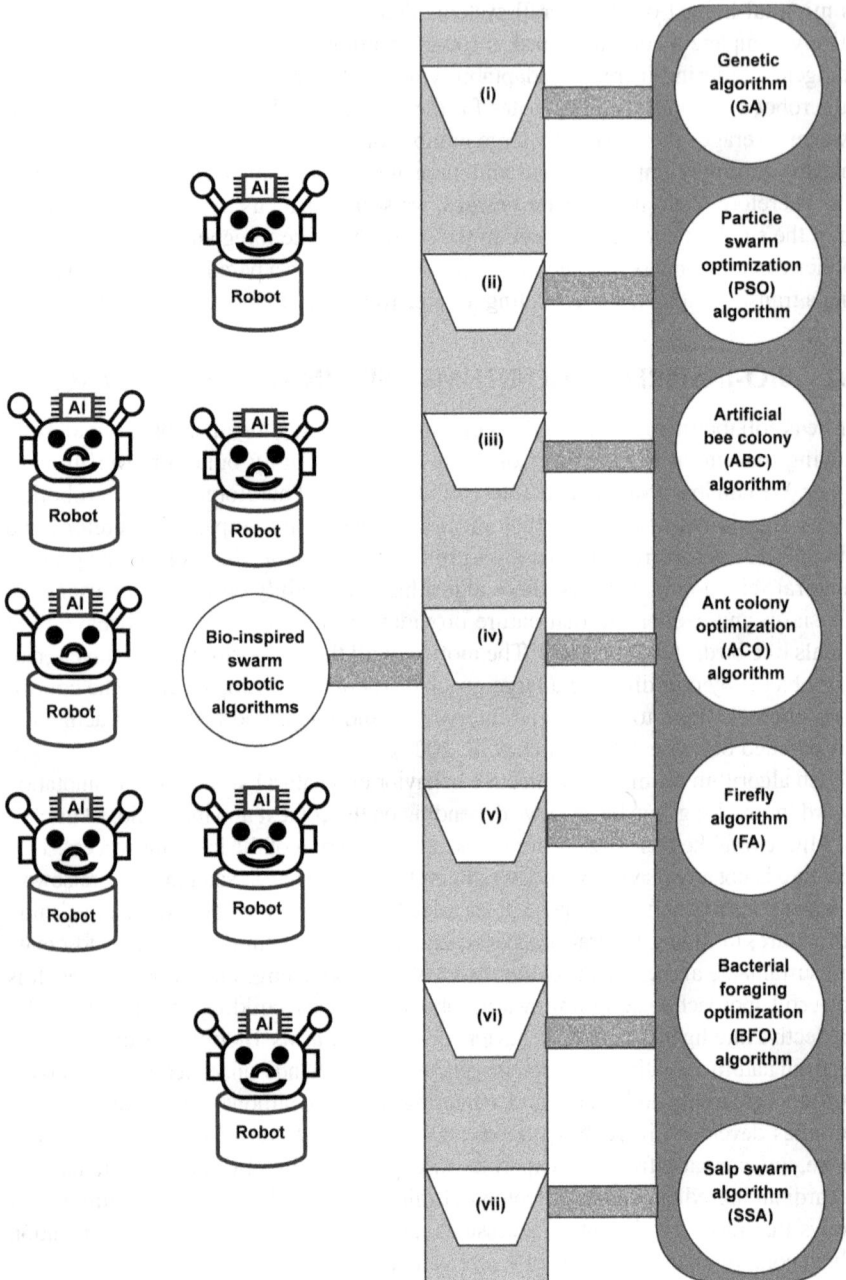

FIGURE 13.1 Examples of algorithms for swarm robotics that are developed by inspiration from biological phenomena.

i. Genetic Algorithm: The GA is a computational method. It is used to optimize the behavior of a group of robots by mimicking the principles of natural selection. In this algorithm, the robot's behaviors are represented as chromosomes containing parameters defining its actions. The actions are movement patterns, decision-making rules, or sensor interpretations. The best robot behaviors are selected, combined, and mutated to produce improved behaviors for the swarm over time. Thus, the robots can collectively solve complex tasks more effectively (Wahab et al. 2024).

ii. Particle Swarm Optimization (PSO) Algorithm: It is a pivotal algorithm for searching for plans of action to achieve the desired results in swarm robotics. The inspiration for this computational model is sparked by bird flocking. The flocking of birds is their instinctive behavior to fly together in formations for finding food and staying safe from predators. In this algorithm, the particles representing robots move toward the best solution found by the swarm (Hereford et al. 2007; Zhang and Wang 2024).

iii. Ant Colony Optimization (ACO) Algorithm: This algorithm is formulated by the inspiration received by humans from observation of the ant colony behavior. Ants are known to lay down pheromone trails to indicate optimal paths between their nest and food sources. Pheromone is a chemical signal. It is secreted by a species to elicit a particular behavioral response from other individuals of the same species. So, subsequently, moving ants use these pheromone trails to tread the paths defined by their predecessors when finding the shortest route. Hence, by following the pheromone trail laid down by previously searching ants, their new colleagues can easily reach the food sources. In the ACO algorithm, the robots are considered as the artificial ants that iteratively build solutions by choosing paths based on pheromone levels (Sharan et al. 2023; Lingkon and Ahmmed 2024).

iv. Artificial Bee Colony (ABC) Algorithm: It emulates the foraging behavior of honeybees to collect nectar and pollens from blooming plants; nectar is a sugary liquid while pollens are powdery substances. The algorithm is modeled after the fascinating world of honeybees, which is depicted as comprising different bee types, namely employed, onlooker, and scout bees. These bee classes search for food sources with distinct roles assigned to different classes. In this distribution of labor, the employed bees exploit local food sources. The onlooker bees choose food sources according to the quality of the material. The scout bees wander about exploring and looking for new areas of food sources (Izaguirre et al. 2021).

v. Firefly Algorithm (FA). It is based on the social behavior of fireflies, the so-called lightning bugs or beetles belonging to the family Lampyridae, which originates from the Greek word 'lampein', purporting 'to shine'. The individual fireflies move toward their brighter associates, representing better solutions in the optimization problem. Drawing an analogy from this movement, the robots mimic the flashing behavior of fireflies, with isolated robots migrating toward brighter fireflies, which represent better solutions. The brighter the firefly, the greater is the emphasis on attraction and convergence toward optimal points (Chaudhary et al. 2024).

    vi. Bacterial Foraging Optimization (BFO) Algorithm: It models the movement and chemotaxis behavior (movement toward nutrient gradients) of bacteria in search of nutrients. Chemotaxis focuses on local search with occasional global exploration through repel and swim phases, allowing for a more thorough search in intricate environments (Hossain and Ferdous 2015). 'Local' relates to a particular place or area, while 'global' refers to a wide area.

    vii. Salp Swarm Algorithm (SSA): It is a computational optimization technique inspired by the natural swarming behavior of barrel-shaped, gelatinous marine animals related to vertebrates. The algorithm closely follows the behavior of salps to form chain-like structures while foraging to solve complex problems by iteratively updating the positions of salp agents within a search space. During this process, a designated leader salp guides the swarm toward the optimal solution. Essentially, it is a swarm intelligence method leveraging the coordinated movement of salps to explore and exploit a solution space effectively (Cheng et al. 2022).

Five noteworthy features of swarm robotics algorithms are underscored. These are their simplicity of approach, scalability to any population size, along with capabilities for decentralization (low-cost communication between agents without the services of a central coordinator), localization (local communication and interaction), and parallelism (breaking down resource-intensive tasks into smaller parts to be executed concurrently for simultaneous solution) (Tan and Zheng 2013).

13.3 GENETIC ALGORITHM

The GA is an algorithm used to optimize the behavior and decision-making of robots in a swarm akin to the evolutionary approach in biology following the natural selection process (Rezk et al. 2014; Bahaidarah et al. 2023; Zhu and Pan 2024). The 'swarm' suggests a population of robots in which each robot has its own set of parameters that can be modified through the GA. Each robot in the swarm represents a potential solution. The parameters of the robots, called genes, are adjusted on the basis of their performance in the environment, favoring the fittest robots to produce better future generations. Through such adjustments, the best solutions for a given task are identified, for example, in robot navigation within a complex environment or robot movement coordination within a swarm. The algorithm proceeds by iteratively improving the parameters called traits of the robots, controlling their actions. Operations like selection, crossover, and mutation in genetics are performed.

13.3.1 OPERATORS IN A GA

These operators, called genetic operators, are defined as (Lamini et al. 2018):

    i. Selection: This operator chooses the best-performing robots to pass on their traits.

    ii. Crossover: This operator combines traits from selected robots to create new offspring.

iii. Mutation: This operator randomly alters specific traits of robots to introduce diversity in the exploration of the solution space.

Figure 13.2a illustrates the vital operations performed in a GA: initialization, selection, crossover, mutation, and replacement. In the initialization step, three samples
are taken: 000, 111, 222, from which two samples are selected: 000, 111. In the
crossover step, 00 from 000 and 1 from 111 are taken to form 001, which is mutated
to 011. This mutated sample is supplied to the replacement step, which feeds it back
to the selection step.

13.3.2 BREAKDOWN OF THE KEY STEPS IN GA

The steps are framed by studying genetics in biology and concepts like heredity,
selection, and mutation. When using a GA to optimize a robot swarm, the key steps
are (Figure 13.2b): initializing a population of robot behaviors by representing
potential solutions as chromosomes, evaluating the fitness of each behavior based
on the swarm's performance in the task, selecting the best-performing behaviors
for reproduction, applying genetic operators like crossover and mutation to generate new behaviors. It is checked whether the newly generated population represents
a satisfactory swarm behavior or meets the stopping criterion. If YES, the process
is stopped. If NO, it reverts to the step of calculation of fitness function, repeating
this process until a satisfactory swarm behavior emerges. These steps essentially
simulate natural evolution to find the most efficient collective behavior for the robot
swarm (McKee 2024).

  i. Initialization of the Robot Population: A diverse initial population of robot
     behaviors is created. Each behavior is represented as a chromosome char-
     acterized by parameters like movement patterns, sensing strategies, com-
     munication protocols, etc. The representation is done randomly or using
     heuristics based on the problem domain, the specific field, phenomenon or
     discipline where the problem exists.
 ii. Evaluation of Fitness: The robot swarm is simulated in the environment
     using the individual behaviors of robots from the population. The perfor-
     mance of the swarm is measured based on the desired task, e.g., coverage,
     foraging efficiency, and obstacle avoidance. A fitness function is used for
     measurement. It is a function that assesses the performance of each robot
     based on the task at hand. A fitness score is assigned to each behavior based
     on performance. The score determines which robots are more likely to be
     selected for reproduction.
iii. Selection (Genetic Operator): The best-performing behaviors having high
     fitness scores are selected from the population to be used for reproduction.
     Methods like roulette wheel selection, tournament selection, or elitism are
     applied. The roulette wheel or fitness-proportionate selection assigns a
     probability of selection to each individual based on the fitness of that indi-
     vidual relative to the total fitness of the population. Tournament selection
     randomly selects a small group of individuals from the population, with

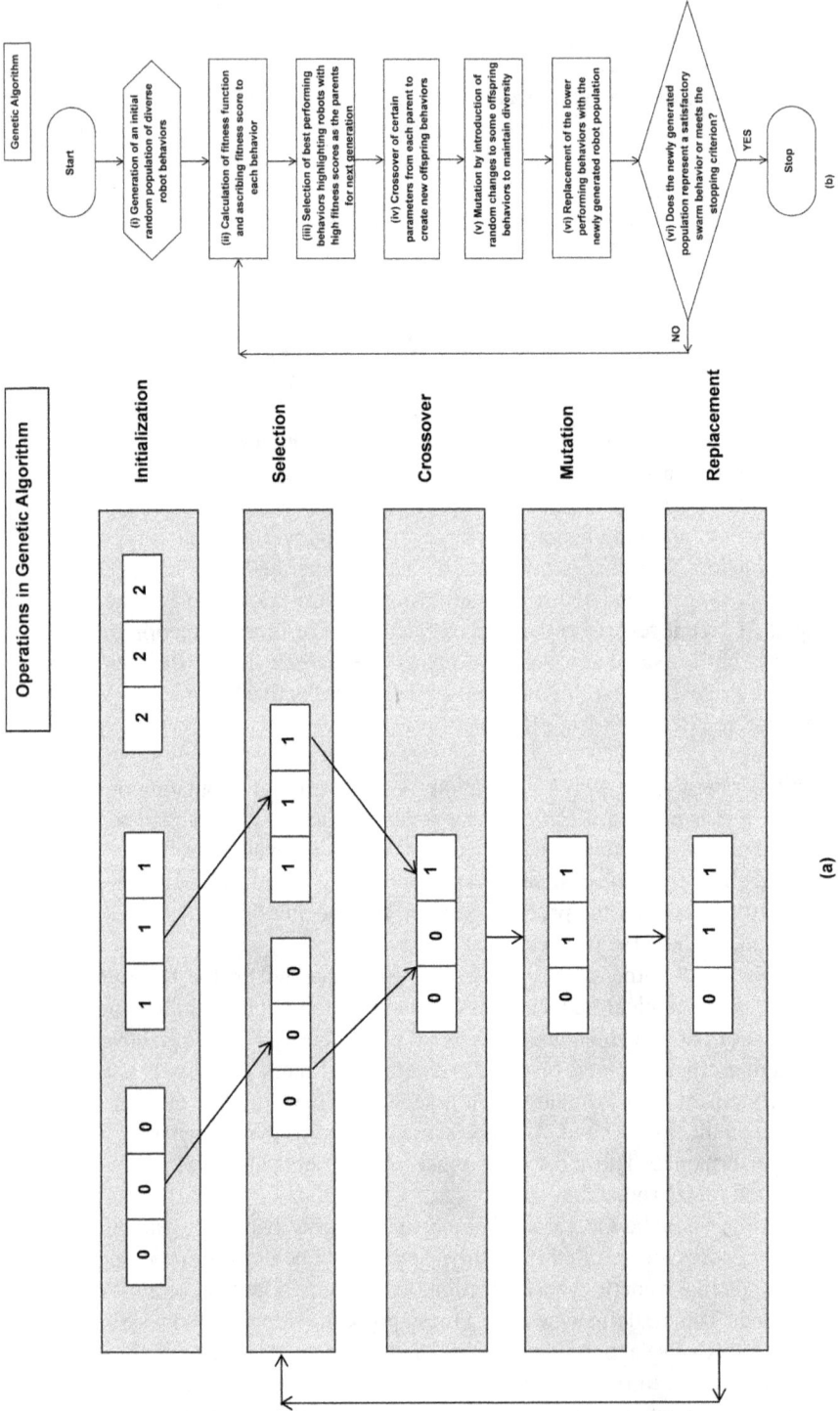

FIGURE 13.2 Genetic algorithm: (a) principal operations and (b) the algorithm.

the fittest individual within that selection chosen for reproduction. Elitism is a scheme in which the best individuals from the current generation are directly copied to the next generation.

iv. Crossover (Genetic Operator): The genetic information from selected parent behaviors is combined to create new offspring behaviors. Variations are generated by swapping parts of chromosomes or mixing parameters.

v. Mutation (Genetic Operator): Certain parameters of the new offspring's behavior are randomly modified to introduce diversity for exploring the solution space.

vi. Replacement: The lower-performing behaviors in the population are replaced with the newly generated offspring.

vii. Iteration: The steps of fitness evaluation, selection, crossover, mutation, and replacement are repeated. The repletion is done until a satisfactory swarm behavior is achieved or a stopping criterion is met.

13.3.3 APPLICATIONS OF GA

To appreciate the significance and relevance of GAs in swarm robotics, we highlight a few applications of the algorithm as follows.

i. Planning of the Robot Paths: The movement paths of robots are optimized within a swarm so that they can navigate vexing environments efficiently.

ii. Controlling the Desired Robot Formation within a Swarm: The desired formations of robots, like a line or a circle, are maintained within a swarm. For their maintenance, the movement patterns of individual robots are adjusted.

iii. Coordination of Multiple Robots for Performing Cooperative Tasks: The actions of multiple robots are coordinated to achieve a collective goal. The goal could be collaborative manipulation of an object.

13.3.4 ADVANTAGES OF GA

When using GAs to control robot swarms, the main advantages include their ability to find near-optimal solutions in entangled environments, explore a wide range of potential behaviors, and adapt to changing conditions. The advantages are:

i. Exploration of Diverse Solutions and Exploitation of Best Solutions: GAs effectively search through a vast space of possible robot behaviors. The search allows for exploration of diverse solutions. It also allows exploitation of the best ones. Both benefits are necessary for involuted swarm tasks.

ii. Adaptation to Changing Environments: The swarm adapts its behavior dynamically to changing environmental conditions. This becomes possible by incorporating evolutionary mechanisms, such as mutation and crossover.

iii. Robustness of Algorithm toward Noise: GAs can withstand noise and tolerate uncertainties in the environment. These advantages make them suitable for real-world robot swarm applications.

iv. Design Flexibility: The design of a GA can be tailored to specific swarm tasks. This is achieved by adjusting the representation of robot behaviors, specifically the chromosome encoding and fitness function.
v. Emergent Behavior: The individual robots are allowed to evolve based on their local information and interactions with the swarm. Hence, complex, collective behaviors can emerge without explicit centralized control.

13.3.5 LIMITATIONS OF GA

Principal limitations of GAs include potential for premature convergence, sensitivity to parameter tuning, and high computational cost associated with large populations, especially in real-time scenarios. These, along with the other limitations, are:

i. Premature Convergence of the Algorithm: In some cases, the GA may converge prematurely to a local optimum. Then it misses potentially better solutions in the search space.
ii. Challenges of Parameter Tuning: The performance of a GA heavily depends on the proper tuning of parameters like population size, mutation rate, and crossover rate. Parameter tuning is often a bothersome activity.
iii. High Computational Cost: Evaluation of the fitness of a large population of robot behaviors is computationally expensive. Real-time applications with many robots are costly.
iv. Difficulty of Interpretation: It is sometimes difficult to comprehend the behaviors evolved by a GA. Interpretation is not straightforward when dealing with complex swarm dynamics.
v. Concerns of Scalability: Management of the communication and computation required for the GA becomes highly complex when there are a large number of robots in a swarm.

13.4 PSO ALGORITHM

From the Darwinian evolution-based GA, we move to the PSO algorithm based on the social behavior of birds/fishes. Both are population-based algorithms. But while GA follows the evolution of species, the PSO algorithm works using swarm intelligence. While GA is good and robust for complex problems, the PSO algorithm is easily implemented and converges fast. Both are prone to local minima. A detailed comparison is deferred for a later discussion.

To introduce the PSO algorithm, we consider an organization comprising a group of robots engaged in a mission. Then the robots in this group constitute a robotic swarm. These robots interact with each other and with their environment using a metaheuristic optimization algorithm as the core mechanism. The algorithm is referred to as the PSO algorithm. The term 'metaheuristic' is a combination of two words. It is a combination of the Greek prefix meta (meaning 'beyond' in the sense of high level) with heuristic (meaning 'search'), which implies a higher-level search procedure. In the PSO algorithm, the robots seek out the best feasible answer to any difficult problem faced by the group by addressing it collaboratively among themselves.

This problem could be regulating and organizing the movements and activities of the individual robots within the robotic cluster. The problem is solved by modifying the relative locations of participating robots and scheduling the actions to be done by them, making use of the collective intellect of the swarm. Figure 13.3a illustrates the particle movements in the PSO algorithm. Four positions of the particle are marked: its current, personal best pbest, global best gbest, and new positions. The three velocities of the particle are represented by respective vectors, which are arrows pointing in their respective directions. These velocities are: its current velocity, velocity based on personal best performance, and velocity based on global best performance. The continuous lines indicate the tendencies of particle motion, while the dashed lines indicate how the particle is carried away under the influence of these tendencies.

13.4.1 PARTICLE REPRESENTATION OF ROBOTS

How does the PSO algorithm represent and treat robots? The PSO algorithm espouses a particle representation of robots wherein each robot is treated as a separate particle. The position of the robot e.g., a robot's location in a navigation task, represents a potential solution to the optimization issue. The PSO is classified as a stochastic search strategy. It is a problem-independent method using randomness for search-space exploration. As opposed to the precise input-based deterministic techniques, it incorporates randomness and uncertainty. It functions on the iterative interaction of each particle that forms the swarm (Hamami and Ismail 2022). Its working mechanism involves regularly updating the relative positions of a swarm of particles from one iteration to another. This process of updating buoys up and supports the PSO algorithm to execute the search in the best possible way (Gad 2022).

13.4.2 SWARM INTELLIGENCE AND THE IDEA OF FITNESS FUNCTION

Swarm intelligence is the fundamental mechanism underlying the PSO algorithm. Duplicating the organized movements of birds in a flock that congregate to forage and travel conjointly; and on similar lines to fish schooling, e.g., a group of fish moving together in the same direction at the same speed; and in consonance with human social behavior of interaction, cooperation, and conflict; each robot in the swarm communicates with other robots in the group to share information about its current position and the best solution. Such inter-robot communication influences the movement of the entire group of robots, leading toward a better overall outcome.

The swarm intelligence enables the robot to change its movement based on its earlier performance as well as the performance of other robots in the swarm. A fitness function is defined keeping an eye on the desired goal for directing the swarm toward the optimal answer. The fitness function is an objective function that defines the objective of the problem in relation to its constraints. It is used as a figure of merit summarizing the closeness of the designed solution to the target. It determines the quality of the potential solutions regarding the position of the robot. The quality ratings are expressed by assigning scores that guide the algorithm on the way to an optimal solution.

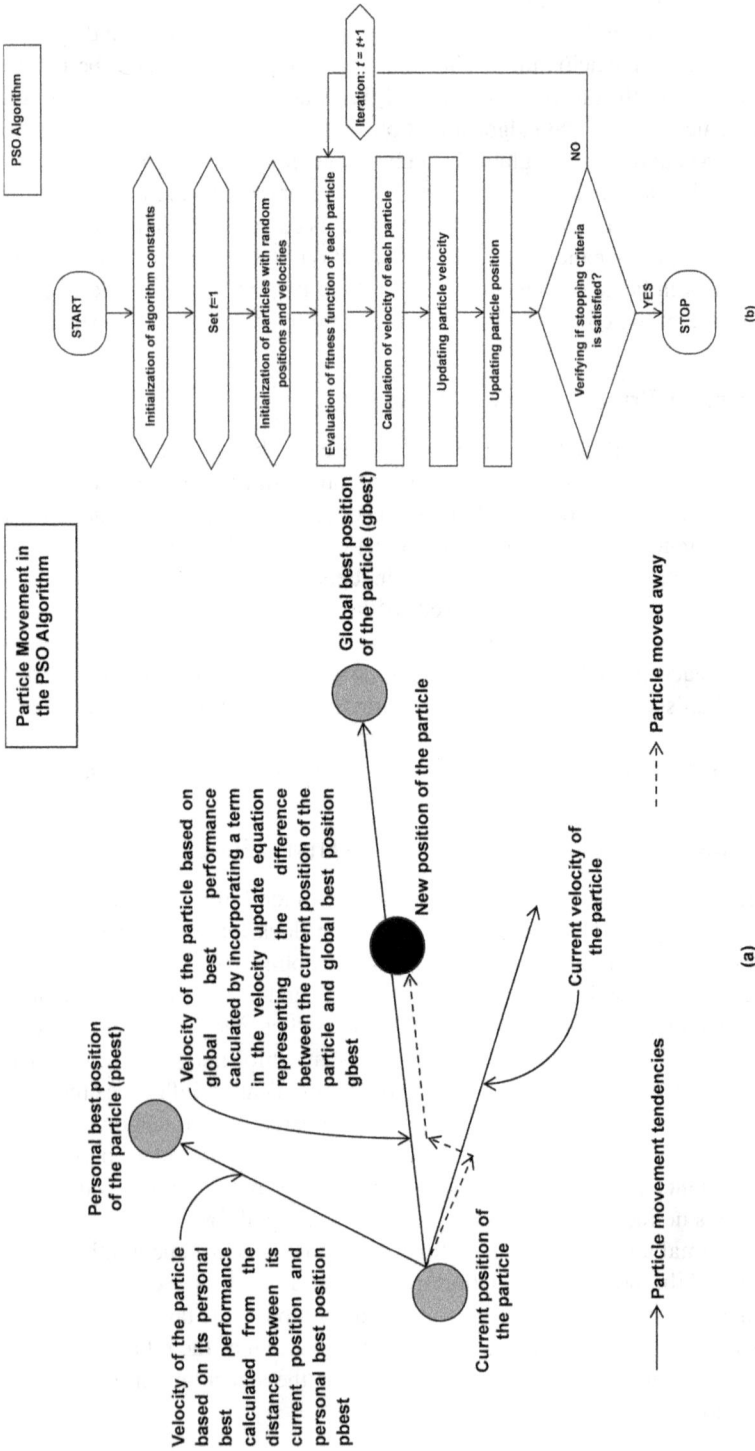

FIGURE 13.3 The PSO algorithm: (a) particle movements and (b) portrayal of the procedures of the particle swarm optimization algorithm by breaking down into discrete steps in the algorithm workflow.

13.4.3 Velocity Updates

The PSO algorithm works by velocity updating. It blends together the local search methods with global search methods. It advances with the notion that the most effective method of conducting the search is to follow the particle that is nearest to the best position, which means the position in the search space that represents the most excellent solution found so far. It is usually called the global best (gbest) position. The 'gbest' attribute indicates that it is the best position discovered by any particle within the complete swarm.

Each particle maintains a record of its own best position encountered thus far. This is known as personal best (pbest) position. For determining the best position, the position of each particle is evaluated as found from the fitness function. Then, the position with the highest fitness value is deemed the best position.

The algorithm utilizes velocity updates during its operation. During each iteration, each robot updates its velocity. The basis of updating is its current position, its best previous position pbest, and the best position found by the whole swarm gbest. The aim is to ascertain the required motion of the robot to make headway toward progressively better solutions.

13.4.4 Steps of the PSO Algorithm

The working procedure of the algorithm becomes evident by learning about its step-wise progress. Figure 13.3b depicts the principal steps of the PSO algorithm. The algorithm begins with the initialization of algorithm constants, setting $t = 1$, and the initialization of particle positions and velocities. Fitness function calculations are followed by velocity calculations for each particle. The particle's velocity and position updating come next. If the stopping criteria are met, the algorithm stops. If not, iteration $t = t + 1$ is done by returning to the fitness function determination. Further details of the workflow of the algorithm are (Market Brew™ 2025):

i. Initialization of Algorithm by Considering a Population of Particles and Randomly Assigning Positions and Velocities to Them: Every particle in the population has its position and velocity initialized with values selected arbitrarily.

ii. Calculation of Fitness Value of Every Particle by Evaluation of Its Objective Function: For each particle, a calculation of the objective function is done to estimate its fitness value at the current position.

iii. Mathematical Determination of Particle Velocity: The velocity of each particle is calculated on the basis of its current position, its best previous position (pbest), and the best position found by the entire swarm (gbest).

iv. Updating the Particle Velocity: The velocity of each particle is modified.

v. Amendment of Particle Position: The position of each particle is revised using its new velocity value.

vi. Checking for Fulfillment of the Stopping Criteria for the Algorithm: A scrupulous comparison of objective functions calculated using updated positions is made with objective functions reckoned through old positions.

In case no noticeable improvements in consecutive objective function values are found, a cessation of the process is warranted.

vii. Process Repetition: If the updated positions improve the objective function values, the process is repeated until arrival at the stopping criterion.

13.4.5 APPLICATIONS OF ROBOTIC PSO ALGORITHM

Robotic PSO is applied to carry out varied responsibilities in dealing with robots engaged in teamwork:

i. Planning and Organization of the Path of Robot's Journey: Since the most effective path must be found while avoiding any obstacles or barriers, moving several individual robots through a complicated environment requires careful optimization to find the most efficient route to the destination.

ii. Cooperative Control and Manipulation of Several Robots: When a multiplicity of robots is involved in grasping and manipulating objects together, their motions and actions must be properly coordinated with accuracy.

iii. Optimization of Sensor Network: In a sensor network for robots, the choice of the positions of placement of sensors is made keeping in view that the largest feasible coverage area is accounted for. At the same time, the network communication expenses must be cut down to the lowest level. Therefore, a trade-off process is performed for achieving the desired outcome by compromising between the sensor positioning plan and the consequent communication expenditure.

iv. Performing Search and Rescue Activities: A squad of robots must be properly coordinated to ferret out designated targets through an expansive area.

13.4.6 ADVANTAGES OF THE PSO ALGORITHM

The primary benefits of the PSO algorithm include its simplicity, ease of use, robustness to parameter changes, computational efficiency, and the ability to search an extremely vast solution space effectively. These assets make it suitable for a broad spectrum of optimization problems. Fewer tuning parameters in the PSO algorithm need to be adjusted compared to other optimization techniques.

i. Simplistic Idea and Easy Execution: The central idea of the PSO algorithm is easily comprehended and implemented, making it accessible to a broader audience.

ii. Fewer Parameters/Settings to Adjust: Rivaled against other optimization algorithms, the PSO method usually requires a smaller number of parameters to be tuned, thereby enormously simplifying the process of setting up the algorithm.

iii. Good Capability of Global Search: The PSO algorithm can effectively explore a vast search space. Premature convergence to local optima is averted.

iv. Rapid Convergence of Algorithm: The PSO algorithm exhibits rapid convergence to a near-optimal solution when the parameters are carefully tuned.

v. Adaptability for Compliance with Different Problem Domains: The PSO algorithm can analyze a variety of optimization problems spanning multiple domains, including those from finance, engineering, and machine learning fields.

vi. Potentiality for Parallel Processing: The particle-based nature of the PSO algorithm makes it capable of easy parallelization to achieve faster computation on multi-core processor systems.

13.4.7 DIFFICULTIES FACED DURING ROBOTIC PSO USAGE

Awareness about the limitations of the algorithm warns us not to venture into areas where we are likely to encounter trouble of some kind or another. The following are the problems encountered when using the PSO algorithm:

i. Tuning of Algorithm Parameters: The three main parameters in the PSO algorithm are the inertia weight (w), the cognitive coefficient (c_1), and the social coefficient (c_2). The inertia weight w-value determines the extent of retention of the previous velocity of a particle. Thus, it strikes a balance between local and global exploration. The coefficient c_1 is a measure of the influence exerted on a particle by its own best position. The coefficient c_2 determines the extent of influence exerted on a particle by the best positions of its neighboring particles. The performance of the PSO algorithm is critically influenced by the choice of these parameters. Therefore, the appropriate selection of these is essential to achieve satisfactory results.

ii. Avoidance of Inter-Robot and Robot-to-Obstacle Collisions: Robot-to-robot collision as well as collisions of robots with any obstacles during their motion in the environment should be unfailingly prevented. In particular, the caution against collision is an issue of paramount importance when dealing with an environment that is highly densely inhabited with robots.

iii. Overhead for Communication: Establishment and maintenance of communications in large swarms of robots is not only a formidable job but demands an exorbitant expenditure in computational overhead.

13.5 ACO ALGORITHM

Like the PSO algorithm, following bird flocking and better suited for continuous optimization, the ACO is also a swarm intelligence technique copying the ant foraging behavior for food, and is suitable for combinatorial problems. The ACO is an important technique for swarm optimization that was introduced in the early 1990s (Blum 2005; Brand et al. 2010). It is used for the planning of robot routes for the purpose of autonomous control and navigation of robot manipulators under dynamic conditions. Let us peek inside the society of ants to get knowledge about its organization.

13.5.1 EUSOCIAL BEHAVIOR OF ANTS

Ants are essentially eusocial insects that live in colonies with only some individuals capable of reproduction. Their primary eusociality traits are:

i. Cooperative Brood Care: This kind of parental care implies a social system in which the offsprings of a colony are attended by ant members other than their biological parents who reproduced them. In this system, individual ants contribute to raising the young of multiple generations. These generations are not necessarily the offspring of the caretaker ants personally.

ii. Overlapping Generations within a Colony of Adults: This social structure features the typical characteristic of the simultaneous coexistence of multiple adult generations within a colony. Hence, young adults are present alongside older adults, resulting in a mixed population.

iii. Division of Labor into Reproductive and Non-reproductive Groups: An ant colony functions as a superorganism partitioned into specialized castes. The castes named as workers, soldiers, queens, etc., perform different roles and undertake various responsibilities. Within a colony, certain individuals, generally a queen, solely bear the burden of reproduction. Other members of the colony are allocated tasks in a dedicated format, such as duties of foraging, defense, and nurturing the young.

13.5.2 THE WORKING PRINCIPLE OF THE ACO ALGORITHM

It is interesting to note that it is a metaheuristic algorithm. As already mentioned, a metaheuristic algorithm is a systematic problem-solving procedure in computer science that proceeds by imitating natural intelligent phenomena, following an instinctive yet methodical approach. For the ACO algorithm, this approach originates from the experience gained through observation of the tiny ants using their pheromone trails to communicate with each other in a self-organizing process. The pheromone trail is a chemical scent left by ants on their paths to food sources, nests, and other stopping places. Leaving this trail is a part of the foraging behavior of ant colonies. A natural consequence of this characteristic is the emergence of a combined, intelligent behavior among ant colonies.

The inception thought underlying the ACO algorithm is to seek the assistance of the pheromone trail laid down by ants during their search for food. Other members of the ant colony use this trail to establish communication among themselves. It marks an opportune path on the ground for other members of the colony to follow, as it is the shortest route to their source of food (Blum 2005; Dorigo et al. 2006; Dorigo and Stützle 2019). Hence, it becomes a path-guiding aid to direct the incoming members of the ant colony. These members have to simply adhere to the pheromone path, preventing the unnecessary repetition of exploratory efforts that their coworkers have already done. Figure 13.4a shows two possible paths for ants to tread upon between their nest and food. Figure 13.4b shows one way in which the ants move from their nest to food along path X and return to their nest along Y. Figure 13.4c shows the

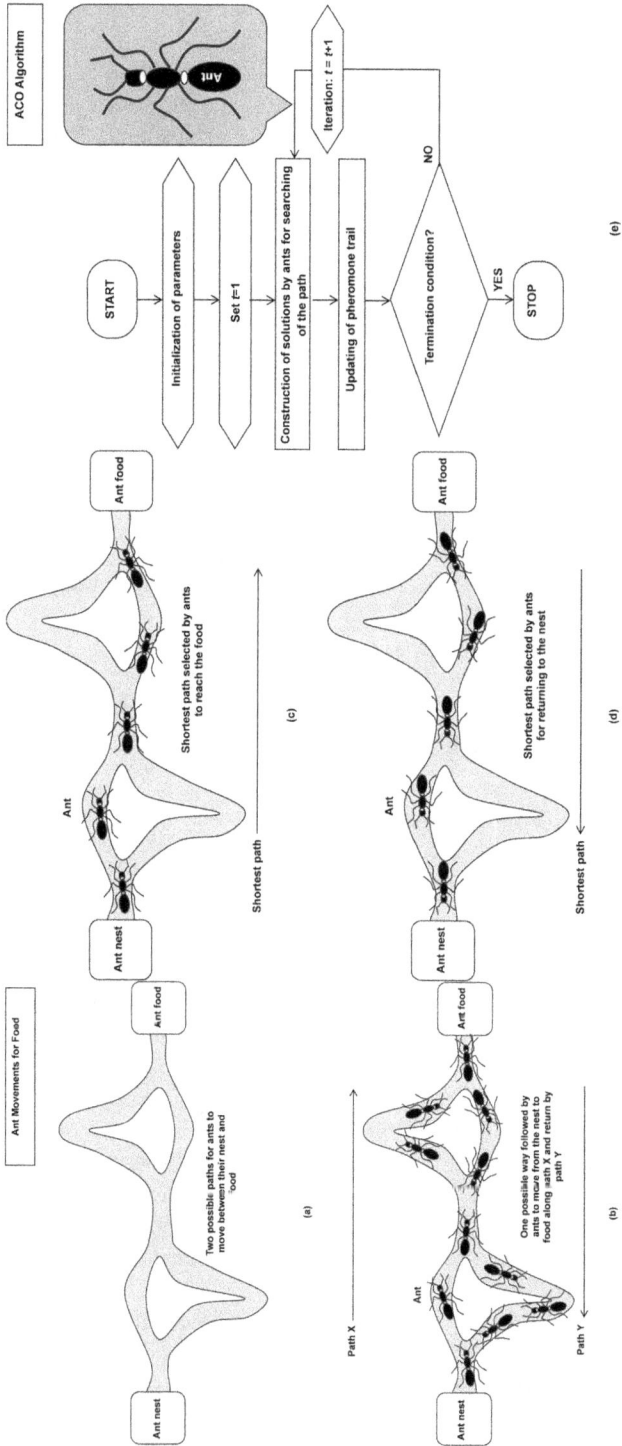

FIGURE 13.4 The ant colony optimization algorithm: (a)–(d) ant movements and (e) the algorithm.

shortest path pursued by ants from their nest to food. Figure 13.4d shows the shortest path followed by ants to return from food to their nest.

The ACO algorithm for robots follows a similar approach to ants in finding the optimal solutions to problems. It is a favorite and eligible choice for solving optimization problems across multiple realms through various fields that were hitherto difficult to decipher by routine methods.

The ACO is a probabilistic technique integrating randomness and uncertainty notions. In this technique, artificial ants are employed to solve computational problems. Good paths are prescribed by taking the help of graphs. The ACO algorithm works as shown in Figure 13.4e along the track: initialization, setting $t = 1$, construction of solutions by ants, and updating of pheromone trail. If the termination condition is satisfied, the algorithm stops. If not, we set $t = t + 1$ and return to constructing solutions by ants.

i. Transformation of the Optimization Problem into a Weighted Graph: The goal of this conversion of the problem into a graphical format is to find the shortest path to the destination.
ii. Construction of a Solution by Each Ant: Each ant randomly constructs a solution to the problem. This solution specifies the order in which the edges of the graph are to be traversed.
iii. Path Comparison: The paths found by the various ants are mutually compared.
iv. Updating the Pheromone Levels: The pheromone levels are made up-to-date on each edge of the graph according to the fresh findings.

13.5.3 STEPS OF THE ACO ALGORITHM

The ACO algorithm is a multi-stage process consisting of the following steps (Fresco Innovation Labs 2023):

i. Initialization of the Algorithm: The algorithm commences its chain of events by generating a colony of artificial ants. These ants have no idea about the problem they are supposed to solve. Quite randomly, they engage themselves in their search for food. In this search process, each ant travels over the solution space. During its movement, the ant creates candidate solutions. A combinatorial optimization problem (COP) ensues. A model of the COP is defined in the form of a triplet (S, Ω, f). In this notation, S denotes a search space, which is defined over a finite set of discrete decision variables. The symbol Ω stands for a set of constraints applied to the variables. The symbol f connotes an objective function that is to be minimized while solving the problem.
ii. Construction of Solutions by Ants: In this step, each ant constructs its own solution. It does so by applying a probabilistic rule. The rule allows the ant to choose the next point in the solution space. Obviously, the ants prefer paths with higher concentrations of pheromone. So, the probability that an

ant moves to the next point is determined by the amount of pheromone dumped by the previous ants on that specific path.

Mathematically speaking, a set of m artificial ants constructs solutions from elements of a finite set of available solution components $C \{c_{ij}\}$. A solution construction starts with an empty partial solution $s^p = \varnothing$. Then, at each step, the current partial solution s^p is extended. The extension is achieved by adding a feasible solution component from the set of feasible neighbors $N(s^p) \subseteq C$. As already stated, the choice of a solution component from the set $N(s^p)$ is done probabilistically at each step. The probability that an ant k located in node i will choose to move to another node j is given by the equation (Blum 2005)

$$p\left(c_{ij} \mid s^p\right) = \frac{\tau_{ij}^{\alpha} \eta_{ij}^{\beta}}{\sum_{c_{il} \in N(s^p)} \tau_{il}^{\alpha} \eta_{il}^{\beta}}, \quad \forall\, c_{ij \in N(s^p)} \tag{13.1}$$

where τ_{ij} is the amount of pheromone deposited for transition from state i to j, $\alpha \geq 0$ is a parameter to control the influence of τ_{ij} on ants, η_{ij} is the desirability of state transition ij, the heuristic value associated with the component c_{ij}; and $\beta \geq 1$ is a parameter to control the influence of η_{ij}. The symbols τ_{il} and η_{il} represent the trail level and attractiveness for the other possible state transitions. The trail level represents the pheromone concentration, while attractiveness is the a priori assessment that represents the extent to which a path is considered appropriate. It is based on factors like distance or cost. The values of positive real parameters α and β determine the relative importance of pheromone versus heuristic information.

iii. Updating the Pheromone Trail: The pheromone trail is updated after all ants have constructed their solutions. The amount of pheromone deposited on a particular edge is a function of the quality of the solution constructed by the corresponding ant. The pheromone level is raised if the solution is good; otherwise, the pheromone level is lowered. In other words, pheromone update increases the pheromone values associated with good solutions and decreases those that are associated with bad ones. The increase or decrease of pheromone is accomplished by:

a. increasing the pheromone levels associated with a chosen set of good solutions, and

b. decreasing all the pheromone values through pheromone evaporation. The equation used is (Dorigo et al. 2006)

$$\tau_{ij} \leftarrow (1 - \rho)\tau_{ij} + \sum_{k=1}^{m} \Delta \tau_{ij}^k \tag{13.2}$$

where τ_{ij} is the amount of pheromone deposited for a state transition ij, ρ is the pheromone evaporation coefficient, m is the number of ants, and $\Delta\tau_{ij}^k$ is the amount of pheromone deposited by kth ant; it is given by

$\Delta \tau_{ij}^k = Q/L_k$ if the ant k uses the curve ij in its journey; $\Delta \tau_{ij}^k = 0$ otherwise

$$(13.3)$$

where L_k is the cost of the kth ant's tour (typically length) and Q is a constant.

iv. Cessation of the Algorithm: The aforesaid steps are cyclic. The calculations are repeated by the ACO algorithm until the optimal solution is found. As the iteration continues, the quality of the solution shows improvement. The iteration terminates when either the optimal solution is found or a predefined number of iterations are reached.

13.5.4 APPLICATIONS OF THE ACO ALGORITHM

The ACO algorithm has found widespread application in solving various complex optimization problems. Its common applications are briefly described below:

i. Robots as Substitutes for Unavoidable Engagements of Workers in Hazardous Conditions: In many dangerous situations, using robots is advisable, e.g., robots are successfully deployed when the environment is dirty, hazardous, likely to cause death or injury to workers as in case of mining, or during detection of leakage in gas pipe, etc. (Joshy and Supriya 2016).

ii. The Traveling Salesman Problem (TSP): A salesman travels to different cities. The TSP problem involves finding the shortest possible path for the traveling salesman to visit all cities and return to the starting city.

iii. The Vehicular Routing Problem: This problem involves determining the optimal route for a vehicle that visits several locations, taking into consideration the constraints, e.g., time windows, vehicle capacity, and related criteria. Numerous investigations have been carried out to solve this problem.

iv. The Knapsack Problem: It is a classical optimization problem. In this problem, a specific weight of objects is placed in a knapsack (backpack) to maximize profit or value. The algorithm is used to solve the knapsack problem subject to different constraints. It is a combinatorial problem of selecting a subset of items having weights and values to fit into the container with a maximum capacity, with the intent to maximize the total value.

13.5.5 ADVANTAGES OF THE ACO ALGORITHM

A few advantages of the ACO algorithm are worth mentioning. These guide us to make a suitable choice from the list of available algorithms that will be most effective for a given swarm robotic problem.

i. Availability of a Fast-Processing Scheme: Optimal solutions for complex problems are effectively searched in a shorter time period using the ACO algorithm than possible by traditional methods.

ii. Pursuit of a Metaheuristic Approach: It is basically a metaheuristic approach applied to solve various optimization problems. The metaheuristic feature makes it an attractive choice for handling any optimization problem.

iii. Easy Implementation and Maintenance Capabilities: It is easily implementable, maintainable, and refinable with updates. Therefore, it demands a comparatively lesser number of iterations to attain convergence in opposition to the orthodox complex methods.

iv. Provision of an Efficient Solution: Good-quality solutions to optimization problems are efficiently determined using the ACO algorithm at a faster speed than other algorithms. The appreciably shorter computation time required to run the algorithm is a significant benefit of the method.

13.5.6 DISADVANTAGES OF THE ACO ALGORITHM

Knowing about the disadvantages of the algorithm is just as important as learning about its advantages, as they inform us about potential sources of errors and situations where algorithmic analysis is prone to failure. The prominent drawbacks of the ACO algorithm are:

i. Instability of Algorithm Performance: The algorithm's performance becomes unstable with an increase in problem size. So, it may not provide the best solution for larger problem sizes. The hindrance to arriving at the best solution occurs when time runs out while trying different combinations.

ii. Dependency on Parameter Fine-Tuning: Several parameters must be fine-tuned in difficult problems to arrive at optimal results. Then, multiple time-consuming iterations of the algorithm are obligatory.

iii. Indispensability of a Large-Scale Memory: Various probabilities are involved in the calculation of the next state. Therefore, a large-scale memory storage is necessary to store the data.

13.6 DISCUSSION AND CONCLUSIONS

Swarm robotics, with its small, agile robots, opens up new possibilities owing to its scalability, robustness, and parallel processing capabilities. Researchers are developing sophisticated, decentralized control algorithms inspired by biological behaviors. These algorithms allow swarms to make collective decisions without relying on a central leader. They make them more robust to failures and adaptable to changing environments. Three ground-breaking algorithms, the GA, the PSO, and ACO algorithms, were treated in this chapter. Table 13.1 gives an overview of the discussions in Chapter 13. The succeeding chapter will present more swarm robotic algorithms to reveal the vast expanse of this field.

TABLE 13.1

Takeaways from This Chapter at a Glance

Sl. No.	Takeaway	Explanation
1	Summary	Swarm robotics, the coordination of many simple robots to work together, is guided by algorithms derived from the observed behaviors of natural swarms, including communication, local interactions, and emergent intelligence. Three important algorithms in swarm robotics were described.
2	GA	In the genetic algorithm, a group of robots optimizes their collective behavior to solve a problem by drawing inspiration from natural selection principles in genetics. The best solutions are evolved through processes such as selection, crossover, and mutation.
3	PSO algorithm	In the particle swarm optimization algorithm, each individual robot is treated as a particle with attributes like position and velocity, and the collective movement of all the robots is determined by information about their own best previous position and the best position found by the entire swarm, allowing them to search for the optimal solution collaboratively (similar to how a flock of birds or a school of fish behaves in nature) for tasks such as path planning, target tracking, and the coordinated movement of multiple robots.
4	ACO algorithm	The ant colony optimization algorithm treats the robots as analogous to artificial ants. Their navigation is based on the pheromone levels on different paths, which are updated based on the quality of previous solutions.
5	GA vs. PSO vs. ACO	GA is better for solving complex problems with diverse constraints, utilizing its crossover and mutation operators, while PSO is favored due to its faster convergence. PSO is typically better suited for continuous optimization problems where solutions can exist across a range of values. At the same time, ACO excels at discrete optimization problems, where solutions are selected from a predefined set of options.
6	Keywords and ideas to remember	Swarm robotics, bio-inspired algorithms, genetic algorithm, particle swarm optimization algorithm, particle representation of robots, fitness function, velocity updates, ant colony optimization algorithm; GA, PSO, and ACO algorithms

REFERENCES AND FURTHER READING

Bahaidarah M., O. Marjanovic, F. Rekabi-Bana and F. Arvin. 2023. An Optimized Robot Swarm Flocking with Genetic Algorithm, *2023 IEEE International Conference on Mechatronics and Automation (ICMA)*, Harbin, Heilongjiang, China, 6–9 August, pp. 1823–1828.

Bhowmick P., S. Das and F. Arvin. 2024. *Bio-Inspired Swarm Robotics and Control: Algorithms, Mechanisms and Strategies*, IGI Global Scientific Publishing, Hershey, PA, USA, 261 pages.

Blum C. 2005. Ant colony optimization: Introduction and recent trends, *Physics of Life Reviews*, Vol. 2, 4, pp. 353–373.

Brand M., M. Masuda, N. Wehner and X.-H. Yu. 2010. Ant Colony Optimization Algorithm for Robot Path Planning, *2010 International Conference on Computer Design and Applications*, Qinhuangdao, China, 25–27 June, pp. V3–436–V3–440.

Chaudhary K., A. Prasad, A. Prasad and B. Sharman. 2024. Robot motion control using stepping ahead firefly algorithm and kinematic equations, *IEEE Access,* Vol. 12, pp. 43078–43088.

Cheng X., L. Zhu, H. Lu, J. Wei and N. Wu. 2022. Robot path planning based on an improved salp swarm algorithm, *Journal of Sensors*, Vol. 2022, Article ID 2559955, pp. 1–16.

Dorigo M., M. Birattari and T. Stutzle. 2006. Ant colony optimization, *IEEE Computational Intelligence Magazine*, Vol. 1, 4, pp. 28–39.

Dorigo M. and T. Stützle. 2019. Ant Colony Optimization: Overview and Recent Advances. In: Gendreau M. and J. Y. Potvin (Eds.), *Handbook of Metaheuristics*. International Series in Operations Research & Management Science, Vol. 272, Springer, Cham, pp. 311–351.

Fresco Innovation Labs. 2023. Ant colony optimization, A guide to swarm intelligence for optimization, https://medium.com/@freskoinnovationlabs/ant-colony-optimization-a-guide-to-swarm-intelligence-for-optimization-5d8842677f5e

Gad A. G. 2022. Particle swarm optimization algorithm and its applications: A systematic review, *Archives of Computational Methods in Engineering*, Vol. 29, pp. 2531–2561.

Hamami M. G. M. and Z. H. A. Ismail. 2022. Systematic review on particle swarm optimization towards target search in the swarm robotics domain, *Archives of Computational Methods in Engineering*, https://doi.org/10.1007/s11831-022-09819-3

Hereford J. M. and M. A. Siebold. 2010. Bio-inspired Search Strategies for Robot Swarms. In: Martin E. M. (Ed.), *Swarm Robotics from Biology to Robotics*, InTech Open, Croatia, pp. 1–26.

Hereford J. M., M. Siebold and S. Nichols. 2007. Using the Particle Swarm Optimization Algorithm for Robotic Search Applications, *2007 IEEE Swarm Intelligence Symposium*, Honolulu, HI, USA, 1–5 April, pp. 53–59.

Hossain M. A. and I. Ferdous. 2015. Autonomous robot path planning in dynamic environment using a new optimization technique inspired by bacterial foraging technique, *Robotics and Autonomous Systems*, Vol. 64, pp. 137–141.

Izaguirre J. M. V., C. Camilo, D. A. Mejia and J. L. Rodriguez-Verduzco. 2021. Intelligent search of values for a controller using the artificial bee colony algorithm to control the velocity of displacement of a robot, *Algorithms*, Vol. 14, 9, 273, pp. 1–12.

Joshy P. and P. Supriya. 2016. Implementation of Robotic Path Planning Using Ant Colony Optimization Algorithm, *2016 International Conference on Inventive Computation Technologies (ICICT)*, 26–27 August, Coimbatore, India, pp. 1–6.

Lamini C., S. Benhlima and A. Elbekri. 2018. Genetic algorithm-based approach for autonomous mobile robot path planning, *Procedia Computer Science*, Vol. 127, pp. 180–189.

Lingkon M. L. R. and M. S. Ahmmed. 2024. Application of an improved ant colony optimization algorithm of hybrid strategies using scheduling for patient management in hospitals, *Heliyon*, Vol. 10, 22, e40134, pp. 1–18.

Market Brew™. 2025. How particle swarm optimization algorithm works: A step-by-step guide, https://marketbrew.ai/how-particle-swarm-optimization-works-a-step-by-step-guide#:~:text=PSO%20is%20an%20iterative%20optimization,solution%20to%20a%20given%20problem

McKee A. 2024. Genetic algorithm: Complete guide with Python implementation, https://www.datacamp.com/tutorial/genetic-algorithm-python

Rezk N. M., Y. Alkabani, H. Bedor and S. Hammad. 2014. A distributed Genetic Algorithm for Swarm Robots Obstacle Avoidance, *2014 9th International Conference on Computer Engineering & Systems (ICCES)*, Cairo, Egypt, 22–23 December, pp. 170–174.

Şahin E. 2005. Swarm Robotics: From Sources of Inspiration to Domains of Application. In: Şahin, E. and W. M. Spears (Eds.), *Swarm Robotics. SR 2004*. Lecture Notes in Computer Science, Vol. 3342, Springer, Berlin, Heidelberg, pp. 10–20.

Sharan S., J. J. Domínguez-Jiménez and P. Nauth. 2023. Development of an Modified Ant Colony Optimization Algorithm for Solving Path Planning Problems of a Robot System, *2023 10th International Conference on Signal Processing and Integrated Networks (SPIN)*, Noida, India, 23–24 March, pp. 52–57.

Tan Y. and Z.-Y. Zheng. 2013. Research advance in swarm robotics, *Defence Technology*, Vol. 9, 1, pp. 18–39.

Wahab M. N. A., A. Nazir, A. Khalil, W. J. Ho, M. F. Akbar, M. H. M. Noor and A. S. A. Mohamed. 2024. Improved genetic algorithm for mobile robot path planning in static environments, *Expert Systems with Applications*, Vol. 249, Part C, p. 123762, https://doi.org/10.1016/j.eswa.2024.123762

Zhang J. and P. Wang. 2024. Improved Particle Swarm Optimization for Trajectory Planning in a Six-Degree-of-Freedom Robotic Arm, *2024 17th International Conference on Advanced Computer Theory and Engineering (ICACTE)*, Hefei, China, 13–15 September, pp. 329–331.

Zhu J. and D. Pan. 2024. Improved genetic algorithm for solving robot path planning based on grid maps, *Mathematics*, Vol. 12, 24, 4017, pp. 1–17.

14 Robotic Swarms
Exploring Additional Avenues and Vistas

14.1 INTRODUCTION

In this chapter, we continue our study of swarm robotic algorithms. As we persevere in our learning efforts, we consider two well-known optimization techniques, the artificial bee colony (ABC) algorithm (Cui et al. 2022, 2024) and the firefly algorithm (FA) (Bisen and Kaundal 2020; Wei et al. 2023), used to solve complex problems concerned with robot path planning, motion control, and obstacle avoidance. These methods make use of swarm intelligence concepts to deal with the problems faced in robotics. They can effectively find solutions across a large solution space.

14.2 ABC ALGORITHM

The ABC algorithm is a much sought-after swarm-based meta-heuristic optimization algorithm in robotics (Li et al. 2018; Xu et al. 2020). As its name suggests, this algorithm functions by simulating the activities of honeybees. It is used in searching for an optimal numerical solution among a large number of alternatives, such as in planning robot paths and solving convoluted robot movement optimization problems. It allows the robots to efficiently explore a space to find the solution that addresses the given problem expertly in the most effective manner. An eloquent example is finding the most optimal route to navigate a labyrinth (Bansal et al. 2013).

We know that the natural activity of honeybees during searching for a food source is based on the distribution of sub-activities among the bees. The sub-activities are related to communication, task allocation, nest site selection, reproduction, mating, floral foraging, pheromone deposition, and patterns of bee navigation. All these traits of the bees are mimicked in the ABC algorithm. Hence, the ABC algorithm is a bio-inspired swarm intelligence optimization technique prompted by the collective foraging behavior of honeybees.

14.2.1 CLASSIFICATION OF BEES INTO THREE GROUPS

Three types of bees participate in the ABC algorithm, each with a disparate assigned role (Zhou et al. 2025).

- i. Scout Bees: The scout bees haphazardly look for new food sources when one area becomes exhausted of food. These food sources are the potential solutions being pursued by bees.

DOI: 10.1201/9781032695266-14

ii. Employed Bees: These bees are entrusted with the exploratory labor in areas near known food sources, which represent the current robot positions. They bring nectar into the hive. They test and evaluate the quality of nectar food sources obtained from the scout bees. They also inform the onlooker bees about the quality of the nectar source.

Incipiently, the employed bees search for new food sources in response to unplanned, sporadic stimuli. A food source is identified as a candidate solution. The suitability or fitness of the same is computed. Subsequently, suppose a new food source is discovered by these bees as a potential candidate solution. Furthermore, suppose that this food source shows a greater suitability than the previous one. In that case, the new source is adopted. Otherwise, the new one is rejected.

iii. Onlooker Bees: These bees are occupied in the evaluation of the quality of food sources. They examine the solutions found by employed bees. They select the best food sources to conduct further exploration. To this end, the onlooker bees obtain the data from the employed bees. The employed bees share the fitness information with the onlooker bees. The onlooker bees select their food sources based on the probability values derived by calculating the ratio of the fitness function of a source to the sum of the fitness functions of all sources. In the circumstance of a failure of the bees to improve the fitness functions of the food sources, their solutions are spurned.

Figure 14.1a shows the two-way interaction between different categories of bees as follows:

Scout and Employed Bees: The scout bees randomly search for food sources, and the employed bees assess the quality of food sources obtained from scout bees.

Employed and Onlooker Bees: Employed bees share the fitness information with onlooker bees, and the onlooker bees evaluate the quality of food sources found by employed bees.

Onlooker and Scout Bees: Onlooker bees abandon non-improved food sources, and scout bees find new food sources.

14.2.2 Phases of the ABC Algorithm

It consists of the rudimentary stages mentioned in Figure 14.1b: start, initialization phase, set $t = 1$, employed bee and onlooker bee phases; two decision steps: Is a scout bee present in the colony? If YES, go to the scout bee phase leading to the termination condition, which is a decision step. If NO, move to checking compliance with the termination condition. If YES, seek the best solution and stop. If NO, set $t = t + 1$ and move to the employed bee phase. More details are given below (Nozohour-leilabady and Fazelabdolabadi 2016):

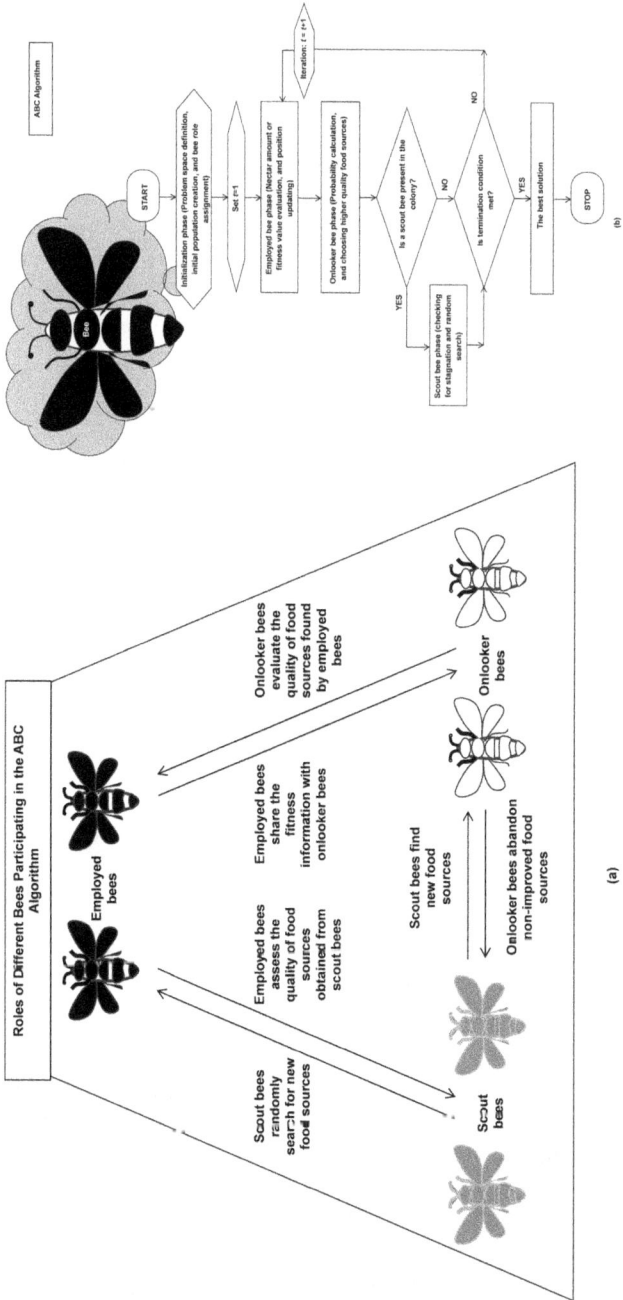

FIGURE 14.1 The artificial bee colony algorithm: (a) different types of bees and (b) the flowchart of the algorithm.

14.2.2.1   Initialization Phase

 i. Definition of the Problem Space: The parameters of the problem are laid down. Principal parameters to be defined are the starting and ending points of the robot, indicating the beginning and conclusion of the robot's journey; the permissible directions of movements of the robot; and the relevant constraints or obstacles likely to be faced by the robot during the course of its movements.

 ii. Creation of Initial Robot Population: A set of potential paths for the robot movements is randomly generated within the search space of the problem. The paths represent sequences of steps taken by the robot to reach food sources. Each food source is described by parameters such as coordinates and movement directions. Thus, an initial set of solutions is created in a random fashion. This randomly distributed set of solutions is given by the equation (Karaboga 2010; Yurtkuran and Emel 2016; Chaudhary 2023)

$$x_{i,j} = x_{\text{minimum},j} + \text{Random number}(0,1)\left(x_{\text{maximum},j} - x_{\text{minimum},j}\right) \qquad (14.1)$$

where $i = 1, 2, 3, \ldots, SN$ (SN is the size of solutions, i.e., food sources), $j = 1, 2, 3, \ldots, D$ (D is the dimension of optimization parameters), x_{ij} is the solution numbered as ith solution with dimension j, $x_{\text{minimum},j}$ is the lower bound for the dimension j and $x_{\text{maximum},j}$ is the upper bound for the dimension j.

 iii. Assignment of Bee Roles: The bee population is divided into three categories:

  a. Employed Bees: These are the bees that are associated with food sources.

  b. Onlooker Bees: These consist of bees that are observing the employed bees.

  c. Scout Bees: These comprise the bees that are exploring unsystematically.

14.2.2.2   Employed Bee Solution Search Phase

After the initialization of the algorithm, the population of food sources or solutions undergoes a series of repeated cycles. The stages in this phase are:

 i. Appraisal of Nectar Amount or Fitness Value: Each employed bee performs a local search around its assigned food source. During this search, the nectar amount or fitness value of the assigned food source is calculated. The chief considerations for this calculation are the distance to the target and collision avoidance. The metrics used in this calculation are the shortest distance and the least collision risk. The quantity of nectar in a food source is a reliable indicator of the quality of the corresponding solution.

 ii. Updating to Change Position: Each employed bee slightly modifies its parameters, such as the current food source position or robot path, based on its nectar amount or fitness value. This is a means of attempting to improve the solution or the path potentially. The new solution is expressed as (Karaboga and Basturk 2007a,b)

$$v_{i,j} = x_{i,j} + \phi_{i,j}\left(x_{i,j} - x_{k,j}\right)$$ (14.2)

where $k = 1, 2, 3, \ldots$, SN, and $j = 1, 2, 3, \ldots$, D. k and j are randomly generated, and k must be different from i; and $\phi_{i,j}$ is a random number in the interval $[-1,1]$.

Briefly stated, each employed bee compares the nectar amount or fitness value of the new source for any randomly selected solution from the swarm with reference to its original value. If the nectar amount of the latest source is higher than that of the previous one in its memory, the employed bee memorizes the new position. It forgets and ignores the old one. If the new source has a lower nectar amount, the employed bee preserves the position of the previous source in its memory.

14.2.2.3 Onlooker Bee Solution Search Phase

The stages in this phase are as follows:

i. Probability Calculation: The onlooker bees search for solutions probabilistically. A technique known as roulette wheel selection is used. Roulette wheel selection, also referred to as fitness proportionate selection, does an impersonation of a casino roulette wheel. Here, individuals are assigned slices proportional to their fitness. The fitter is an individual, the higher its probability of being selected for reproduction.

The onlooker bees conduct their search based on better solutions. A probability is calculated for each employed bee to be selected by an onlooker bee. This is obtained from the nectar amount or fitness value of each food source. Therefore, each solution in the swarm is associated with a selection probability calculated by an onlooker bee. This onlooker bee evaluates the nectar information taken from all employed bees and calculates a probability related to its nectar amount as (Huang and Chuang 2020)

$$p_i = \frac{\text{fit}_i}{\sum_{j=1}^{SN} \text{fit}_j}$$ (14.3)

where fit_i denotes the fitness value of solution X_i. The fitness value fit_i is defined as follows:

$$\text{fit}_i = \frac{1}{1 + f(X_i)} \quad \text{if } f(X_i) \geq 0$$ (14.4)

$$\text{fit}_i = 1 + |f(X_i)| \quad \text{if } f(X_i) < 0$$ (14.5)

where $f(X_i)$ represents the objective function value of the decision vector X_i.

ii. Selection and Updating of Positions of Food Sources: As in the case of the employed bee, the onlooker bee checks the nectar amount of the candidate source and produces a modification of the source position in its memory. Based on the fitness values, onlooker bees choose food sources with higher quality, suggestive of better robot paths, and therefore with higher probability. The onlooker bees perform similar position updates as employed bees, further refining the good solutions. Similar to employed bees, the onlooker bees perform local search around the selected food sources.

14.2.2.4 Scout Bee Solution Search Phase

The stages under this phase are as follows:

i. Stagnation Assessment Check: An important control parameter in the ABC algorithm is the limit or abandonment criteria. It is stipulated as a predetermined number of cycles or trials. When a solution cannot be improved after reaching this limit, i.e., if the food source or robot path of an employed bee does not improve after the permissible number of iterations defined in the limit, then that food source or path is treated as stagnant. It is relinquished and substituted by a new one in the scout phase. This means that the corresponding employed bee that is assigned to that solution assumes the role of a scout bee.

ii. Random Search by Scout Bee: This freshly produced scout bee is sent to randomly explore the search space to find a potentially new promising path, and a new food source is generated. All other solutions in the swarm follow the same process.

14.2.2.5 Repetition and Updating Food Sources

The steps in Sections 14.2.2.2–14.2.2.4 are repeated. After each phase, the food sources are updated based on the best solutions found by the bees. A gradual improvement of the overall path quality is thereby achieved. Iterations are continued through the employed, onlooker, and scout bee phases until a termination condition is met, such as reaching a maximum number of iterations or finding a satisfactory solution, as mentioned above.

14.2.3 Objective Function for Guiding the Bees in the ABC Algorithm

The objective function holds crucial significance in the ABC algorithm because it serves as the yardstick that determines the quality of potential solutions to the problem, which is represented by the positions of food sources. The objective function is used to calculate the fitness or quality of that solution. It allows the bees to compare different options offered.

How is the objective function calculation utilized in the algorithm? The ABC algorithm uses the objective function values to guide the movement of bees. The bees tend to explore areas with better objective function values, which leads them toward the optimal solution. The information from the objective function is used by the onlooker bees to decide which food sources or solutions to focus on. Food sources with higher quality or better objective function values are prioritized. The algorithm frequently reaches a conclusion and halts when the improvement in the objective function value becomes negligible, signifying that a near-optimal solution has been found.

14.2.3.1 Objective Function as a Measure of the Robot Swarm Performance

In an ABC algorithm applied to a robot swarm, the objective function represents a measure of the proficiency with which the swarm is performing its designated task. The measurement of the proficiency is inclusive of operations such as minimizing the total distance traveled to reach a target, maximizing coverage area in an exploration scenario, or optimizing the collective decision-making process, depending on the specific application. It quantifies the quality of the current configuration or behavior of the swarm. Therefore, it provides supervisory recommendations to the algorithm for adjusting robot positions and actions to improve the overall performance.

14.2.3.2 Considerations about Choosing the Objective Function in a Robot Swarm

Vital considerations to be kept in mind during the selection of the objective function in a robot swarm ABC algorithm are:

i. Foundation of the Objective Function Formula: The specific formula for the objective function is directly related to the desired outcome of the swarm. Possible outcomes are minimizing the average distance to a target, maximizing the number of points covered in a search operation, or balancing resource allocation among robots.
ii. Modeling and Simulation of Collective Swarm Behavior: As each robot within the swarm contributes to the overall objective function depending on its current position, actions, and sensory data, the algorithm evaluates the collective behavior of the swarm, and its values portray the same.
iii. Assessment of Fitness Score: The objective function bears an analogy to the concept of nectar in a bee colony. Its value acts as a fitness score for each potential solution or robot configuration. A higher value of the objective function for a solution reflects better performance for that solution.

14.2.3.3 Dependence of Objective Functions on Robot Swarm Goals

For a robot swarm using ABC, the objective function is chosen in accordance with the goal to be reached:

i. For Area Coverage Responsibility: Aiming to distribute robots evenly across the area, the sum of the distances between each robot and its nearest neighbors is taken as the objective function.
ii. For Target Search Job: Focusing on the target, the minimum distance between any robot in the swarm and the target location is preferred as the objective function.
iii. For Task Allocation Duty: Good counseling to the swarm is provided by a combination of factors like completion time, efficiency, and workload distribution among robots. So, the objective function is defined with these issues in mind.

14.2.3.4 Objective Function for ABC Algorithm Applied to Organizing Robot Navigation

In the ABC algorithm applied to robot navigation, the objective function typically represents the shortest path distance between the starting point of the robot and its goal. At the same time, it considers obstacles and other constraints in the environment. So, it smartly aims to minimize this distance to find the optimal path.

Prime features of the objective function in the ABC algorithm for robots are

i. Emphasis on Minimization of Objective Function Value: The objective function is usually designed to be minimized. This statement means that the algorithm seeks the path that yields the lowest total distance traveled.

ii. Incorporation of Path Parameters: The objective function incorporates parameters such as the coordinates of locations or landmarks along the robot's path. Then the ABC algorithm works for the adjustment of these parameters to optimize the route.

iii. Penalty Term Inclusion for Obstacle Avoidance: When navigating through obstacles, the objective function includes penalty terms for reaching in close vicinity of obstacles, encouraging the robot to find a safe path, e.g.,

$$
\begin{array}{c}
\text{Objective function for} \\
\text{obstacle avoidance}
\end{array}
=
\begin{array}{c}
\text{Distance between consecutive} \\
\text{points on the path of the robot}
\end{array}
+
\begin{array}{c}
\text{Penalty value if the robot} \\
\text{reaches very close to} \\
\text{an obstacle}
\end{array}
$$

$$(14.6)$$

iv. Multi-Objective Optimization in Complex Scenarios: The objective function takes multiple factors into account. These factors are the travel time, energy consumption, or smoothness of the path. As a result, the end task becomes the optimization of a multi-objective problem.

14.2.4 APPLICATIONS OF THE ABC ALGORITHM IN ROBOTICS

The ABC algorithm finds applications not only in robot path planning and multi-robot coordination but also in robot arm manipulation. Significant areas where it makes an impact are:

i. Robot Path Planning and Navigation: The ABC algorithm is widely used in robotics for robot path planning. It helps in finding the best, optimal, or near-optimal collision-free routes for robots. Mirroring the eating habits of honeybees, it navigates through cluttered environments by optimizing the sequence of waypoints and movement parameters. It takes factors like distance, time taken, obstacles, terrain, and other constraints into consideration.

ii. Multi-robot Coordination: The ABC algorithm aids in coordinating the movement of multiple robots to achieve a collective goal. These could be assigning tasks and optimizing their paths to avoid collisions and maximize efficiency. The algorithm is particularly useful in scenarios with obstacles.

iii. Robot Arm Manipulation: The ABC algorithm is used for optimizing the trajectory of a robotic arm to precisely reach a target position while circumventing the limits imposed by the joints of the robot and the obstacles on its way. This activity of the robot is based on:

a. Optimization of Joint trajectory: The movement of robot joints is adjusted to achieve smooth and efficient motion.

b. Grasp Planning: The best hand configuration to grasp an object is determined.

c. Object Maneuvering: The movements of the robot are controlled to manipulate objects precisely.

iv. Adaptive Robot Control: The ABC algorithm is applied to dynamically adjust the parameters of a robotic control based on changing environmental conditions, thereby enhancing its performance in real-time situations.

14.2.5 Advantages of the ABC Algorithm in Robotics

The ABC algorithm offers several benefits in robotics. These include its simplicity, ease of use, robust exploration skills, the ability to strike a proper equilibrium between exploration and exploitation, the capacity to manage intricate, high-dimensional search spaces, and the capability for fast convergence. These recompenses make it suitable for solving various optimization problems. These include robot path planning and motion control, particularly when talking about non-linear or multimodal scenarios. The advantages are enumerated below:

i. Simple in Understanding and Easy in Implementation: The ABC algorithm is based on the foraging behavior of bees. It translates to a relatively straightforward concept, making it easier to implement compared to other complex optimization methods. The straightforward structure of the ABC algorithm with a few parameters makes it easy to integrate into robotic systems.

ii. Effectiveness of Exploration and Exploitation: The ABC algorithm balances exploration (searching new areas of the solution space) and exploitation (focusing on promising areas and refining favorable solutions) through its distinct bee types: employed bees, onlooker bees, and scout bees. This balancing allows it to find good solutions in complex environments.

iii. Capability of Handling Complex Problems: Due to its swarm intelligence nature, the ABC algorithm can efficiently tackle multifaceted optimization problems with many variables and constraints. It can effectively handle multiple constraints in robotic issues, such as joint limits, obstacle avoidance, and energy consumption. These kinds of problems are common in robotic applications.

iv. Fast Convergence to a Solution: The ABC algorithm usually converges quickly to a near-optimal solution. The quick convergence is crucial for real-time robotic decision-making. However, quick convergence sometimes evades thorny problems.

v. Adaptability: The algorithm is easily modified and hybridized with other optimization techniques to suit specific needs in robotics.

vi. Parallelization Potential: The independent behavior of bees allows for parallel processing. The parallelism approach can significantly improve computation speed in multifarious robotic scenarios.

14.2.6 LIMITATIONS OF THE ABC ALGORITHM IN ROBOTICS

When applied to robotics, the ABC algorithm suffers from several limitations, including: slow speed of convergence with increasing problem complexity, vulnerability to local optima, poor exploitation ability, and anticipated difficulty in handling complex, high-dimensional robotic problems. These shortcomings lead to suboptimal solutions and inefficient path planning in real-time scenarios.

i. Weakness in Exploitation: ABC excels in exploration. This means it is able to search a wide range of solutions but struggles to refine solutions near the optimal point. Every so often, it gets stuck in local minima due to its basic search mechanism.

ii. Slow Convergence in Solving Complex Problems: The algorithm requires a substantial number of iterations to reach a near-optimal solution for intricate problems. The large number of iterations is problematic in time-critical robotic applications.

iii. Parameter Tuning Sensitivity: The performance of the ABC algorithm is sensitive to the selection of appropriate parameters, such as the population size and the number of iterations (i.e., the iteration count). It requires careful parameter tuning for solving specific robotic problems. A casual approach to tuning can severely impact algorithm performance.

iv. Limited Dimensionality Handling Capability: The standard ABC algorithm struggles with high-dimensional search spaces commonly encountered in complex robotic tasks. In these cases, many variables must be optimized simultaneously.

v. Likelihood of Premature Convergence: In certain situations, especially when dealing with inexplicable, non-convex optimization landscapes, the ABC algorithm converges too quickly to a local minimum, suboptimal solution. The potential for unduly hasty, untimely convergence necessitates modifications to enhance exploration.

14.2.7 ADDRESSING THE ABC ALGORITHM LIMITATIONS

After knowing the limitations of the ABC algorithm, one can devise suitable strategies to overcome its shortcomings. To cope with these limitations, the principal strategies that evolved are:

i. Hybrid Approaches: The ABC algorithm is combined with other optimization algorithms like the particle swarm optimization (PSO) algorithm. This unification of algorithms is beneficial because it enables leveraging the strengths of both methods to improve exploration and exploitation capabilities.

ii. Adaptive Parameter Tuning: The algorithm's parameters are dynamically adjusted, taking the search progress into account. The dynamic adjustment procedure helps in addressing the sensitivity of the algorithm to the setting of parameters.

iii. Improved Search Operators: The core search equation of the ABC algorithm is modified with a view to enhancing the exploitation phase and navigating weird search spaces in a better way.

iv. Multi-objective Optimization: It involves the utilization of multi-objective variants of ABC algorithm. This allows handling of multiple competing objectives in robotic tasks. Robot path planning is conducted while considering factors such as distance, safety, and energy consumption.

14.3 FIREFLY ALGORITHM

Like the algorithms treated in foregoing sections, the FA is a nature-inspired optimization technique (Fister et al. 2013; Patle et al. 2017, 2018, 2023). It facsimiles the social behavior of fireflies for the coordination and guidance of a group of robots toward a desired goal. It helps robots find the best path through a messy and chaotic environment by sidestepping obstacles.

In this algorithm, each robot is considered a firefly. Each firefly represents a potential solution to the problem. The intensity of light emitted by it indicates the quality of that solution. The movements of fireflies are determined by their attraction to each other. The attraction depends on the luminous intensity, or brightness, of the fireflies. The less bright fireflies move toward the better-performing brighter ones. So, each firefly migrates toward the brighter fireflies in the swarm (Figure 14.2a). Figure 14.2a shows a Firefly 1 in its initial state. It has the lowest brightness value. The brightness of fireflies increases in the order 1, 2, 3, 4, with Firefly 1 at the minimum and Firefly 4 at the maximum level. So, the fireflies move in the sequence 1, 2, 3, 4. Firefly 5, which is less bright than Firefly 4, is also attracted toward Firefly 4, after which the goal state is reached.

Each firefly or robot starts with a random position. At the starting point, it evaluates the quality of its initial path. The algorithm effectively allows the robots to explore and find optimal solutions to complex problems collaboratively. Robot path planning, target tracking, and environmental mapping are the types of issues that are resolved.

14.3.1 Essential Points about the FA

Some vital features and ideas related to the FA warrant the reader's attention:

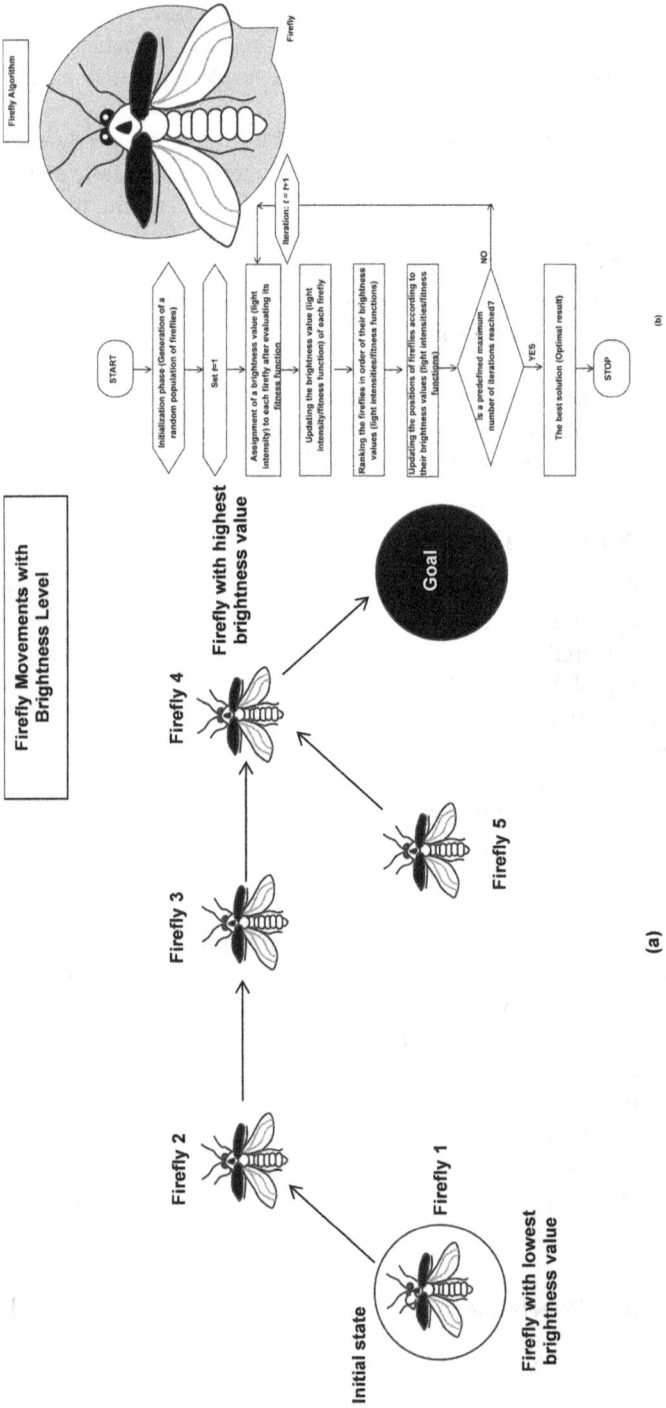

FIGURE 14.2   The firefly algorithm: (a) firefly movements and (b) the algorithm flowchart.

i. Origination of Algorithm from Bio-Inspiration: The algorithm does an impersonation of the flashing behavior of fireflies to solve optimization problems. A brighter firefly attracts a less brilliant one. The brighter firefly represents a robot that is moving toward a better solution, dependent on a fitness function correlated to the task at hand.
ii. Dynamics of Robot Motion: Each robot calculates its fitness value, viz., the brightness. Based on the fitness value, it moves toward the brighter robots in its proximity. As a consequence, a simulation of the attraction between fireflies is performed.
iii. Exploration and Exploitation: The algorithm balances two activities. These activities are exploration and exploitation. Exploration is concerned with searching a wide area. Exploitation focuses on promising regions. The balancing of activities is done by adjusting parameters such as attractiveness and randomness. This process of balancing enables the swarm to stumble upon diverse solutions and then converge on the best solution among the discovered ones.

14.3.2 IMPORTANT PARAMETERS OF THE FA

Fundamental parameters specific to the FA must be defined. These are as follows:

i. Brightness Function: It is a mathematical function determining the brightness of a robot. The brightness depends on the current state of the robot or its performance on the given task. The brightness of a robot firefly is directly proportional to the intensity of light radiated by it.

Light Intensity: It represents the fitness value of a firefly and governs its attractiveness to other fireflies. The light intensity $I(x)$ of a firefly is related to the objective function by the equation (Mashhour et al. 2020)

$$I(x) \propto f(x) \tag{14.7}$$

where $f(x)$ is the value of the objective function.
ii. Attractiveness Parameter: It controls the strength with which a firefly is attracted to other brighter fireflies, i.e., how strongly a robot is attracted to a brighter neighboring robot. It is usually measured by a parameter that depends on the distance between the robots.
iii. Randomness Parameter: It introduces random movement of robots to prevent premature convergence of the algorithm. Hence, the search space is explored more effectively.

14.3.3 MAIN STEPS OF THE FA

The principal steps of the FA in swarm robotics are shown in Figure 14.2b: start, initializing a population of fireflies by placing the fireflies at random positions, setting $t = 1$, determining the brightness of each firefly from a fitness function, updating the brightness of fireflies according to their fitness, checking if the prefixed maximum number of iterations is reached, if NO setting $t = t + 1$, iteratively moving each firefly toward brighter fireflies in the population, and repeating the process until a suitable solution is found. If YES, record the best solution and stop. The algorithmic process is essentially a representation of the social behavior of fireflies. In their society, the fireflies are attracted to brighter individuals. The attraction of fireflies allows the swarm to converge toward an optimal solution. Further specifics about the algorithm stages are declared as follows (Banerjee et al. 2022):

i. Initialization: A random population of fireflies is generated. This population represents potential solutions in the search space. A brightness value is assigned to each firefly based on its fitness function. The higher the fitness, the brighter the firefly.
   Firefly Movement and Attraction: For each firefly:
   a. The distance to every other firefly in the population is calculated.
   b. If another firefly is brighter, the firefly under consideration moves toward it based on the distance and brightness difference between fireflies.
   c. From the calculated movement of the current firefly, the position of the current firefly is updated.
ii. Brightness Update: After each movement, the brightness of each firefly is re-evaluated by feeding its updated position and the fitness function.
iii. Iteration and Termination: The firefly movement and brightness update steps are repeated for a predefined number of iterations. The best solution is usually the one having the brightest firefly at the end of the iterations.
iv. Convergence: In order to avoid getting stuck in local optima, the algorithm should balance between the activities of exploration (searching a large area) and exploitation (focusing on promising regions).

14.3.4 APPLICATIONS OF FA

Besides robot path planning, the FA helps in target localization and tracking, cooperative exploration and decision-making in a group of robots, and in many other ways.

i. Robot Path Planning: The algorithm uses fireflies to find optimal routes for locomotion of mobile robots in complex environments, taking cognizance of the obstacles and terrain variations, and maximizing energy efficiency. The robotic navigation is rendered possible by collectively finding the most efficient route. As the brighter fireflies represent better paths, each robot is attracted to the brightness-steered best path discovered by other robots in the swarm.
ii. Target Localization and Tracking by Robots: A swarm of robots collaboratively track a moving target. They do so by adjusting their positions based on the brightness signal received from the target, i.e., by moving toward

brighter signals emitted from the target. In this way, they are collectively able to locate a target.

iii. Cooperative Exploration and Decision-Making in a Robotic Swarm: A robotic swarm collectively makes decisions by simulating firefly behavior. The best solution emerges based on the brightness of different alternative options. Along these lines, the robots can explore a large area by moving toward regions with the highest brightness, which represent the most interesting features.

iv. Facilitating Swarm Robotic Jobs: The movements and task allocations of multiple robots in a swarm are coordinated to achieve collective goals like coverage or exploration

v. Sensor Deployment and Fusion: Robots collectively build a more accurate picture of the environment by sharing information about their brightness values indicated by sensor readings. This pictorial representation of the environment is used to optimize the placement of sensors in a given area, maximizing coverage and efficiency.

vi. Robot Arm Manipulation: Robot arm trajectories are optimized for carrying out precise manipulation tasks satisfactorily.

14.3.5 Advantages of the FA

The FA offers several advantages in robotics. Among these, the benefits that are most worthy of attention include its ability to handle complex, multi-dimensional optimization problems and find near-optimal solutions in dynamic environments. Additionally, it enables efficient path planning, effective multi-robot coordination, and generally provides a good balance between exploration and exploitation. These capabilities make the algorithm suitable for multifaceted robotic applications like swarm robotics and sensor deployment strategies.

The advantages of the FA in robotics are:

i. Global Optimization Capability: FA can effectively search for near-global optima in complex problem spaces. This benefit offered by FA is crucial for finding optimal robot paths or coordinating a swarm of robots in challenging environments.

ii. Adaptability to Dynamic Environments: The algorithm adapts to changing conditions by adjusting the movement of fireflies representing robots in response to real-time information. Flexible path planning in dynamic environments can therefore be made.

iii. Multi-robot Coordination: The FA can facilitate coordinated movement and task allocation among multiple robots in a swarm by simulating the attraction behavior of fireflies. This facilitation makes efficient collaborative behaviors possible.

iv. Easy Implementation: The FA is relatively simple to implement and understand. Hence, it can be rapidly prototyped to carry out experiments in robotic applications.

   v. Non-requirement of a Good Initial Solution: FA does not require precise initial guesses, unlike some other optimization algorithms. Abdication of guesswork makes it suitable for scenarios where the initial state of a robot is uncertain.

  vi. Ability to Handle Complex Constraints: The algorithm is adaptable to incorporate various constraints related to robot movement. Examples of constraints are obstacle avoidance or energy consumption, while searching for optimal solutions.

14.3.6 Disadvantages of the FA

In robotics, the main disadvantages of the FA include its tendency to get fastened to local optima and provide a slow convergence speed. Besides these shortfalls, sensitivity to parameter tuning and potential for premature convergence hinder its ability to find optimal solutions in complex robotic navigation scenarios. The issues are especially aggravated when dealing with multimodal problems having multiple solutions, including one or more global solutions.

The FA disadvantages in robotics are as follows:

   i. Local Optima Trap: The algorithm follows the principle of movement toward brighter fireflies. As it progresses, it easily gets stuck in a local optimum. This happens whenever a firefly encounters a seemingly better solution at an early stage. In such a situation, the exploration of the wider search space is thwarted.

  ii. Slow Convergence of the Algorithm: In certain scenarios, FA makes a large number of iterations to reach a satisfactory solution. This slow convergence process makes it less efficient for real-time robotic applications.

 iii. Premature Convergence of the Algorithm: In some cases, the algorithm converges too quickly to a suboptimal solution. This is likely especially when dealing with high dimensionality or complex environments.

  iv. Parameter Sensitivity: The performance of FA heavily depends on the correct selection of parameters like attractiveness and randomness coefficients. Such over-reliance has unfavorable repercussions. Optimization of these parameters is often an uphill task in complicated robotic problems.

   v. Limited Exploration Ability: The movement of fireflies is primarily directed toward brighter ones. This approach restricts the exploration of diverse areas in the search space, often resulting in the missing of better solutions.

14.3.7 Possible Solutions to Mitigate Disadvantages of FA

Knowledge of the drawbacks of the FA aids in developing methods to alleviate the deficiencies. Among these methods, the following are regarded as exigent and demanding:

   i. Hybrid approaches: Exploration ability is improved along with the possibility to escape from local minima when FA is combined with other optimization algorithms, e.g., genetic algorithms or simulated annealing.

ii. Adaptive Parameter Tuning: Mechanisms can be implemented to adjust parameters based on the optimization progress dynamically. These adjustments can enhance the performance of the algorithm.

iii. Improved Attractiveness Function: The attractiveness function is modified to better represent the problem and encourage more varied exploration.

iv. Algorithm Based on Leader Strategy: The problem of unbalanced exploration and exploitation, and the insufficient diversity of the algorithm result in a firefly search algorithm based on the leader and follower population model being proposed (Zhang and Wang 2023).

14.4   DISCUSSION AND CONCLUSIONS

Table 14.1 gives a presentation of the discussions in this chapter in a terse tabular format. An important consideration when using the ABC and FAs in robotics is the correct formulation of the problem because successful optimization depends on the accurate definition of the objective function. No less significant is the incorporation of constraints of robot kinematics and dynamics into the optimization process. As robotic systems require real-time decision-making, a high computational efficiency of the algorithm used is yearned for.

TABLE 14.1
Takeaways from This Chapter at a Glance

Sl. No.	Takeaway	Explanation
1	Summary	Two popular swarm intelligence optimization techniques used for robot path planning, motion control, obstacle avoidance, and multi-robot control were described, namely, the artificial bee colony (ABC) and the firefly algorithm (FA).
2	ABC algorithm	The ABC algorithm mimics the foraging behavior of honeybees, where bees work in different roles, such as employed, onlooker, and scout bees, collaborating to find the best food source. The algorithm iteratively updates potential solutions (representing robot movements) based on the quality of the food source (fitness function), allowing for both exploration (searching new areas) and exploitation (refining good solutions).
3	FA	The FA is based on the flashing behavior of fireflies, where brighter fireflies attract the less bright ones. Each firefly represents a potential solution, and the fireflies move toward brighter (better) solutions based on their relative brightness, gradually converging toward the optimal solution.
4	ABC algorithm vs FA	The ABC algorithm is better suited for solving problems that involve exploring a large search space, thanks to its diverse bee roles. FA is more suitable for solving complex, high-dimensional problems and fine-tuning solutions due to its attraction mechanism.
5	Keywords and ideas to remember	Artificial bee colony algorithm; employed bee, onlooker bee, and scout bee solution search phases; objective function, firefly algorithm, comparison of ABC algorithm, and FA

Swarm robotics utilizes a variety of algorithms, allowing the robots to coordinate and solve entwined tasks by exploring and exploiting search spaces. As a multitude of swarm robotic algorithms have come into the limelight and continue to do so (Nayak et al. 2020), the discussion of algorithms will be continued in the ensuing chapter.

REFERENCES AND FURTHER READING

Banerjee A., D. Singh, S. Sahana and I. Nath. 2022. Impacts of Metaheuristic and Swarm Intelligence Approach in Optimization. In: Mishra S., H. K. Tripathy, P. K. Mallick, A. K. Sangaiah and G.-S. Chae (Eds.), *Cognitive Data Science in Sustainable Computing, Cognitive Big Data Intelligence with a Metaheuristic Approach*, Academic Press, pp. 71–99.

Bansal J. C., H. Sharma and S. S. Jadon. 2013. Artificial bee colony algorithm: A survey, *International Journal of Advanced Intelligence Paradigms*, Vol. 5, 1/2, pp. 123–159.

Bisen A. S. and V. Kaundal. 2020. Mobile Robot for Path Planning Using Firefly Algorithm, *2020 Research, Innovation, Knowledge Management and Technology Application for Business Sustainability (INBUSH)*, Greater Noida, India, 19–21 February, pp. 232–235.

Chaudhary K. C. 2023. A modified version of the ABC algorithm and evaluation of its performance, *Heliyon*, Vol. 9, 5, e16086, pp. 1–19.

Cui Y., W. Hu and A. Rahmani. 2022. A reinforcement learning based artificial bee colony algorithm with application in robot path planning, *Expert Systems with Applications*, Vol. 203, p. 117389, https://doi.org/10.1016/j.eswa.2022.117389

Cui Y., W. Hu and A. Rahmani. 2024. Multi-robot path planning using learning-based Artificial Bee Colony algorithm, *Engineering Applications of Artificial Intelligence*, Vol. 129, p. 107579, https://doi.org/10.1016/j.engappai.2023.107579

Fister I., I. Fister Jr., X.-S. Yang and J. Brest. 2013. A comprehensive review of firefly algorithms, *Swarm and Evolutionary Computation*, Vol. 13, pp. 34–46.

Huang H.-C. and C.-C. Chuang. 2020. Artificial bee colony optimization algorithm incorporated with fuzzy theory for real-time machine learning control of articulated robotic manipulators, *IEEE Access,* Vol. 8, pp. 192481–192492.

Karaboga D. 2010. Artificial bee colony algorithm. *Scholarpedia*, Vol. 5, Article No. 6915, https://doi.org/10.4249/scholarpedia.6915

Karaboga D. and B. Basturk. 2007a. Artificial Bee Colony (ABC) Optimization Algorithm for Solving Constrained Optimization Problems. In: Melin, P., O. Castillo, L. T. Aguilar, J. Kacprzyk and W. Pedrycz (Eds.), *Foundations of Fuzzy Logic and Soft Computing. IFSA 2007*. Lecture Notes in Computer Science, Vol. 4529, Springer, Berlin, Heidelberg, pp. 789–798.

Karaboga D. and B. Basturk. 2007b. A powerful and efficient algorithm for numerical function optimization: Artificial bee colony (ABC) algorithm, *Journal of Global Optimization*, Vol. 39, 3, pp. 459–471, https://doi.org/10.1007/s10898-007-9149-x

Li X., Y. Huang, Y. Zhou and X. Zhu. 2018. Robot Path Planning Using Improved Artificial Bee Colony Algorithm, *2018 IEEE 3rd Advanced Information Technology, Electronic and Automation Control Conference (IAEAC)*, Chongqing, China, 12-14 October, pp. 603–607.

Mashhour E. M., E. M. F. El Houby, K. T. Wassif and A. I. Salah. 2020. A novel classifier based on firefly algorithm, *Journal of King Saud University – Computer and Information Sciences*, Vol. 32, 10, pp. 1173–1181.

Nayak J., B. Naik, D. Pelusi and A. V. Krishna. 2020. A Comprehensive Review and Performance Analysis of Firefly Algorithm for Artificial Neural Networks. In: Yang X. S. and X. S. He (Eds.), *Nature-Inspired Computation in Data Mining and Machine Learning*. Studies in Computational Intelligence, Vol. 855, Springer, Cham, pp. 137–159.

Nozohour-leilabady B. and B. Fazelabdolabadi. 2016. On the application of artificial bee colony (ABC) algorithm for optimization of well placements in fractured reservoirs; efficiency comparison with the particle swarm optimization (PSO) methodology, *Petroleum*, Vol. 2, 1, pp. 79–89.

Patle B. K., A. Pandey, A. Jagadeesh and D. R. Parhi. 2018. Path planning in uncertain environment by using firefly algorithm, *Defence Technology*, Vol. 14, 6, pp. 691–701.

Patle B. K., D. R. Parhi, A. Jagadeesh and S. K. Kashyap. 2017. On firefly algorithm: Optimization and application in mobile robot navigation, *World Journal of Engineering*, Vol. 14, pp. 65–76.

Patle B. K., B. Patel, A. Jha and S. K. Kashyap. 2023. Self-directed mobile robot navigation based on functional firefly algorithm (FFA), *Eng*, Vol. 4, 4, pp. 2656–2681.

Wei Q., C. Chen, X. Pang and M. Huang. 2023. Robot Dynamic Path Planning Based on Improved Firefly Algorithm, *2023 3rd International Symposium on Artificial Intelligence and Intelligent Manufacturing (AIIM)*, Chengdu, China, 27–29 October, pp. 122–125.

Xu F., H. Li, C.-M. Pun, H. Hu, Y. Li, Y. Song and H. Gao. 2020. A new global best guided artificial bee colony algorithm with application in robot path planning, *Applied Soft Computing*, Vol. 88, p. 106037, https://doi.org/10.1016/j.asoc.2019.106037

Yurtkuran A. and E. Emel. 2016. An enhanced artificial bee colony algorithm with solution acceptance rule and probabilistic multisearch, *Computational Intelligence and Neuroscience*, Vol. 2016, Article ID 8085953, pp. 1–13.

Zhang X. and S. Wang. 2023. Firefly search algorithm based on leader strategy, *Engineering Applications of Artificial Intelligence*, Vol. 123, Part B, 106328, https://doi.org/10.1016/j.engappai.2023.106328

Zhou X., G. Tan, H. Wang, Y. Ma and S. Wu. 2025. Artificial bee colony algorithm based on multi-neighbor guidance, *Expert Systems with Applications*, Vol. 259, p. 125283, https://doi.org/10.1016/j.eswa.2024.125283

15 Robotic Swarms
Expanding Horizons

15.1 INTRODUCTION

This chapter further extends our coverage of swarm robotics, continuing the discussion from the previous chapter. Herein, we undertake the study of bacterial foraging optimization (BFO) algorithm (Yang et al. 2014; Majumder et al. 2019) and salp swarm algorithm (SSA) (Romeh and Mirjalili 2023; Yang et al. 2024). These algorithms are used to determine the optimal paths for robots to navigate through complex environments, taking into account obstacles on the route and variations in terrain. They escort robots to avoid collisions with obstacles during navigation. The algorithms are also utilized in multi-robot collaboration, facilitating the coordinated movement and decision-making of a team of robots to work cooperatively and complete challenging tasks. Hence, they are indispensable components in the toolkit of robotic algorithms.

Advancing still further in this chapter, a comparative analysis of the various swarm robotic algorithms discussed in Chapters 13–15 will be performed. It is predicated on the optimality of the path, the complexity of computation, adaptability to various environments, real-time performance, and suitability for specific robotic activities. The advantages and disadvantages of the algorithms are emphasized to determine the best choice for a particular situation. Comparing the pros and cons of each option provides a structured and logical approach to making a decision.

15.2 BFO ALGORITHM

The BFO is a swarm intelligence optimization algorithm. It borrows its root working idea from the collective food searching behavior of bacteria like *Escherichia coli* (*E. coli*) (Guo et al. 2021; Wang et al. 2022). It treats each robot as a bacterium. Bacterial actions, such as chemotaxis, reproduction, and elimination-dispersal, serve as mechanisms that can be applied to solve robotic problems. In these mechanisms, the bacterium moves toward nutrient-rich areas by swimming and tumbling, eventually falling head over heels (Figure 15.1a and b). Figure 15.1a shows an *E. coli* bacterium moving straight in the running mode, while Figure 15.1b shows the bacterium in the clockwise tumbled position. The algorithm essentially simulates the process of searching for the best solution in a complex problem space. Indeed, it is used to find optimal solutions to Byzantine optimization problems across various fields of engineering, data science, and robotics.

15.2.1 ESSENTIAL FEATURES OF THE BFO ALGORITHM

The four prominent characteristics of the BFO algorithm are (Fiveable 2024):

DOI: 10.1201/9781032695266-15

i. Chemotaxis: It concerns the directed movements of cells, organisms, or their parts in response to a chemical stimulus. It is the primary behavior of bacteria as they move toward a nutrient source. In this behavioral style, the bacteria adjust their swimming direction based on the concentration gradient of the nutrient. Chemotaxis essentially represents the algorithm's hunt for better solutions.

ii. Swarming: This occurs when bacteria gather around a high concentration of nutrients, representing a local search phase. Here, the algorithm focuses on refining promising solutions within a specific area.

iii. Reproduction: High-fitness bacteria or good solutions are allowed to replicate. The propagation of better solutions is thereby proliferated.

iv. Elimination-Dispersal: Low-fitness bacteria or poor solutions are eliminated from the population. Their stamping out prevents stagnation and encourages exploration and investigation of new areas in the search space.

15.2.2 MAIN STEPS OF THE BFO ALGORITHM

Figure 15.1c depicts the course of actions followed in the execution of the BFO algorithm. The stages in the BFO algorithm are: start, initialization phase, evaluation of the fitness function, beginning and ending of the chemotaxis phase, commencing and closing of the reproduction phase, start and end of the elimination-dispersal phase, optimization of values, the best solution phase, and then the algorithm comes to a halt. The ends of chemotaxis, reproduction, and elimination-dispersal phases mark decision steps after which further progress is made through self-examination of the status of calculations and, accordingly, determining what to do next. If YES, the algorithm moves forward. If NO, it moves back to the step of evaluation of the fitness function. The output from the step of optimization of values is fed back to the initialization phase. We explore these stages in depth as follows to uncover more details (Gan and Xiao 2020):

i. Initialization: A randomly distributed population of bacteria is dispersed within the search space. This random distribution represents the potential solutions or fixes to the problem.

ii. Chemotaxis Phase: Each bacterium moves toward a better solution. During movement, a bacterium freely adjusts its position based on the fitness function. A mechanism of tumbling or randomly changing direction acts as a means of escaping precarious episodes. It helps to avoid situations in which the bacterium becomes immobilized in local optima.

iii. Reproduction Phase: Bacteria with higher fitness values are allowed to replicate. By replication, new bacteria are created, possessing similar characteristics.

iv. Elimination-Dispersal Phase: Bacteria with low fitness are eliminated from the population. In reciprocation, new bacteria are randomly introduced into the population to maintain diversity and heterogeneity.

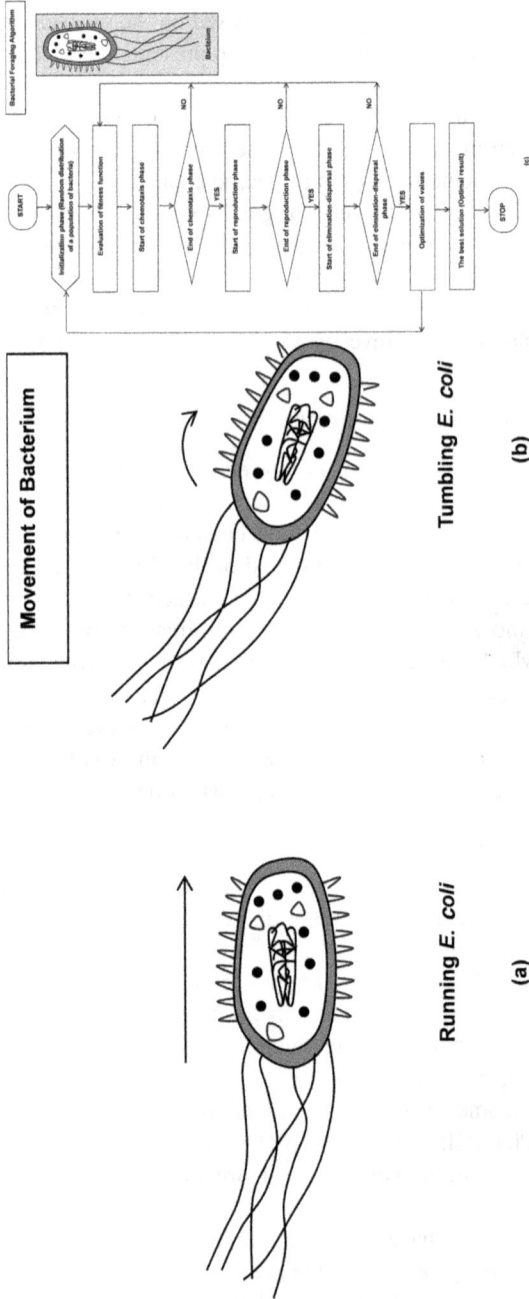

FIGURE 15.1 The swarm robotic algorithm enthused by the mechanisms of the natural process of bacterial foraging optimization: (a) running mode of the bacterium, (b) tumbling mode of the bacterium, and (c) the algorithm.

15.2.3 APPLICATIONS OF THE BFO ALGORITHM

The BFO algorithm finds applications in swarm robotics primarily for efficient allocation of tasks that require decision-making in a decentralized manner with a dispersed approach, and coordinated movements among a group of robots. It capitalizes on the natural bacterial properties to optimize collective behavior, such as area coverage, path planning, and resource search, within a swarm environment.

i. Area Coverage by Robots: The BFO algorithm effectively distributes robots across a given area by simulating the bacteria's chemotaxis behavior. Each robot navigates toward nutrients, representing target points. Vigilance is maintained to avoid collisions and optimize coverage efficiency.

ii. Robot Path Planning: The BFO algorithm is used to plan optimal paths for individual robots within a swarm to navigate complex environments by simulating the movements of bacteria toward food sources. The bacteria move by avoiding obstacles to reach the targeted destinations efficiently.

iii. Resource Search by Robots: The BFO algorithm guides robots to search different areas based on chemical gradients representing resource concentrations. Faster and more comprehensive search operations are rendered possible by this algorithm in scenarios where a swarm needs to locate disseminated and scattered resources.

iv. Task Allocation to Robots: Robots dynamically adapt their roles within the swarm by adjusting the reproduction and elimination steps in the BFO algorithm. These steps are designed for allocating tasks based on their current position, capabilities, and environmental conditions.

v. Robot Swarm Coordination: The decentralized nature of BFO allows each robot to make local decisions. The decision made by a robot is based on its immediate entourage or neighborhood. The robot's decision facilitates emergent behaviors, such as the aggregation of the swarm, its splitting, and re-grouping when needed.

15.2.4 ADVANTAGES OF THE BFO ALGORITHM

Calling attention to the benefits of the BFO algorithm, mention may be made of:

i. Easy Comprehension and Implementation: The basic BFO algorithm is relatively simple to understand in principle and easy to apply in practice. These favorable features make it suitable for real-time applications on robot swarms.

ii. Effective for Complex Optimization Issues: The BFO algorithm offers diverse search mechanisms. Therefore, it is capable of handling problems with multiple local optima.

iii. Good Balance between Exploration and Exploitation: The tumbling behavior in chemotaxis allows for exploration of the search space. At the same time, focus on refining promising solutions is maintained.

iv. Global Search Capability: BFO effectively explores a large search space. This ability of the BFO algorithm ensures that the swarm can find optimal solutions even in complex and tangled environments.

v. Environmental Adaptability: The BFO algorithm can adapt to dynamic environments and changing task requirements. This adaptation is achieved by adjusting parameters like the size of the chemotaxis step.

vi. Parallel Processing: The independent behavior of individual bacteria or robots allows for efficient parallel processing. Parallel operation is crucial for robotic large-scale swarm systems.

15.2.5 LIMITATIONS OF THE BFO ALGORITHM

The BFO algorithm has several limitations, primarily its slow convergence speed and susceptibility to becoming trapped in local optima due to a fixed chemotactic step size. Difficulty is experienced in balancing exploration with exploitation. Exploration is trying new things. Exploitation is utilizing what is known. Potentially weak connections between bacteria incite suboptimal solutions. Such limitations hamper its efficacy in real-time robotic applications. In these cases, making a fast and accurate decision is a mandatory requirement.

i. Fixed Step Size: The standard BFO algorithm uses a constant chemotaxis step size. The constant size of the step is the cause of the algorithm's poor performance in complex environments. In such situations, different levels of exploration are required, depending on the circumstances. The action will vary with the specific context.

ii. Local Optima Trapping: Due to the fixed step size, the BFO algorithm readily gets trapped in local optima. When so trapped, it delivers a suboptimal solution instead of the global best solution.

iii. Slow Convergence of the Algorithm: In certain scenarios, the BFO algorithm takes a long time to reach a satisfactory solution. The slow convergence makes it less appropriate for real-time robotic applications that require rapid response.

iv. Need for Algorithm Parameter Tuning: For specific robotic problems, it has been found that optimizing the parameters of the BFO algorithm, such as the chemotaxis step size and reproduction rate, becomes difficult, requiring precise execution, and thus needs to be carefully performed.

v. Limited Applicability of the Algorithm: The BFO algorithm is not the best choice, especially when dealing with highly dynamic or unpredictable situations, which depend on the task and environmental complexity.

15.2.6 POTENTIAL SOLUTIONS TO ADDRESS THE BFO ALGORITHM LIMITATIONS

Several alternative methodologies are suggested to deal with the limitations of the BFO algorithm:

i. Making Chemotaxis Self-Adaptive: A self-adaptive chemotaxis step size is adopted. The size of the step is adjusted based on the search process. This adjustment improves exploration and exploitation capabilities.

ii. Hybrid Approaches: The BFO algorithm is often combined with other optimization algorithms, such as evolutionary algorithms. This combination helps in overcoming its limitations. Furthermore, it offers improved performance in specific robotic tasks.

iii. Improvement in Population Diversity: Strategies to maintain population diversity prevent premature convergence of the algorithm and improve the search process. Introduction of random perturbations and incorporation of diversity measures are examples of such strategies.

15.3   SALP SWARM ALGORITHM

From the foraging behavior of bacteria, we transition to the swarming behavior of salps moving in a chain-like structure. The SSA is a computational optimization technique in swarm robotics. It intimately parallels the collective behavior of marine creatures called salps to solve optimization problems (Faris et al. 2019; Houssein et al. 2020; Castelli et al. 2022). A group of robotic agents, copying the chain-like formation of salps, work together to solve Gordian problems. The problems are solved by iteratively updating the positions of salps based on the position of the leader salp. Efficient exploration and exploitation of the search space can therefore be made within a given environment. It is time and again used for path planning and organizing coordinated movement in robotic swarms.

15.3.1   SALIENT POINTS ABOUT THE SSA ALGORITHM

Before proceeding further, a few words are in order regarding the biological inspiration, search mechanism, and suitability for application of the SSA algorithm.

(i) Biological Inspiration: Salps are known to form chain-like formations while swimming (Figure 15.2a and b). Figure 15.2a shows the leader salp, while Figure 15.2b shows the follower salp; the leader salp is drawn with thick lines to distinguish it from the follower salp, made in thin lines. The basis for the algorithm structure is a single leader guiding the group. This leader salp directs the movement of the other follower salps. The primary objective is to identify the source of food within the search space. Figure 15.2c shows a salp chain with the leader salp (thick lines) in the front reaching near the food, and several follower salps (thin lines) behind it forming a circular ring.

(ii) Search Mechanism: It utilizes a balance between exploration and exploitation. A wormhole mechanism is employed; a wormhole is a theoretical passageway connecting two points in spacetime. Salps move in different directions depending on their positions in the chain.

(iii) Application Suitability: It is especially helpful for solving complicated optimization problems involving high dimensionality owing to its diverse search capabilities.

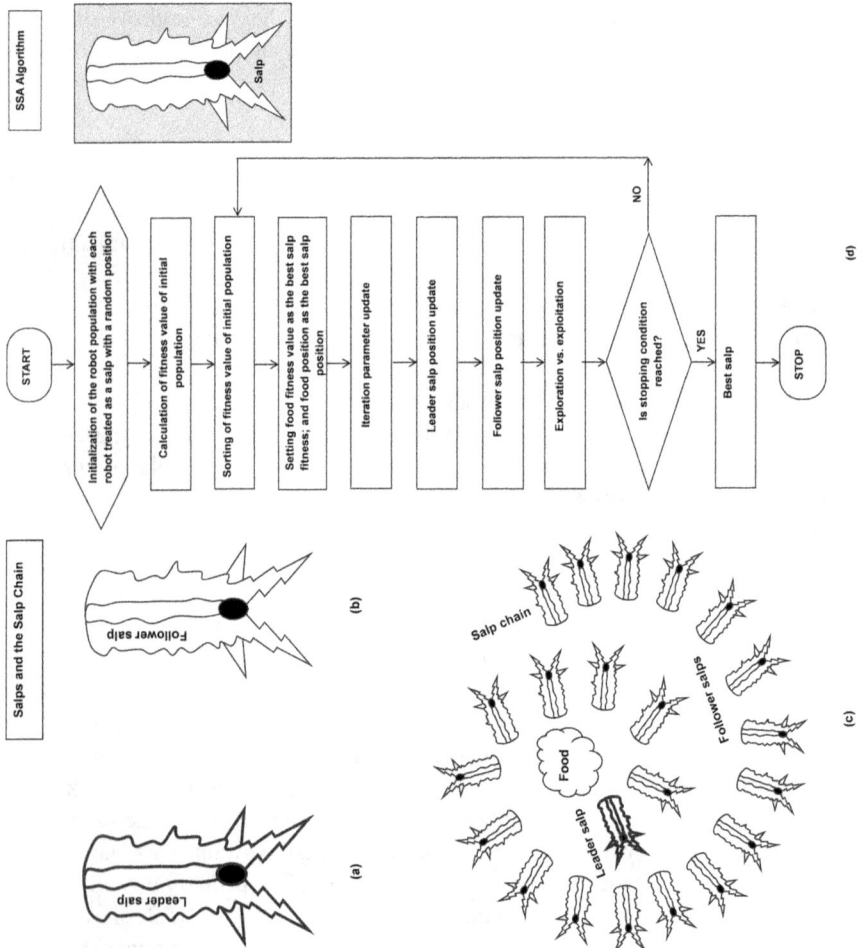

FIGURE 15.2 Quest for food by marine creature salps, and the corresponding swarm robotic algorithm: (a) leader salp, (b) follower salp, (c) salp chain, and (d) the algorithm formulated by adopting the collective activities of marine creature salps.

15.3.2 MAIN STEPS OF THE SSA ALGORITHM

The steps of this algorithm are shown in Figure 15.2d. The stages in the salp algorithm are: start, initialization of the robot population, calculation and sorting of fitness values, setting the food fitness value and food position as the best parameters, updating the iteration parameter, leader and follower salp positions, followed by exploration versus exploitation. If the stopping condition is reached, the best salp is obtained and the process stops. Otherwise, the process goes back to the stage of sorting fitness values. The algorithmic procedures are clarified below (Hegazy et al. 2020):

   i. Initialization of Robot Population: Each robot in the swarm is initialized as a salp. To get started, it is assigned a random position within the search space.
  ii. Fitness Value Calculation and Sorting: The fitness value of the initial salp population is determined. The fitness values obtained are sorted.
 iii. Setting the Food Fitness and Position: The food fitness is set as the best salp fitness. The food position is considered the optimal salp position.
  iv. Iteration Parameter Update: The iteration parameter represents the current number of cycles or loops that the algorithm has undergone during the optimization process. This number is updated.
   v. Leader Update: The leader salp position is updated. The updating is based on the best solution found so far, attracting the other salps toward the optimal area.
  vi. Follower Update: The follower salps update their positions by following the movement of the salp in front of them. Hence, the chain-like structure of salps is maintained.
 vii. Exploration vs. Exploitation: The algorithm dynamically balances between searching an extensive area (exploration) against concentrating on promising regions (exploitation). Balancing is achieved by adjusting the parameters based on the progress of the iteration.

15.3.3 APPLICATIONS OF THE SSA ALGORITHM

Apart from its use in robot path planning, and cooperative manipulation, the SSA algorithm is also useful for the optimized placement of sensors in a network.

   i. Robot Path Planning: The SSA algorithm optimizes the route for efficient movement and collision avoidance of robots. It thus helps a group of robots to navigate through a baffling and convoluted environment. In order to optimize the search for food sources in a distributed robot system, the SSA algorithm simulates the foraging behavior of salps.
  ii. Cooperative Manipulation: Multiple robots are coordinated to manipulate an object collaboratively. For this purpose, a chain-like formation is utilized to maintain stability and adjust positions.

iii. Sensor Network Deployment: The placement of sensors in a network is optimized to achieve maximization of coverage area and minimization of energy consumption.

15.3.4 ADVANTAGES OF THE SSA ALGORITHM

Special characteristics of the SSA algorithm can be beneficially exploited to get meaningful results in swarm robotics.

   i. Simplicity of the Basic Algorithm Concept. Conceptually, the SSA algorithm is relatively easy to understand. Putting the concept into practice also requires minimal effort.
  ii. Flexibility of the Algorithm: The algorithm is adaptable to different swarm robotics tasks. The parameters of the algorithm are varied, and additional constraints are incorporated to facilitate adaptation.
 iii. Global Search Capability: Salps form chain-like structures called aggregates. The chains comprise many identical salps. This structure allows for efficient exploration of a large search space.

15.3.5 LIMITATIONS OF THE SSA ALGORITHM

To prevent failures during the application of the algorithm and minimize the likelihood of obtaining erratic results, the limitations of the SSA algorithm must not be overlooked.

   i. Parameter Tuning: The performance of the SSA algorithm is sensitive to parameter values. Careful tuning of parameters is essential for specific applications.
  ii. Potential for Local Optima: In certain scenarios, the algorithm gets stuck in a local optimum. This happens when it is not properly designed.

15.4 COMPARISON OF GA WITH PSO ALGORITHM

From this section onward, we attempt to make a series of comparisons among the swarm robotic algorithms that we have discussed so far (Warnakulasooriya and Segev 2025). Comparison is a powerful learning technique that can be leveraged to discover a new breadth of view and develop a correlational vocabulary. By making comparisons, one can acquire a better overall view of the entire landscape of these algorithms. Comparisons encourage analysis of information and bring to the forefront the subtleties and nuanced differences among algorithms. They foster critical thinking, thereby preparing us to make the most suitable choices compatible with the needs and priorities of the situation, while also providing a clear, cogent, and articulate explanation for the decisions made.

We start by comparing Genetic Algorithm (GA) with the PSO algorithm (Wihartiko et al. 2018). Both GA and PSO algorithm are population-based optimization algorithms. The key difference between them lies in their approach to exploring the search space.

TABLE 15.1

Genetic Algorithm and PSO Algorithm

Sl. No.	Specific Feature/ Aspect Considered	Genetic Algorithm	PSO Algorithm
1	Evolutionary inspiration	GA draws inspiration from biological evolution, utilizing concepts such as selection, crossover, and mutation to generate new solutions.	The PSO algorithm is based on the collective behavior of bird flocks or fish schools.
2	Exploration vs. exploitation	GA generally has a stronger exploration capability due to its diverse genetic operations. The stronger exploration capability allows it to search a wider range of solutions.	PSO tends to focus more on exploitation. Although it rapidly converges toward a solution, it potentially gets stuck in a local optimum.
3	Complexity in implementation	GA can sometimes be considered more complex to implement due to the need to design appropriate crossover and mutation operators.	The PSO algorithm typically has a simpler structure with fewer parameters to adjust.
4	Suitable situations in which particular algorithms are preferred	GA is used in situations where: (i) A large, complex search space with diverse solutions is to be explored (ii) The problem involves discrete variables or constraints. (iii) Premature convergence to a local optimum is desired to be avoided.	PSO is used in cases where: (i) A fast convergence to a solution and high computational efficiency are needed. (ii) There are continuous variables and a well-defined search space in the problem. (iii) A simple implementation with fewer tunable parameters is desirable.

GA relies more on survival of the fittest through crossover and mutation operations. The PSO algorithm mimics the collective behavior of a swarm. In the PSO algorithm, the individuals adjust their movements based on their own best position and the best position in the swarm. The PSO algorithm generally leads to faster convergence but is potentially more prone to getting stuck in local optima compared to GA. GA is better for complex problems with diverse solutions. The PSO algorithm is more suitable for faster convergence in continuous optimization problems. Table 15.1 presents a comparison between the GA and the PSO algorithm with respect to specific points.

15.5 COMPARISON OF PSO, ABC, AND ACO ALGORITHMS

When comparing PSO, artificial bee colony (ABC), and ant colony optimization (ACO) algorithms, the significant difference lies in their inspiration from natural phenomena: PSO mimics bird flocking behavior, ABC simulates honeybee foraging, and ACO models the foraging patterns of ant colonies (Selvi and Umarani 2010; Arora et al. 2023). Each algorithm has distinct strengths and weaknesses in various optimization scenarios. Generally, the PSO algorithm is better for continuous optimization problems. The ACO algorithm excels in combinatorial optimization tasks due to its path-building nature. The ABC algorithm is effective in both domains, depending on the complexity of the problem.

The differences among these algorithms are pointed out in the context of their search mechanisms, strengths, weaknesses, and applications, as outlined in Table 15.2.

15.6   COMPARISON OF FIREFLY, ABC, AND ACO ALGORITHMS

When comparing firefly, ABC, and ACO algorithms, it is found that the main differences lie in their biological inspiration: The Firefly algorithm (FA) shadows the flashing behavior of fireflies. Brighter fireflies attract less bright ones. The ABC algorithm simulates the foraging behavior of honeybees. The ACO algorithm models the way ants find food using pheromone trails (Lazarowska 2023). Each algorithm has its own lustiness and frailty depending on the optimization problem at hand. The comparisons among the three algorithms are made in Table 15.3 with reference to the mechanism of movement of individuals, strengths/weaknesses, and from an all-inclusive viewpoint.

15.7   COMPARISON OF BFO, ABC, AND ACO ALGORITHMS

BFO, ABC, and ACO are all swarm intelligence algorithms, but they differ in their ways of drawing inspiration from nature. The BFO algorithm takes the lesson from bacteria searching for food. The ABC algorithm simulates the foraging behavior of honeybees. The ACO algorithm replicates the pathfinding behavior of ants. Distinct propitious and unpropitious aspects are observed in their optimization approaches.

The important differences are brought out in Table 15.4.

15.8   COMPARISON OF FIREFLY AND BFO ALGORITHMS

Both the firefly and BFO algorithms are swarm intelligence algorithms that are used to control robot swarms. Both these algorithms gain insight from different biological behaviors. Generally, the FA is more suitable for swarm robotic tasks where precise localization and coordinated movement are given a high priority. The BFO algorithm excels in handling dynamic environments with uncertain conditions. Its ability to adapt and explore a wider search space effectively is beneficial in these circumstances. Distinct favorable and unfavorable traits of these algorithms are noticed depending on the intended application.

The differences between these algorithms are given in Table 15.5.

15.9   COMPARISON OF SSA, ABC, AND ACO ALGORITHMS

All three algorithms are swarm intelligence algorithms used for optimization problems. They differ in the stimulus for encouragement they receive from nature. The salp algorithm follows the movement of salps in the ocean. The ABC algorithm

TABLE 15.2
PSO, ABC, and ACO Algorithms

Sl. No.	Specific Feature/ Aspect Considered	PSO Algorithm	ABC Algorithm	ACO Algorithm
1	Basic search mechanisms of the algorithms	The particles update their positions by considering their own best position and the best position discovered by the swarm, adjusting their velocities accordingly to navigate toward the optimal solution within the search space.	The food sources are updated based on the quality of their nectar or the fitness function.	The ants probabilistically choose paths. The path selection is based on pheromone levels. The best paths accumulate more pheromone.
2	Advantages and strengths of algorithms	Easy in implementation, fast in convergence, and performs well in continuous optimization problems.	Effective in solving complex problems, balancing exploration and exploitation.	Excellent for combinatorial optimization problems, and capable of finding near-optimal solutions in complex constraint scenarios.
3	Disadvantages and weaknesses of algorithms	The algorithm becomes trapped in local optima, necessitating careful parameter tuning to solve complex problems.	The algorithm is slower than other algorithms for certain problems. Also, it is sensitive to the selection of parameters.	The algorithm has a low efficiency for solving high-dimensional problems. It gets stuck in suboptimal solutions if the pheromone update is not properly managed.
4	Typical application examples of algorithms	Processing of images, optimization of engineering designs, optimization of functions.	Clustering of data, scheduling problems, and selection of features.	Traveling Salesman Problem, Route Optimization, and Resource Allocation.

TABLE 15.3
Firefly, ABC, and ACO Algorithms

Sl. No.	Specific Feature/ Aspect Considered	Firefly Algorithm	ABC Algorithm	ACO Algorithm
1	Movement mechanisms of individuals	Individual fireflies move toward brighter or better solutions. The strength of attraction between individuals decreases as the distance between them increases.	Bees explore the search space. They update their positions based on the quality of their current food source.	Ants update pheromone levels on potential paths. The pheromone levels influence the probability that other ants will choose those paths.
2	Advantages and strengths of algorithms	Problems with complex search spaces are easily solved due to the attraction mechanism, which serves as the basis of the algorithm. The algorithm can also handle continuous optimization problems well.	The algorithm is efficient in local search. Additionally, it is able to balance exploration and exploitation, leveraging its diverse bee types.	The algorithm is effective in finding optimal paths in graph-based problems. Combinatorial optimization issues are efficiently handled.
3	Disadvantages and weaknesses of algorithms	The algorithm gets stuck in local optima whenever parameter tuning lacks optimality.	The algorithm is sensitive to the selection of parameters. It may not perform well in high-dimensional spaces.	The algorithm suffers from stagnation if the pheromone updates are not carefully managed.
4	Overall comparison of algorithms	For problems requiring global exploration with a focus on finding better solutions based on attractiveness, like feature selection, the firefly algorithm is a good choice.	For problems with well-defined local search requirements, where exploitation of good solutions is crucial, the ABC algorithm is more suitable.	If a problem involves finding optimal paths in a graph structure, the ACO algorithm is preferable.

TABLE 15.4

BFO, ABC, and ACO Algorithms

Sl. No.	Specific Feature/ Aspect Considered	BFO Algorithm	ABC Algorithm	ACO Algorithm
1	Search mechanisms of algorithms	This algorithm utilizes chemotaxis, a process by which cells move in a directed manner toward a chemical gradient. Occasional random perturbations occur to prevent stagnation.	This algorithm employs a combination of local search, performed by employed bees, and global search, carried out by onlooker bees, with scout bees responsible for diversifying the search space.	It iteratively updates pheromone levels on potential solutions. The updating of pheromone levels guides subsequent ants to follow better paths, benefiting from the accumulated pheromone strength.
2	Advantages and strengths of algorithms	This algorithm is effective in solving problems with complex search spaces. Its diverse local search mechanisms and ability to escape local optima are helpful in these situations.	This algorithm strikes a good balance between exploration and exploitation. The distinct roles of bee types serve as a blessing in this regard. It performs correctly in multimodal optimization problems.	It demonstrates efficiency in handling combinatorial optimization problems, such as routing. In routing problems, the graph structure is well-defined. Therefore, a good solution quality is achieved with a relatively simple implementation.
3	Disadvantages and weaknesses of algorithms	This algorithm is sensitive to parameter tuning. So, a careful adjustment of parameters is necessary for specific problems.	This algorithm suffers from premature convergence if not properly managed. High-dimensional problems are afflicted by such difficulties.	It gets stuck in local optima if pheromone updates are not carefully designed. Large-scale problems experience this setback.
4	Summary	The BFO algorithm displays strength in its ability to explore diverse areas of the search space.	The ABC algorithm excels at balancing local and global search.	The ACO algorithm is suited for graph-based optimization problems. In these problems, the path selection is a crucial activity.

TABLE 15.5
Firefly and BFO Algorithms

Sl. No.	Specific Feature/ Aspect Considered	Firefly Algorithm	BFO Algorithm
1	Advantages and strengths of algorithms	It is efficient in finding local optima due to the attraction mechanism built into this algorithm. It performs well in tasks requiring precise coordination and movement patterns.	It works well in dynamic environments where changing conditions need to be addressed. It is capable of exploring a wider search space. Its random movement and chemotaxis behavior help in this exploration.
2	Applications of algorithms	Planning of path, avoidance of obstacles, localization of target, and aggregation of swarm.	Problems of optimization in complex environments, deployment of sensor networks, and dynamic task allocation.

simulates the behavior of honeybees looking for food. The ACO algorithm mirrors the pathfinding procedure of ant colonies. Distinct optimism and pessimism are seen when applying these algorithms in different scenarios. Generally, the SSA algorithm is better for complex, high-dimensional problems. Its diverse search strategy is the main cause for this superiority. The ABC algorithm is effective in facing problems with well-defined search spaces. The ACO algorithm outshines other algorithms in finding optimal paths in graph-based problems.

The chief differences among these algorithms are listed in Table 15.6.

15.10 DISCUSSION AND CONCLUSIONS

Progress in BFO and salp swarm robotic algorithms was reviewed in this chapter. Table 15.7 presents a brief sum-up of Chapter 15. In swarm robotics, several robots are organized for search and rescue missions. Each robot in the swarm has its own sensing, processing, and communication capabilities. Coordination of a large number of tasks among robots engaged in teamwork demands efficient multi-robot task allocation or MRTA methods (Khamis et al. 2015; Chakraa et al. 2023). Several challenges are posed by real-life MRTA applications, e.g., simulating fleets of robots in a congested shopping center (Surma et al. 2021). Disaster response, environment monitoring, and reconnaissance operations deserve special mention. These challenges are encountered in the form of dynamically occurring tasks that have deadlines. Robots with payload capacity and ferry range constraints are involved. Such combinatorial optimization problems have been solved by several approaches (Park et al. 2022).

TABLE 15.6
The SSA, ABC, and ACO Algorithms

Sl. No.	Specific Feature/ Aspect Considered	SSA Algorithm	ABC Algorithm	ACO Algorithm
1	Inspiration of algorithms	This algorithm is created by closely watching the movement patterns of salps in the ocean. It applies their chain-like formations during navigation and foraging.	This algorithm is developed by copying the foraging behavior of honeybees. The roles played by employed bees, onlooker bees, and scout bees contribute to the common aim.	It moves in the way ants communicate through pheromone trails to find the shortest path to a food source.
2	Search mechanisms of algorithms	This algorithm utilizes a balance between exploration and exploitation through a wormhole mechanism. The salps move in different directions depending on their position in the chain.	This algorithm employs a combination of local search, using employed bees, and global search, via onlooker bees, to explore the solution space. Local and global searches unite to form an efficient search strategy.	The ants iteratively update pheromone levels on potential paths. By such updating, valuable guidance is provided to subsequent ants, leading to better solutions and preventing the wastage of time in further efforts required to find the path if it has not been previously decided and marked with pheromone by predecessor ants.
3	Application suitability of algorithms	It is particularly useful for solving complex optimization problems with high dimensionality. Its diverse search capabilities are beneficially utilized to address these cases.	It is well suited for dealing with problems having clearly defined search spaces. Also, it is befitting to tackle issues where balancing between exploration and exploitation is compulsory.	It is commonly used for solving path planning problems. When dealing with graphs and network optimization, it supersedes other algorithms in many respects.

TABLE 15.7

Takeaways from This Chapter at a Glance

Sl. No.	Takeaway	Explanation
1	Summary	This chapter described the bacterial foraging optimization (BFO) and salp swarm algorithm (SSA) used in robotic path planning and navigation. Detailed comparisons were made among the GA and PSO algorithms; the PSO, ABC, and ACO algorithms; the firefly, ABC, and ACO algorithms; the BFO, ABC, and ACO algorithms; the firefly and BFO algorithms; and the SSA, ABC, and ACO algorithms.
2	BFO algorithm	The BFO algorithm mimics the foraging behavior of *E. coli* bacteria, including chemotaxis (movement toward nutrients), swarming (clustering around food sources), and a reproduction, elimination, and dispersal mechanism. A population of virtual bacteria navigates the search space, updating their positions based on local information about the nutrient (the optimal solution) and adjusting their movement according to chemotaxis, swarming, and reproduction steps.
3	Salp algorithm	The SSA is motivated by the schooling behavior of salps, where individuals follow a leader and maintain a certain distance from each other while moving in a coordinated manner. A population of salps is represented as points in the search space, with a leader salp guiding the movement of the others. The salps update their positions based on a balance between exploration (random movement) and exploitation (moving toward the best solution found so far).
4	BFO *vs* Salp algorithm	The BFO algorithm tends to excel in exploration due to its random movement during chemotaxis and dispersal phases, while SSA strikes a good balance between exploration and exploitation by adjusting the influence of the leader salp. The BFO is relatively simpler to implement, while SSA requires more fine-tuning of parameters due to its leader-follower structure.
5	Keywords and ideas to remember	Bacterial foraging optimization algorithm, salp swarm algorithm, comparison of genetic algorithm and PSO algorithm; PSO, ABC, and ACO algorithms; firefly, ABC, and ACO algorithms; BFO, ABC, and ACO algorithms; firefly and BFO algorithms; SSA, ABC, and ACO algorithms.

THE ROBOTIC RESCUE TEAM

The robotic team
Is held in high esteem
On accomplishing the rescue scheme
Very dreadful it may have seemed
But when the distressed screamed
Robots brought a hopeful gleam
Miraculous, real, and supreme.

Although the swarm robotic technology has a promising future, it must be ardently and categorically declared that the technology is relatively new and there is ample scope for improvement before it is widely accepted and becomes commonplace in solving practical, real-life problems. The technology is considered to be in its infancy. Currently, most swarm robotic demonstrations are limited to controlled laboratory conditions. Moreover, the research work is primarily focused on developing foundational concepts. In this early childhood stage of the technology, the design of control algorithms and suitable hardware platforms for swarm robots is a hot topic. The technical hurdles that need to be addressed include managing interactions among a large number of robots, ensuring decentralized decision-making, and overcoming the uncertainties of the environment. The development of AI techniques, particularly distributed learning algorithms in AI, which require limited computation and can operate with CPUs and AI-optimized processors in small, reasonably priced robots, will enable robot swarms to gradually increase their autonomy (Dorigo et al. 2021). An emerging field of interest concerns swarms of flying robots. This burgeoning field offers unique capabilities, including aerial mobility, rapid maneuverability, and the ability to cover large areas quickly, thereby providing advantages such as affordability, multitasking, scalability, resilience, and flexibility (Alqudsi 2024; Alqudsi and Makaraci 2025).

WELCOMING THE ERA OF AI ROBOTICS

Robots working in labor-intensive jobs, robots working in hazardous
*   conditions*
Robots guiding vehicles on roads and assisting doctors in critical
*   surgical operations*
Affable, cordial, and well-behaved robots are working everywhere
To look after human security and welfare
Robots working side by side with humans
With the enthusiasm of a mechanical acumen
To evolve a happy robot-cum-man society
Filled with joy and gaiety
Let's unify manly and robot efforts
To bring a new quality of life and comforts
Robots are good friends, sharing emotions
In moments of stress and commotion
Welcome to the age of Robotics and AI

Gazing from the horizon in the blue, expansive blue sky
Let our thoughts fly into dreams and soar high
And think of a beautiful, peaceful earth
Drenched in merriment and mirth!

REFERENCES AND FURTHER READING

Alqudsi Y. 2024. Coordinated Formation Control for Swarm Flying Robots, *2024 1st International Conference on Emerging Technologies for Dependable Internet of Things (ICETI)*, Sana'a, Yemen, 25–26 November, pp. 1–8.

Alqudsi Y. and M. Makaraci. 2025. Exploring advancements and emerging trends in robotic swarm coordination and control of swarm flying robots: A review, Proceedings of the Institution of Mechanical Engineers, *Part C: Journal of Mechanical Engineering Science*, Vol. 239, 1, pp. 180–204.

Arora G. K. Bala, H. Emadifar and M. Khademi. 2023. A comparative study of particle swarm optimization and artificial bee colony algorithm for numerical analysis of Fisher's equation, *Discrete Dynamics in Nature and Society*, Vol. 2023, Article ID 9964744, 10 pages.

Castelli M., L. Manzoni, L. Mariot, M. S. Nobile and A. Tangherloni. 2022. Salp swarm optimization: A critical review, *Expert Systems with Applications*, Vol. 189, p. 116029, https://doi.org/10.1016/j.eswa.2021.116029

Chakraa H., F. Guérin, E. Leclercq and D. Lefebvre. 2023. Optimization techniques for multi-robot task allocation problems: Review on the state-of-the-art, *Robotics and Autonomous Systems*, Vol. 168, p. 104492, https://doi.org/10.1016/j.robot.2023.104492

Dorigo M., G. Theraulaz and V. Trianni. 2021. Swarm robotics: Past, present, and future [Point of View], *Proceedings of the IEEE*, Vol. 109, 7, pp. 1152–1165.

Faris H., S. Mirjalili, I. Aljarah, M. Mafarja and A. A. Heidari. 2019. Salp Swarm Algorithm: Theory, Literature Review, and Application in Extreme Learning Machines. In: Mirjalili S., et al. (Eds.), *Nature-Inspired Optimizers, Studies in Computational*, Springer Nature, Switzerland, pp. 1–15.

Fiveable. 2024. Swarm intelligence and robotics review: 3.5 bacterial foraging optimization, https://library.fiveable.me/swarm-intelligence-and-robotics/unit-3/bacterial-foraging-optimization/study-guide/BgBC9E9Htxm3GdWT

Gan X. and B. Xiao. 2020. Improved bacterial foraging optimization algorithm with comprehensive swarm learning strategies, *Advances in Swarm Intelligence*, Vol. 12145, pp. 325–334.

Guo C., H. Tang, B. Niu and C. B. P. Lee. 2021. A survey of bacterial foraging optimization, *Neurocomputing*, Vol. 452, pp. 728–746.

Hegazy A. E., M. A. Makhlouf and G. S. El-Tawel. 2020. Improved salp swarm algorithm for feature selection, *Journal of King Saud University – Computer and Information Sciences*, Vol. 32, 3, pp. 335–344.

Houssein E. H., I. E. Mohamed and Y. M. Wazery. 2020. Salp Swarm Algorithm: A Comprehensive Review. In: Oliva D. and S. Hinojosa (Eds.), *Applications of Hybrid Metaheuristic Algorithms for Image Processing*. Studies in Computational Intelligence, Vol. 890, Springer, Cham, pp. 285–308.

Khamis A., A. Hussein and A. Elmogy. 2015. Multi-robot Task Allocation: A Review of the State-of-the-Art. In: Koubâa A. and J. R. Martínez-de Dios (Eds.), *Cooperative Robots and Sensor Networks 2015*, Studies in Computational Intelligence, Vol. 604, Springer International Publishing, Switzerland, pp. 31–51.

Lazarowska A. 2023. A comparison of the firefly algorithm and the ant colony optimization for ship collision avoidance, *Procedia Computer Science*, Vol. 225, pp. 2037–2046.

Majumder A., D. Laha and P. N. Suganthan. 2019. Bacterial foraging optimization algorithm in robotic cells with sequence-dependent setup times, *Knowledge-Based Systems*, Vol. 172, pp. 104–122.

Park B., C. Kang and J. Choi. 2022. Cooperative multi-robot task allocation with reinforcement learning, *Applied Sciences*, Vol. 12, 272, pp. 1–19.

Romeh A. E. and S. Mirjalili. 2023. Multi-robot exploration of unknown space using combined meta-heuristic salp swarm algorithm and deterministic coordinated multi-robot exploration, *Sensors*, Vol. 23, 4, 2156, pp. 1–23.

Selvi V. and R. Umarani. 2010. Comparative analysis of ant colony and particle swarm optimization techniques, *International Journal of Computer Applications*, Vol. 5, 4, pp. 1–6.

Surma F., T. P. Kucner and M. Mansouri. 2021. Multiple Robots Avoid Humans to Get the Jobs Done: An Approach to Human-Aware Task Allocation, *2021 European Conference on Mobile Robots (ECMR)*, Bonn, Germany, 31 August to 3 September, pp. 1–6.

Wang Z., J. Peng and S. Ding. 2022. A bio-inspired trajectory planning method for robotic manipulators based on improved bacteria foraging optimization algorithm and tau theory, *Mathematical Biosciences and Engineering*, Vol. 19, 1, pp. 643–662.

Warnakulasooriya K. and A. Segev. 2025. Comparative analysis of accuracy and computational complexity across 21 swarm intelligence algorithms, *Evolutionary Intelligence*, Vol. 18, Article Number 18, pp. 1–37.

Wihartiko F. D., H. Wijayanti and F. Virgantari. 2018. Performance comparison of genetic algorithms and particle swarm optimization for model integer programming bus timetabling problem, *IOP Conf. Series: Materials Science and Engineering*, Vol. 332, 012020, pp. 1–6.

Yang B., Y. Ding and K. Hao. 2014. Target Searching and Trapping for Swarm Robots with Modified Bacterial Foraging Optimization Algorithm, *Proceeding of the 11th World Congress on Intelligent Control and Automation*, Shenyang, 29 June to 4 July, pp. 1348–1353.

Yang Z., Y. Jiang and W. C. Yeh. 2024. Self-learning salp swarm algorithm for global optimization and its application in multi-layer perceptron model training, *Scientific Reports*, Vol. 14, 27401, 29 pages.

Appendix

Interactive Mini Glossary of AI Algorithms and Related Terms for Robotics in Question–Answer Format

AlexNet: It is a deep learning algorithm. *What specifically does it represent?* It is a convolutional neural network architecture. *What is the purpose of this architectural design?* It is designed for image recognition and classification. *What are the applications of AlexNet in robotics?* It is utilized in robotics for object detection, scene understanding, and autonomous navigation. It is also used to guide robotic arms in picking objects or performing complex manipulations.

Ant Colony Optimization: It is a pathfinding algorithm inspired by the foraging behavior of ants. *What is its use in robotics?* It is a popular choice for robot path planning. *How does it work?* In this algorithm, the robots find optimal paths by simulating pheromone trails. *Is the algorithm adaptable?* Yes, the algorithm is adaptable to dynamic environments. For adaptation, pheromone levels are updated based on the current state of the environment.

Artificial Bee Colony: It is a swarm intelligence optimization technique. *How many types of bees are differentiated in this technique?* In this technique, three types of artificial bees—employed, onlooker, and scout—are used to search for food sources. *How do the bees find the best path?* The bees communicate their findings to other bees to find the best path. *How is the method applied in robotics?* The technique is particularly effective for robot path planning. It optimizes both the robot's path length and the smoothness of the path. It is also used for multi-robot path planning, ensuring their collision-free movement.

Artificial Potential Field Algorithm: It is a path-planning technique. *How is the robot's environment simulated?* The robot's environment is simulated as a potential field. *How does the algorithm guide the robot?* It guides a robot toward a goal by an attraction-repulsion mechanism, attracting it toward the goal and repelling it away from obstacles.

A* Search Algorithm: It is a pathfinding algorithm used to find the shortest path between two points in a graph. *Are obstacles and constraints considered?* Yes, the barriers and limitations on the robot's path are duly taken into consideration. *What special tool does the algorithm use?* The algorithm works by leveraging a heuristic function to estimate the cost of reaching the goal from any given node.

Backpropagation: It is a core component of neural networks. *Where is it used in robotics?* It is used to train robots to learn and make decisions. *How does the robot learn?* The robot learns by adjusting network weights to minimize errors between the predicted and actual outputs.

Bacterial Foraging Optimization: It is an algorithm designed to solve optimization problems. *What are these problems?* These problems include robot path planning, obstacle avoidance, and the coordination of a swarm of robots. *What does the algorithm do?* The algorithm enables robots to search for optimal solutions in dynamic environments efficiently. *What natural phenomenon does the algorithm mimic?* It operates by mimicking the foraging behavior of *E. coli* bacteria.

Bayesian Inference: It is a statistical method that utilizes Bayes' theorem as the guiding principle for operation. *How is Bayes' theorem applied?* Bayes' theorem is used to update the probability of a hypothesis or belief by combining prior knowledge with new sensor data received from the robot. The probability updating yields an estimate of the posterior probability, which represents the updated belief about the robot's state. *In what ways does Bayesian inference help robots?* Bayesian inference allows robots to learn, adapt, and perform tasks in changing environments.

Branch-and-Bound Scheme: It is a general algorithm design paradigm. *What is it used for?* It is used for solving discrete or combinatorial optimization problems. *How does it operate?* Its operation is based on the systematic exploration of a tree of candidate solutions. As it progresses, it prunes branches that cannot contain the optimal solution. For pruning the branches, it uses bounds, namely the upper and lower estimates. Thereby, it reduces the search space. The reduction of the search space improves the efficiency of finding the optimal solution.

Bug Algorithms: These are a class of simple, sensor-based path-planning techniques. *How are they exploited in robotics?* They are applied by robots for navigating unknown environments. By using these algorithms, the robots can move in an organized manner without needing a map of the environment. *What are the principal measurements and data on which robots depend?* For their motion, the robots rely on sensor measurements for guidance. Local sensor data, such as contact or range sensors, are used by robots to determine their positions and the presence of obstacles.

CNN: It is an artificial neural network for analyzing visual data. *In what fields does it give excellent performance?* It excels at recognizing patterns and features in images. *What are the main tasks at which CNN is highly skilled, and where are its capabilities used in robotics?* Three primary image-related tasks are expertly handled by it: object detection, image classification, and scene understanding. Due to these beneficial features, it is used in robotics for obstacle avoidance based on real-time camera images, enabling robots to navigate and plan their paths.

Convolutional Neural Network: See CNN.

Decision Tree: It is a supervised machine learning technique. *Where does the name 'decision tree' come from?* The name 'decision tree' originates from its tree-like structure. *What are the parts of the decision tree?* The tree has a root node, internal nodes, branches, and leaf nodes. *What are the functions of the different parts of the decision tree?* Each internal node of the tree represents a test carried out on an attribute. Each branch represents the outcome of the test. Each leaf node represents the final outcome or prediction. *What types of tasks is the decision tree used for?* The decision tree is used by robots for executing both classification and regression tasks. It enables robots to classify situations or predict actions. Classification predicts categories. Regression predicts continuous values.

Dijkstra's Algorithm: It is a pathfinding algorithm used in robot navigation. *What does the algorithm give to robots?* It provides them with efficient and collision-free path planning. *How is the pathfinding implemented?* The pathfinding is done by transforming the robot's environment into a graph. The shortest path between two points is easily found once a graph is drawn.

Firefly Algorithm: It is an optimization algorithm used in robotics for path planning. *How are the robots represented in this algorithm?* In this algorithm, the robots are described as fireflies. *What rule do the robot fireflies follow, and what do they achieve?* The less bright robot fireflies move toward the brighter ones. Following this brightness rule, the robot fireflies move toward brighter or better locations in the environment to establish a suitable path.

Generalized Voronoi Diagram: It is a roadmap that provides a comprehensive global overview of the robot's environment. *What does a global overview show?* The global overview shows all possible paths in an environment containing obstacles. *How is the depiction of paths helpful?* An inspection of all feasible paths facilitates efficient path planning by robots. They can focus on the free space or areas of maximum clearance from obstacles. *What other facilities and services does the generalized Voronoi diagram provide?* Besides path planning, the Voronoi diagram enables stealthy navigation. The robots move surreptitiously, minimizing their visibility. The generalized Voronoi diagram is also used for surveillance and area coverage.

Genetic Algorithm: It is a type of evolutionary algorithm that simulates the biological process of natural selection to find optimal solutions to problems. *How does it find optimal solutions to problems, such as robot path planning?* It finds optimal solutions by simulating this natural selection process. *How does the algorithm work?* The algorithm operates by iteratively evolving a population of potential solutions, known as chromosomes. It advances through selection, crossover, and mutation, always aiming for the fittest individuals.

Hidden Markov Model: It is a probabilistic model. In this model, the observed data are generated by a sequence of hidden states. *What is meant by hidden states?* The hidden states refer to the robot's position or environmental conditions, as determined by its sensors. These hidden states are inferred based on the observed data. *In what ways does the hidden Markov model assist robots?* The hidden Markov model enables a robot to estimate its position or location in an unknown environment based on the sensor readings it has acquired.

Image-of-Interest Detection: It is an algorithm that uses computer vision techniques. *What is its main purpose?* Its purpose is to identify and locate specific objects or regions of interest within images or videos. *What does it do for robots?* It enables robots to understand their surroundings. Hence, robots navigate, manipulate objects, and perform tasks that require visual perception.

ISODATA: It is an algorithm to find compact clusters. *How does it work?* It works by grouping data points into clusters based on similarity. During operation, it iteratively updates cluster representatives. The updating is based on the mean of the vectors assigned to each cluster. *How is data clustering utilized in robotics?* In robotics, this clustering technique is used for object recognition, scene segmentation, and data analysis.

***k*-Means Clustering:** It is an unsupervised machine learning algorithm. *What is it used for?* It is used to partition a given dataset into k clusters. *What is the basic partitioning approach of the algorithm?* The approach for partitioning involves minimizing the sum of squared distances between each data point and its assigned centroid. *What does the symbol k in this algorithm stand for?* The symbol k represents a user-defined parameter, denoting the desired number of clusters. *What is the relevance of the algorithm in robotics?* In robotics, it groups similar data points, such as robot locations or sensor readings, into clusters. Thus, it facilitates the coordinated movement of a group of robots.

NAS: It is an automated approach to designing efficient and high-performing neural network architectures. *What is its specialty?* It represents a departure from the manual, trial-and-error approach. It leverages learning algorithms and deep learning techniques to find optimal architectures without manual intervention. *How is the use of the NAS approach beneficial?* Using NAS can lead to the discovery of architectures that outperform hand-crafted designs. *How are the designs evaluated?* The designs are evaluated in terms of performance, efficiency, and resource utilization.

Neural Architecture Search: See NAS.

Neural Networks: These are algorithms inspired by the functioning of the human brain. *What is the unique feature of these algorithms?* Their exceptional feature is the ability of learning and making predictions from data. *What role do these algorithms play in robotics?* These algorithms enable robots to perceive and understand their environment using techniques such as computer vision. They allow robots to make autonomous decisions based on learned patterns and real-time data. In this way, they aid in developing sophisticated control systems for robot movement. These include robot path planning and manipulation.

Particle Swarm Optimization: It is a population-based stochastic algorithm. *What natural phenomenon does this algorithm impersonate?* It mimics the social behavior of birds flocking or fish schooling. *How is the solution to a problem found?* Each potential solution to the problem is represented as a particle with a certain velocity. Particles move through the search space, updating their velocity and position. The update is based on two criteria: their own best position and the best position of the swarm. The global optimum is found by iteratively updating the positions of the particles. *Where is this algorithm applied in robotics?* The algorithm is used to find the optimal path for robots in a given environment. It is also used to optimize various aspects of robot behavior, including task allocation and resource management. Furthermore, it is used for target tracking where robots need to find and track specific objects or locations. It is particularly well-suited for coordinating the behavior of multiple robots in a swarm to achieve a common goal.

PID Algorithm: It is a feedback control mechanism. *How does this mechanism operate?* It operates within a closed-loop system to regulate the robotic system's output. It works by continuously adjusting a control variable based on the error between the desired set point and the actual value. During this work, it utilizes the proportional, integral, and derivative terms. The proportional (P) term responds to the current error. The integral (I) term accounts for accumulated error over time. The derivative (D) term responds to the rate of change of the error. *Where are PID controllers used in robotics?* PID controllers are used to control the speed, position, and orientation of robotic arms, wheels, and other actuators, enabling motion control. They are also used to maintain a robot's path, speed, and heading, enabling autonomous navigation. Another use of PID controllers is to control the force exerted by a robotic gripper or other actuators during manipulation. A further use of these controllers is to maintain a specific temperature in a robotic system for temperature control.

PNAS: It is a method for automatically designing convolutional neural networks. *What is its functional approach?* It works by sequentially searching the space of cell structures in a step-by-step manner. In the course of work, it optimizes cell structures. These structures are the building blocks of larger networks. This approach begins with simple models and progresses to more complex ones. It often leads to the discovery of CNN architectures that outperform those designed manually.

Probabilistic Roadmap: It is a path-planning technique for robots. *What is its governing principle?* It constructs a graph of possible paths by randomly sampling nodes in free space and connecting them. Thus, it enables efficient pathfinding between a start and goal configuration while avoiding obstacles. It starts by randomly sampling points or nodes within the robot's free space (sampling). In this space, the robot can move without colliding. Then the algorithm determines if a robot can move safely and smoothly between them (connectivity). If two nodes are connected, an edge or connection is created between them, forming a roadmap through graph construction. Then, the algorithm can efficiently find paths between a start and goal configuration by searching the graph, a process known as pathfinding.

Progressive Neural Architecture Search: See PNAS.

Proportional-Integral-Derivative Algorithm: See PID algorithm.

Rapidly Exploring Random Tree Algorithm: It is a probabilistic motion planning algorithm. *What is its basic working principle?* The algorithm works by iteratively expanding a tree of possible paths from a starting point, known as the root of the tree. Then it iteratively generates random samples (points) within the search space. For each new sample, the algorithm identifies the nearest node in the existing tree, called the nearest neighbor. A new node is added to the tree and connected to the closest neighbor. Thereby, the tree is effectively extended toward the randomly sampled point (tree extension). The process continues until the tree reaches the goal or a desired number of iterations are completed. At this point, a path is extracted from the tree by tracing back from the goal to the starting point. This process is known as pathfinding. RRT handles obstacles by ensuring that the tree nodes and paths do not collide with obstacles (obstacle avoidance). *How is this algorithm used in robotics?* It is used in robotics for finding paths through complex, high-dimensional spaces. It is particularly suitable for spaces with obstacles and non-holonomic constraints. Such constraints restrict the velocities of a system, rather than its positions. They are not integrable into the position constraint.

R-CNN: It is a type of machine learning model. *What are its primary features?* R-CNN excels at identifying and localizing objects within images or videos. It leverages the power of CNNs, which are adept at extracting features from visual data. *What special technique does R-CNN use?* The R-CNN uses a method called region proposals to identify potential object regions in an image. Once these regions are proposed, CNNs extract features from them. These features are then used to classify the objects within those regions. R-CNN also refines the bounding boxes around the detected objects. Thus, it ensures accurate localization of objects. *What are the uses of R-CNN?:* It is used for object detection and localization in computer vision. It is used in robotics for object recognition and scene understanding. R-CNNs are utilized in self-driving cars for object detection and lane-keeping purposes. Warehouses or hospitals use R-CNNs to identify and pick up objects. In industrial automation, R-CNNs are used for quality control and inspection tasks in manufacturing.

Region-Based Convolutional Neural Network: See R-CNN.

Salp: It is a population-based optimization algorithm. *What biological phenomenon does it imitate?* It mimics the swarming behavior of salps. *What are salps?* The salps are marine organisms that move in chains to forage for food. *What is the working approach of the algorithm?* The algorithm employs a leader-follower approach. The leader salp explores the search space, and the follower salps follow the leader's path. The salps iteratively refine their positions to find the optimal solution. *What role does this algorithm play in robotics?* The algorithm is used to find optimal paths for robot movements in dynamic and complex environments, known as path planning. It is used to assign tasks to robots in a multi-robot system. It is also used in optimization engagements to achieve efficiency and optimize resource utilization, including task assignment. Another use is to optimize various robotics-related parameters, such as robot arm movements, sensor placement, and control algorithms, in robotics optimization.

Scale-Invariant Feature Transformation: See SIFT.

Self-Organizing Map It is a type of artificial neural network. *What is its primary objective?* It helps to visualize and cluster high-dimensional data. *How does it function?* It functions by mapping the data onto a lower-dimensional grid, typically a 2D grid. During data mapping, relationships between data points are preserved. It is a type of unsupervised learning algorithm. *How is a self-organizing map utilized in robotics?* The algorithm is used to create maps of the environment and assist robots in navigation by identifying regions and paths, thereby facilitating navigation and path planning. It clusters similar objects together, making it easier to identify and classify objects in the robot's environment. This clustering is called object recognition. It detects unusual patterns or behaviors in the robot's sensor data, alerting the robot to potential problems through anomaly detection. It helps visualize complex sensor data, making it easier for humans to understand the robot's environment and behavior through data visualization.

Semantic Parsing: It is the process of mapping natural language, like spoken or written commands, into a formal representation of its meaning. *How is this representation done, and what is its use?* This representation is done in a machine-understandable format. Thus, it allows computers to interpret and act upon the meaning of the input. *What are the roles of semantic parsing in robotics?* It facilitates human-robot interaction, robotic task execution, navigation, and manipulation by enabling robots to understand and execute human instructions. It thus makes them more intuitive and easier to control. Hence, it enables robots to perform tasks based on natural language instructions. It helps robots to navigate to specific locations, manipulate objects, or perform actions based on natural language commands.

SIFT: It is a powerful algorithm for extracting distinctive features from images. *What main jobs does it accomplish?* It enables robust object recognition and matching across different scales, rotations, and partial occlusions. It helps robots identify objects in their environment. Identification is possible even when objects are viewed from different angles or at varying scales, demonstrating object recognition and scale invariance. *How is the SIFT algorithm applied in robotics?* It is used to create maps of an environment and assist robots in navigation by recognizing landmarks, enabling robotic mapping and navigation. It is used to stitch together multiple images of a scene to create a panoramic view, a process known as image stitching. It is used to reconstruct 3D models of objects from 2D images, a process called 3D modeling.

Simultaneous Localization and Mapping: See SLAM.

SLAM: It is an algorithm that enables a robot to build a map of an unknown environment (mapping) while simultaneously determining its own position (location tracking) within that map. *What help does mapping and localization provide to robots?* It enables robots to navigate and interact with unfamiliar environments.

Sliding Window Algorithm: It is an analysis and pattern recognition technique for processing streams of data or sequences. *How does it function?* It functions by examining the data through a fixed-size or variable-size window. The window moves across data to find patterns, subarrays, or subsequences that meet certain criteria within a larger dataset. *For what purposes is the algorithm used in robotics?* The algorithm is used for performing image processing and path mapping in robotics.

Support Vector Machine: It is a supervised machine learning algorithm. *What are its main uses?* It is used for classification and regression tasks. *How does it perform these tasks?* It performs these tasks by finding optimal decision boundaries or hyperplanes that maximize the margin between data classes. *Where is it applied?* It is particularly effective for object recognition, path planning, and robot control.

Transformer Network: It is a type of neural network architecture that excels at processing sequential data, like text or sensor data. It accomplishes the data processing by using a mechanism called self-attention. *How do robots use transformer networks?* Robots utilize transformer networks to comprehend and respond to spoken natural language commands, as well as for speech generation. These networks enable robots to communicate with humans through more intuitive interactions, including speech recognition, generation, and natural language processing. They also help predict the optimal sequence of actions for a robot to perform a task, based on the current situation and goals, facilitating action planning and sequencing. They are applied to interpret sensor data (e.g., images, sounds) and understand the surroundings of the robots. In this way, they enable more autonomous navigation and manipulation, including perception and understanding of the environment. Furthermore, they are used to predict the optimal movements for a robot's arm or other actuators to perform specific tasks, such as grasping an object or assembling parts (robotic manipulation).

VCG-16: It is a convolutional neural network architecture. It works as a powerful type of deep learning model used for processing image data. *Who developed this model?* The Visual Geometry Group developed it at the University of Oxford. *Where is it used in robotics?* It is used in robotics for image classification, object detection, and feature extraction. *How does it perform its duties?* It does so by leveraging its pretrained capabilities for such tasks.

Vector Field Histogram: It is a real-time motion planning technique in robotics. *How does it use sensor data?* It utilizes range sensor data to compute obstacle-free steering directions, taking into account the robot's dynamics and shape. *What is its working principle?* It works by creating a polar histogram of obstacle density. The areas devoid of obstacles, called valleys, are identified. The valley closest to the target direction is selected. Thus, a computationally efficient and robust method is provided for mobile robots to navigate and avoid obstacles while moving toward a target.

Velocity Updates: It is the process of adjusting or modifying a robot's speed and direction in response to real-time information about its environment or task requirements. *How are velocity updates applied in practice?* Sensors and controllers are used to ensure that the robot moves efficiently and safely. Velocity updates allow robots to avoid collisions. The robots maintain stability, and any accidents are prevented. Thus, the robots adapt to obstacles, changes in terrain, and dynamic situations.

Visual Geometry Group-16: See VGG-16.

Index

For Product Safety Concerns and Information please contact our EU
representative GPSR@taylorandfrancis.com
Taylor & Francis Verlag GmbH, Kaufingerstraße 24, 80331 München, Germany

9 781032 695198